MASS SUPPLY AND FLOWS IN THE SOLAR CORONA

MASS SUPPLY AND FLOWS IN THE SOLAR CORONA

The 2nd SOHO Workshop

edited by

B. FLECK

Space Science Department of ESA, ESTEC, Noordwijk, The Netherlands

G. NOCI

Dipartimento di Astronomia e Scienza dello Spazio, Firenze, Italy

and

G. POLETTO

Osservatorio Astrofisico di Arcetri, Firenze, Italy

Reprinted from Space Science Reviews, Vol. 70, Nos. 1–2, 1994

SPRINGER-SCIENCE+BUSINESS MEDIA, B.V.

A C.I.P. Catalogue record for this book is available from the Library of Congress.

ISBN 978-94-010-4401-1 ISBN 978-94-011-0930-7 (eBook)
DOI 10.1007/978-94-011-0930-7

Printed on acid-free paper

TABLE OF CONTENTS

III CORONAL STREAMERS

IV CORONAL HOLES AND SOLAR WIND

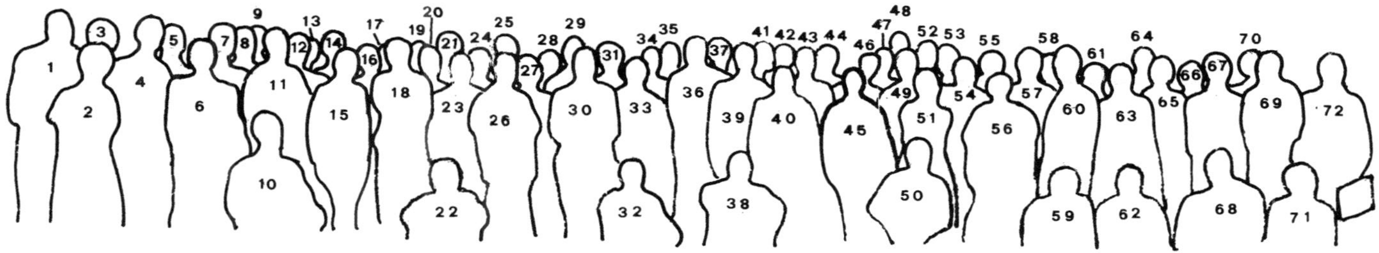

1 S. Antiochos **2** Mrs. Gulyaev **3** S. Koutchmy **4** B. Fleck **5** A. Sterling **6** L. Sanchez **7** P. Martens **8** M. Laming
9 H. Kunow **10** B. Foing **11** A. Pedersen **12** B. Somov **13** ? **14** G. Simnett **15** B. Schmieder **16** ? **17** ?
18 C. Chiuderi **19** ? **20** C. Schultz **21** J. Kohl **22** S. Blood **23** E. Hiei **24** C. Cheng **25** ? **26** R. Gulyaev
27 I. Kim **28** G. Timothy **29** V. Hansteen **30** E. Priest **31** G. Brueckner **32** V. Domingo **33** F. Drago
34 X. Zhao **35** ? **36** M. Huber **37** R. Harrison **38** P. Lemaire **39** E. Marsch **40** S. Sahal-Brechot **41** G. Noci
42 ? **43** K. Dere **44** P. Brekke **45** G. Poletto **46** E. Antonucci **47** K. Strong **48** M. Romoli **49** G. Peres
50 R. Falciani **51** M. Guhathakurta **52** L. Kulčár **53** O. Kjeldseth-Moe **54** A. Kosovichev **55** D. Hassler
56 T. Stepanova **57** R. Kopp **58** D. Spadaro **59** R. Ventura **60** S. Fineschi **61** J.-P. Meyer **62** F. Reale
63 L. Maccari **64** ? **65** M. Neugebauer **66** ? **67** S. Suess **68** S. Pohjolainen **69** R. Schwenn **70** M. Velli
71 H. Mason **72** Y. Yermolaev

List of Participants

Aboudarham, Jean	Institut d'Astrophysique Spatiale, Orsay, France
Antiochos, Spiro	Naval Research Laboratory, Washington, USA
Antonucci, Ester	Università di Torino, Italy
Axford, Ian	Max Planck Inst. für Aeronomie, Lindau, Germany
Bel, Nicole	Observatoire de Paris, Meudon, France
Benna, Carlo	Smithsonian Astrophys. Observ., Cambridge, USA
Blood, Steven	Williams College, Williamstown MA, USA
Bonnet, Roger	ESA Headquarters, Paris, France
Bothmer, Volker	Max Planck Inst. für Aeronomie, Lindau, Germany
Brekke, Pal	Institute of Theoretical Astrophys. , Oslo, Norway
Brueckner, Günther	Naval Research Laboratory, Washington, USA
Celnikier, L.	Observatoire de Paris, Meudon, F
Cheng, Chung-Chieh	Naval Research Laboratory, Washington, USA
Chiuderi, Claudio	Dipart. di Astronomia, Univ. di Firenze, Italy
Chiuderi Drago, Franca	Dipart. di Astronomia, Univ. di Firenze, Italy
Dahlburg, Russel	Naval Research Laboratory, Washington, USA
Dere, Ken	Naval Research Laboratory, Washington, USA
Domingo, Vicente	ESA Space Science Department, ESTEC, NL
Dryer, Murray	NOAA Space Environment Lab., Boulder, USA
Efimov, Anatoly	Institute of Radio Engineering, Moscow, Russia
Esser, Ruth	Smithsonian Astrophys. Observ., Cambridge, USA
Falciani, Roberto	Dipart. di Astronomia, Univ. di Firenze, Italy
Fineschi, Silvano	Smithsonian Astrophys. Observ., Cambridge, USA
Fleck, Bernhard	ESA Space Science Department, ESTEC, NL
Foing, Bernard	ESA Space Science Department, ESTEC, NL
Garcia, Howard	NOAA Space Environment Lab., Boulder, USA
Gardner, Larry	Smithsonian Astrophys. Observ., Cambridge, USA
Gibson, Sarah	University of Colorado, Boulder, USA
Gloeckler, George	University of Maryland, College Park, USA
Gulyaev, Rudolf	IZMIRAN, Troitsk, Russia
Guhathakurta, M.	NASA/GSFC, Greenbelt, USA
Habbal, Shadia	Smithsonian Astrophys. Observ., Cambridge, USA
Hansteen, Viggo	Institute of Theoretical Astrophys. , Oslo, Norway
Harrison, Richard	Rutherford Appleton Laboratory, Chilton, UK
Hassler, Don	Smithsonian Astrophys. Observ., Cambridge, USA
Hiei, Eijiro	National Astronomical Observatory, Mitaka, Japan
Hildner, Ernest	NOAA Space Environment Lab., Boulder, USA
Hoeksema, Todd	CSSA, Stanford University, USA

Huber, Martin	ESA Space Science Department, ESTEC, NL
Kalkofen, Wolfgang	NASA Headquarters, Washington, USA
Kallenbach, Reinald	University of Bern, Switzerland
Kim, Iraida	Sternberg Institute, Moscow University, Russia
Kjeldseth-Moe, Olav	Institute of Theoretical Astrophys. , Oslo, Norway
Kohl, John	Smithsonian Astrophys. Observ., Cambridge, USA
Kopp, Roger	Los Alamos National Lab., Los Alamos, USA
Korzhavin, A.N.	Special Astrophysical Observatory, Pulkovo, Russia
Kosovichev, Alexander	Institute of Astronomy, Cambridge University, UK
Kotrč, Pavel	Astronomical Institute, Ondřejov, Czech Republic
Koutchmy, Serge	Institut d'Astrophysique, CNRS, Paris, France
Kulčár, Ladislav	Universita M. Bela, Banská Bystrica, Slovakia
Kunow, Horst	Institut fur Kernphysik, Universität Kiel, Germany
Kyrie, Natalya	General Physics Institute, Moscow, Russia
Laming, Martin	Naval Research Laboratory, Washington, USA
Landini, Massimo	Dipart. di Astronomia, Univ. di Firenze, Italy
Lemaire, Philippe	Institut d'Astrophysique Spatiale, Orsay, France
Lumme, Mikael	Space Research Lab., Univ. of Turku, Finland
Maccari, Lisa	Dipart. di Astronomia, Univ. di Firenze, Italy
Marsch, Eckart	Max Planck Inst. für Aeronomie, Lindau, Germany
Martens, Piet	ESA Space Science Department, ESTEC, NL
Mason, Helen	Cambridge University, UK
Meyer, Jean-Paul	Service d'Astrophysique, CE-Saclay, France
Michels, Don	Naval Research Laboratory, Washington, USA
Monsignori-Fossi, B.	Osservatorio di Arcetri, Firenze, Italy
Moses, Dan	Naval Research Laboratory, Washington, USA
Neugebauer, Marcia	Jet Propulsion Laboratory, Pasadena, USA
Noci, Giancarlo	Dipart. di Astronomia, Univ. di Firenze, Italy
Pedersen, Arne	ESA Space Science Department, ESTEC, NL
Penn, Matthew	National Solar Observatory, Sunspot, USA
Peres, Giovanni	Osservatorio Astrofisico di Catania, Italy
Pohjolainen, Silja	Metsaehovi Radio Res. Station, Kymaelae, Finland
Poletto, Giannina	Osservatorio di Arcetri, Firenze, Italy
Priest, Eric	University of St. Andrews, Scotland
Rami, Vaimio	Metsaehovi Radio Res. Station, Kymaelae, Finland
Reale, Fabio	Osservatorio di Palermo, Italy
Romoli, Marco	Smithsonian Astrophys. Observ., Cambridge, USA
Sahal-Brechot, Sylvie	Observatoire de Paris, Meudon, France
Salakhutdinov, Rafik	Institute of Solar-Terrestrial Phys., Irkutsk, Russia
Sanchez-Duarte, Luis	ESA Space Science Department, ESTEC, NL

Schmieder, Brigitte	Observatoire de Paris, Meudon, France
Schultz, Goran	University of Turku, Finland
Schwenn, Rainer	Max Planck Inst. für Aeronomie, Lindau, Germany
Simnett, George	University of Birmingham, UK
Sizemore, Kenneth	NASA/GSFC, Greenbelt, USA
Somov, Boris	Sternberg Institute, Moscow University, Russia
Spadaro, Daniele	Osservatorio Astrofisico di Catania, Italy
Steinolfson, Richard	Southwest Research Institute, San Antonio, USA
Stepanova, Tatiana	General Physics Institute, Moscow, Russia
Sterling, Alphonse	Institute for Space and Astronomical Sciences, Japan
Strachan, Leonard	Smithsonian Astrophys. Observ., Cambridge, USA
Strong, Keith	Lockheed Solar and Astrophys. Lab., Palo Alto, USA
Suess, Steven	NASA Marshall Space Flight Center, Huntsville, USA
Timothy, Gethyn	CSSA, Stanford University, USA
Tondello, Giuseppe	Dipartimento di Elettronica, Univ. di Padova, Italy
Torsti, Jarmo	Space Research Laboratory, Univ. of Turku, Finland
Urpo, Seppo	Metsaehovi Radio Res. Station, Kymaelae, Finland
van Ballegoijen, Aad	Smithsonian Astrophys. Observ., Cambridge, USA
Vekstein, Gregory	University of St. Andrews, Scotland
Velli, Marco	Dipart. di Astronomia, Univ. di Firenze, Italy
Ventura, Rita	Osservatorio Astrofisico di Catania, Italy
Waljeski, Katrina	Naval Research Laboratory, Washington, USA
Wang, Yi-Ming	Naval Research Laboratory, Washington, USA
Widing, Kenneth	Naval Research Laboratory, Washington, USA
Wu, Shi Tsan	University of Alabama, Huntsville, USA
Yermolaev, Yuri	Space Research Institute, Moscow, Russia
Zhao, Xuepu	CSSA, Stanford University, USA

Foreword

The 2nd SOHO Workshop on "Mass Supply and Flows in the Solar Corona" was held in Marciana Mariana on the island of Elba, Italy, in the week September 27 to October 1, 1993, as part of a series of workshops planned by the Solar Corona and Particles Working Group of the SOHO Science Working Team (SWT). The purpose of this workshop series is to acquaint the solar community with the capabilities of SOHO, and prepare scientific projects and observing plans for the mission. This Workshop, which was attended by more than one hundred scientists from different countries, focused on the following topics:

1) Fine Scale Structures
2) Loops and Prominences
3) Coronal Streamers
4) Coronal Holes and Solar Wind

Each of these four topics was introduced by an observational and a theoretical overview highlighting the most recent advances in their area. A third review illustrated how SOHO might help in solving open problems. Oral papers and poster presentations were followed by Working Group sessions. On the last day of the Workshop the Group Leaders reported on the activities of their Working Group.

These proceedings include most of the papers presented at the Workshop, including the poster papers and Working Group reports by the Group Leaders as well as two overview papers of the SOHO mission. All papers have been refereed.

It is our pleasure to thank all those who contributed to the success of this Workshop. In particular, we appreciate the cooperation of the Group Leaders during the Workshop and the efforts of the referees who worked to a very tight schedule. Further, we would like to thank Mrs. Clare Bingham whose efficient secretarial support allowed the timely publication of this volume.

B. Fleck, G. Noci, and G. Poletto

SOHO Overview

SOHO AND THE SOLAR COMMUNITY

R.M. BONNET
ESA Science Directorate, 8-10 Rue Mario-Nikis, 75739 Paris Cedex 15, France

I am very pleased to be here, among friends on the beautiful island of Elba, to participate in the Second SOHO Workshop, even though the constraints of the SOHO preparation make it difficult sometimes to maintain as serene a relationship as one might wish.

SOHO will probably be the last major solar observatory before the end of this decade and will therefore have a central position in solar physics. It is the first quiet Sun space observatory since the OSOs and SKYLAB, and the next similar mission (high resolution observations, solar probe, etc.) will not be in space before the next decade is well advanced.

To give a little of the historical background of SOHO, it had its origins in the DISCO proposal, which was concerned with helioseismology. I myself was deeply involved in DISCO, but it was not selected by ESA for Phase A Study. Then came SOHO which had the same objectives as DISCO, as well as chromosphere and corona instruments. I was very surprised when I learned that some of the instruments on SOHO might be as large as telephone booths. I was also surprised when I saw the mock-up of SOHO at the Paris Air Show last June. However, it came as no surprise when I learned of the problems which some of you are currently facing, due to the complexity of the instruments. In spite of this, SOHO is not doing too badly and will be the next and most complete solar observatory in space.

Concerning future missions, the Solar Corona Probe was proposed in response to the Call for M3 Mission Proposals issued in November 1992. It was not selected for assessment study but, in view of the international interest already expressed by three other agencies through IACG (Inter-Agency Consultative Group for Space Science), it was recommended as a possible component of the post Horizon 2000 exercise.

The SIMURIS proposal (a Solar Inteferometric Mission for Ultrahigh Resolution Imaging and Spectroscopy) was not retained since it was considered to be too complex and too expensive.

I should also like to remind you of the Workshop on "Future Scientific Requirements for Future Solar-Physics Space Missions" organised by Per Maltby in January of this year, and which addressed the question of the planning of the next generation of instruments.

To predict the next step is a delicate and somewhat perilous exercise. Will it come from higher angular- and spectral-resolution observations? But

Space Science Reviews **70**, 3–5.

here there is obviously a physical as well as an instrumental limitation. Will it concern the X-ray or the ultraviolet, or the visible, or the far-infrared part of the spectrum? Will it come from more refined neutrino detectors? Or will it be from new instruments like those of Ulysses, SOHO, Yohkoh, HRTS or Coronas? Clearly, many hopes are riding on the capabilities of these missions/instruments and they will probably contribute new discoveries. Should we plan the next missions now? If we agree to start planning now, then the key question is: What to plan? I will merely enumerate a few criteria or necessary conditions for planning with some chance of success:

1. The science we address must be first class, and even outstanding, in order to win in the unavoidable competition with other disciplines.
2. We must prove that space is a unique asset for the mission(s) and that similar progress cannot be accomplished from the ground.
3. Considerations of prestige and of originality which may attract the public's attention should not be disregarded. They play an important role in the minds of the politicians who sit on our Committees.
4. Similarly, considerations of what technological benefits might be derived from the mission(s) are important and can also play an important role in raising the interest of industry or of the hardware institutes.

At ESA, we are starting the process now with our Space Science Advisory Committee. We have called on the science community to contribute new mission concepts, new ideas on which to elaborate the 'Horizon 2000 Plus Programme'. In the course of 1994, our advisory bodies and our ESTEC engineers will have to evaluate these missions and establish priorities, which will then form the basis for this future programme. It will also be the time when we should discuss potential cooperation with our usual partners: the Japanese, the Russians and the Americans. I encourage you all to respond - before 15 October - to this Call for Ideas.

Turning to today's Workshop, I believe we should have an open attitude towards the larger solar community. Therefore, the SOHO community has the responsibility to make the most out the of the SOHO investigations. It must attract the cooperation of the solar community in order to be able to extract as much science as possible from the SOHO data.

The aims of the Workshop series are:

- to summarise the experimental and theoretical status in the selected topic area;
- to acquaint the scientific co-investigators and potential guest investigators with the instrumental capabilities of SOHO;
- to obtain the widest possible scientific output for coordinating the SOHO observing programme.

The need for coordinated observations and cooperative data analysis is important. SOHO has been conceived with a complementary set of instruments, such that their measurements would produce a complementary set of data for the study of the phenomena in the solar atmosphere, in the solar wind and in the solar interior. Therefore, SOHO will only achieve its aims by coordinated operation of its instruments and cooperative analysis of the data.

It must not be forgotten that SOHO is part of the Solar-Terrestrial Science Programme and as such must attract not only the cooperation of the other space-based (Yohkoh, Coronas, etc.) and ground-based solar observers, but also exploit the fact that SOHO forms part of STSP because there is a large fraction of space science - plasma physics - that may benefit from a coordinated approach between "in situ" observers such as Cluster and remote and "in situ" observers as those found in SOHO. All of these are good reasons to make this series of Workshops extremely important. It is essential that they do not end up by being just good conferences where we hear again about the status of the field, but that they are followed by the development of coordinated observing sequences, concrete plans for observing campaigns, formation of teams for the development of missing theoretical work, development of appropriate data analysis tools, etc.

This being said, I would like to wish you a very successful Workshop.

THE SCIENTIFIC PAYLOAD OF THE SPACE-BASED SOLAR AND HELIOSPHERIC OBSERVATORY (SOHO)

V. DOMINGO AND B. FLECK
Space Science Department of ESA, ESTEC, 2200 AG Noordwijk, The Netherlands

and

A.I. POLAND
Laboratory for Astronomy and Solar Physics, NASA/Goddard Space Flight Center, Greenbelt, MD 20771 USA

Abstract. The space-based Solar and Heliospheric Observatory (SOHO) is a joint venture of ESA and NASA within the frame of the Solar Terrestrial Science Programme (STSP), the first "Cornerstone" of ESA's long-term programme "Space Science — Horizon 2000". The principal scientific objectives of the SOHO mission are: a) a better understanding of the structure and dynamics of the solar interior using techniques of helioseismology, and b) a better insight into the physical processes that form and heat the Sun's corona, maintain it and give rise to its acceleration into the solar wind. To achieve these goals, SOHO carries a payload consisting of 12 sets of complementary instruments which are briefly described here.

Key words: Solar Physics – Solar Interior – Solar Corona – Solar Wind – SOHO

1. Mission Overview

The SOHO spacecraft (Figure 1) is scheduled for launch on an Atlas II-AS in July 1995. After a transfer phase of ca. 4 months it will be injected into a halo orbit around the Sun-Earth L1 Lagrangian point, about 1.5×10^6 km sunward from the Earth, where it will be continuously pointing to Sun center with an accuracy of 10". Pointing stability will be better than 1" over 15 min time intervals. The halo orbit will have a period of 180 days and has been chosen because, 1) it provides a smooth Sun-spacecraft velocity change throughout the orbit, appropriate for helioseismology, 2) is permanently outside of the magnetosphere, appropriate for the "in situ" sampling of the solar wind and particles, and 3) allows permanent observation of the Sun, appropriate for all the investigations. The Sun-spacecraft velocity will be measured with an accuracy better than 0.5 cm/s. SOHO is being designed for a nominal operational lifetime of two years, but on-board consumables will be sized for up to six years.

Planning, coordination and operation of the scientific payload will be conducted from the Experiment Operations Facility (EOF) at NASA's Goddard Space Flight Center (GSFC). The SOHO payload produces a continuous science data stream of 40 kbits/s which will be increased by 160 kbits/s whenever the solar oscillations imager (SOI) is operated in its high rate mode. For telemetry the Deep Space Network (DSN) will be used. There are 3

Space Science Reviews **70**, 7–12.

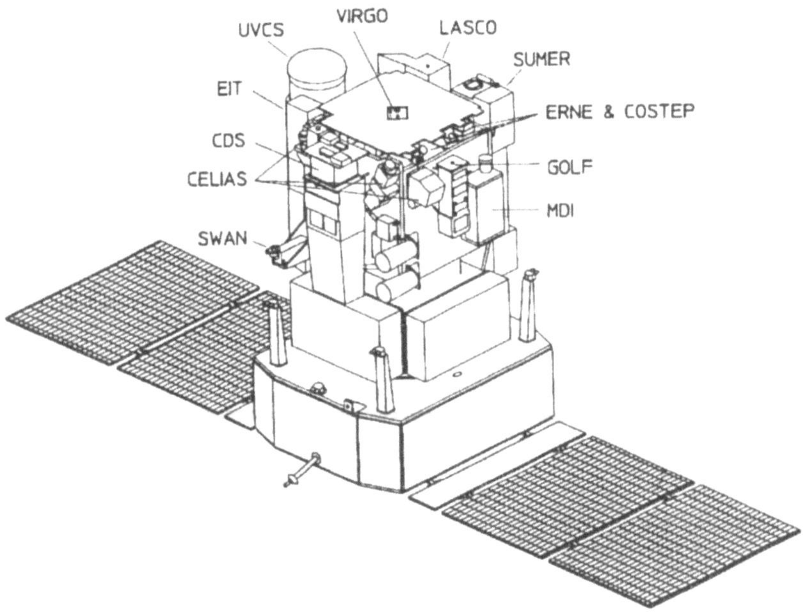

Fig. 1. SOHO spacecraft schematic view.

short (1.6 hours) and one long (8 hours) contacts. The short passes are used to transmit "play-back" data stored on-board during periods without DSN coverage. During the 8-hr passes SOHO will be operated in real-time in an observatory type mode.

2. The Scientific Payload

The scientific payload of SOHO comprises 12 sets of instruments which can be divided into three main groups, according to their area of research: helioseismology instruments, solar corona instruments, and solar wind "in-situ" instruments. Detailed instrument descriptions by the consortia involved can be found in the ESA volumes edited by Domingo and Guyenne (1989) or Mattok (1992).

2.1 Helioseismology instruments

GOLF (Global Oscillations at Low Frequencies), using a very stable sodium vapour resonance scattering spectrometer, aims to obtain observations of the global solar velocity field ($l \leq 3$) with a sensitivity of better than 1 mm/s over the whole frequency range from 0.1 μHz to 6 mHz (periods from 3 min to 100 days). It will also measure the long-term variations of the global average of the line-of-sight magnetic field with a precision of 1 mG. VIRGO (Variability

TABLE I
The SOHO Payload

Investigation	Measurements	Technique	Bit Rate (kb/s)
HELIOSEISMOLOGY			
GOLF	Global Sun velocity and oscillations (ℓ=0-3)	Na-vapour resonant scattering cell, Doppler shift and circular polarization	0.160
VIRGO	Low degree (ℓ=0-7) irradiance oscillations and solar constant	Global Sun and low resolution (12 pixels) imaging, active cavity radiometers	0.1
MDI/SOI	Velocity oscillations, harmonic degree up to 4500	Fourier tachometer, angular resolution: 4 and 1.5"	5 (+160)
SOLAR ATMOSPHERE REMOTE SENSING			
SUMER	Plasma flow characteristics (T, density, velocity) chrom. through corona	Normal incidence spectrometer, 50-160 nm, spectral res. 20000-40000, angular res. 1.5"	10.5 (+21)
CDS	Temperature and density: transition region and corona	Normal and grazing incidence spectrometers, 15-80 nm, spectr. res. 1000-10000, angular res.≈3"	12 (+22.5)
EIT	Evolution of chromospheric and coronal structures	Images (1024 x 1024 pixels in 42' x 42') in the lines of He I, Fe IX, Fe XII and Fe XV	1 (+26.2)
UVCS	Electron and ion temp. densities, velocities in corona (1.3-10 $R_\odot$)	Profiles and/or intensity of several EUV lines between 1.3 and 10 $R_\odot$	5
LASCO	Structures evolution, mass, momentum and energy trans. in corona (1.1-30 $R_\odot$)	1 internal and 2 externally occulted coronagraphs Spectrometer for 1.1-3 $R_\odot$	4.2 (+26.2)
SWAN	Solar wind mass flux anisotropies. Temporal variations	Scanning telescopes with H absorption cell for Ly-α light	0.2
SOLAR WIND 'IN SITU'			
CELIAS	Energy distribution and composition (mass, charge) of ions (0.1-1000 keV/e)	Electrostatic deflection, time-of-flight measurements, solid state detectors	1.5
COSTEP	Energy distribution of ions (p, He) 0.04-53 MeV/n and electrons 0.04-5 MeV	Solid state and plastic scintillator detector telescopes	1.01
ERNE	Energy distribution and isotopic composition of ions (p - Ni) 1.4-540 MeV/n and electrons 5-60 MeV	Solid state and scintillator crystal detector telescopes	1.01

of solar IRradiance and Gravity Oscillations) will perform high sensitivity observations of *p*- and *g*-mode solar intensity oscillations with a 3-channel sun-photometer measuring the spectral irradiance at 402, 500, and 862 nm, and with the 12 resolution elements Luminosity Oscillations Imager (LOI) ($l \leq 7$). The relative accuracy of these data will be better than 1 ppm (for 10 s integration time). VIRGO will also measure the solar constant with an absolute accuracy of better than 0.15% using two different types of absolute radiometers (PMO6 and CROM). Both GOLF and VIRGO lay particular emphasis on the very low frequency domain of low order *p*- and *g*-modes which penetrate deep into the solar core and which are difficult to observe from the Earth due to unavoidable day/night gaps in ground-based observing sequences and the disturbing influence of the earth's atmosphere. The Solar Oscillations Imager (SOI) focusses on the intermediate to very high degree *p*-modes. By sampling the Ni I 676.8 nm line with a wide-field tunable Michelson interferometer, SOI will provide high precision solar images (1024×1024 pixels) of the line-of-sight velocity, line intensity, continuum intensity, longitudinal magnetic field components, and limb position. It can be operated in a full disk mode (2" equivalent pixel size) to resolve modes in the range $3 \leq l \leq 1500$ or in a high-resolution mode (0.62" pixel size) to resolve modes as high as $l = 5000$. SOI will run 4 different observing programs: The "structure program" provides a continuous 5 kbits/s data stream of various spatial and temporal averages of the full disk velocity and intensity images. It runs at *all times*. The "dynamics program" operates during 60 consecutive days each year with continuous high rate telemetry (+ 160 kbits/s). Each minute, a full disk velocity image and either a full disk intensity image or a high resolution velocity image will be transmitted. The "campaign program" will be conducted during 8 consecutive hours each day when the high rate telemetry is available. This is a very flexible operations mode to perform a variety of more narrowly focussed scientific investigations (e.g. studies of meso- and supergranulation, active region seismology, etc.). Finally, the "magnetic field program" will provide several real-time magnetograms per day for planning purposes and correlative studies.

2.2 Coronal instruments

The solar atmosphere remote sensing investigations are carried out with a set of telescopes and spectrometers that will produce the data necessary to study the dynamic phenomena that take place in the solar atmosphere at and above the chromosphere. The plasma will be studied by spectroscopic measurements and high resolution images at different levels of the solar atmosphere. Plasma diagnostics obtained with these instruments will provide temperature, density, and velocity measurements of the material in the outer solar atmosphere.

SUMER (Solar Ultraviolet Measurements of Emitted Radiation) is an

UV telescope equipped with a normal incidence spectrometer to study plasma flows, temperatures, densities, and wave motions in the upper chromosphere, transition region and lower corona with high spatial (1") and high time resolution (typically 10 s) by measuring line profiles and intensities of UV lines in the range from 500 to 1600 Å. The spectral coverage varies between 20 and 44 Å with a spectral resolving power of $\lambda/\Delta\lambda = 18.800 - 40.000$. With SUMER, it should be possible to measure velocity fields in the transition region and corona down to 1 km/s. At shorter wavelengths (150 to 800 Å) CDS (Coronal Diagnostic Spectrometer), a Walter II grazing incidence telescope equipped with both a normal incidence and a grazing incidence spectrometer, will measure absolute and relative intensities of selected EUV lines to determine temperatures and densities of various coronal structures. The spatial resolution of CDS is about 3", the spectral resolution varies between 2000 and 10000. EIT (Extreme-ultraviolet Imaging Telescope) will obtain full sun high resolution EUV images in 4 emission lines (Fe IX 171 Å, Fe XII 195 Å, Fe XV 284 Å, and He II 304 Å) corresponding to 4 different temperature regimes. The wavelength separation is achieved by multilayer reflecting coatings deposited on the four quadrants of the telescope mirrors and a rotatable mask to select the quadrant illuminated by the Sun. A 1024×1024 CCD camera with an effective pixel size of 2.6" is used as detector. UVCS (UltraViolet Coronagraph Spectrometer) is an occulted telescope equipped with high resolution spectrometers to perform spectroscopic observations of the solar corona out to 10 solar radii to locate and characterize the coronal source regions of the solar wind, to identify and understand the dominant physical processes that accelerate the solar wind, and to understand how the coronal plasma is heated. One of the gratings is optimized for line profile measurements of Ly-α, another one for line intensity measurements in the range 944 to 1070 Å. LASCO (Large Angle and Spectrometric COronagraph) is a triple coronagraph having nested, concentric annular fields of view with progressively larger included angles. The fields of view of the three coronagraphs C1, C2 and C3 are 1.1-3, 1.5-6 and 3-30 solar radii, respectively. C1 will not only be the first spaceborne "mirror coronagraph" in the visible, but it will also be the first spaceborne coronagraph with spectroscopic capabilities for visible radiation. It is equipped with a Fabry-Perot interferometer to perform spectroscopic measurements with a spectral resolution of $\approx$ 700 mÅ in the lines Fe XIV 5303 Å, Fe X 6374 Å, Ca XV 5964 Å, Na D_2, and Hα. SWAN (Solar Wind ANisotropies) will measure the latitude distribution of the solar wind mass flux from the equator to pole by mapping the emissivity of the interplanetary Ly-α light.

2.3 Solar wind "in-situ" instruments

The instruments to measure "in situ" the composition of the solar wind and energetic particles will determine the elemental and isotopic abundances, the ionic charge states and velocity distributions of ions originating in the solar atmosphere. The energy ranges covered will allow the study of the processes of ion acceleration and fractionation under the various conditions that cause their acceleration from the "slow" solar wind through solar flares.

CELIAS (Charge, ELement and Isotope Analysis System) consists of three mass- and charge-discriminating sensors based on the time-of-flight technique, making use of electrostatic deflection, post-acceleration and residual energy measurements. It will measure the mass, ionic charge and energy of the low and high speed solar wind, of suprathermal ions, and of low energy flare particles. It also carries SEM (Solar Extreme-ultraviolet Monitor), a very stable photodiode spectrometer which will continuously measure the full disk solar flux in the He II 304 Å line as well as the absolute integral flux between 170 and 700 Å. To study the energy release and particle acceleration processes in the solar atmosphere as well as particle propagation in the interplanetary medium, COSTEP (COmprehensive SupraThermal and Energetic Particle analyser) will measure energy spectra of electrons (up to 5 MeV), protons and He nuclei (up to 53 MeV/nuc). ERNE (Energetic and Relativistic Nuclei and Electron experiment), having the same scientific objectives as COSTEP, will measure energy spectra of elements in the range Z=1–30 (up to 540 MeV/nuc), abundance ratios of isotopes as well as the anisotropy of the particle flux.

3. Coordinated Research

The SOHO payload has been conceived as an integrated package of complementary instruments. Therefore, to achieve the scientific aims of the mission, it is essential to operate the instruments in a coordinated programme and analyse the data in a correlative and cooperative spirit. Further, to reach an adequate and sound understanding of the dynamics of the solar corona and its mass and energy balance, it is essential to know the magnetic field topology and the velocity field (both flows and waves) in the photosphere and chromosphere with the highest spatial and temporal resolution achievable. At present, these data can only be supplied by ground-based observatories. Therefore, to have a successful mission, it is vital to have a close cooperation and coordination between SOHO and ground-based solar observatories (both optical and radio).

References

Domingo, V., Guyenne, T.D. (eds.): 1989, The SOHO mission — scientific and technical aspects of the instruments, ESA SP-1104

Mattok, C. (prep.): 1992, Coronal streamers, coronal loops, and coronal and solar wind composition, ESA SP-348

SOHO OPERATIONS AND GROUND SYSTEM

A.I. POLAND
Laboratory for Astronomy and Solar Physics, NASA/Goddard Space Flight Center, Greenbelt, MD 20771 USA

and

V. DOMINGO AND B. FLECK
Space Science Department of ESA, ESTEC, 2200 AG Noordwijk, The Netherlands

Abstract. SOHO is a joint ESA/NASA mission to study the sun from its interior to, and including, the solar wind in interplanetary space. It is currently scheduled for launch in 1995. After launch SOHO will be operated from the Experiment Operations Facility (EOF) at Goddard Space Flight Center (GSFC). The EOF will consist of facilities for instrument commanding, data reception, data reduction and data analysis. In this paper the operations concepts including instrument ground commanding from the EOF and communications capabilities between the EOF and ground observatories and the public networks in general will be described.

Key words: Sun – Corona – SOHO

1. Science Operations Overview

The diagram in Figure 1 shows the basic data connections that will be present for the SOHO science operations. The SOHO Experimenter Operations Facility (EOF), to be located at NASA's Goddard Space Flight Center, will serve as the focal point for mission science planning and instrument operations. At the EOF, experiment PI representatives will receive real-time and playback flight telemetry data, process these data to determine instrument commands, and send commands to their instruments, both in near real-time and on a delayed execution basis. They will be able to perform data reduction and analysis, and have capabilities for data storage. To accomplish these ends, the appropriate experiment teams will use workstations (IWS) that will be connected to an EOF Local Area Network (LAN). Additional workstations and X-terminals will be used to support the Project Scientists and for SOHO planning and operations support staff in the EOF. There will be connections from the EOF to external facilities to allow transfer of incoming data from GSFC support elements, remote investigator institutes, other solar observatories, and ESA facilities. There will also be connections for the EOF to interact with the SOHO Mission Operations Control Center (SMOCC) and other required elements at GSFC for scheduling and commanding the SOHO flight experiments. Short term and long term data storage will be either within the EOF or at an external facility with electronic communication access from the EOF.

The Deep Space Network (DSN) will receive S/C telemetry during three

Space Science Reviews **70**, 13–18.

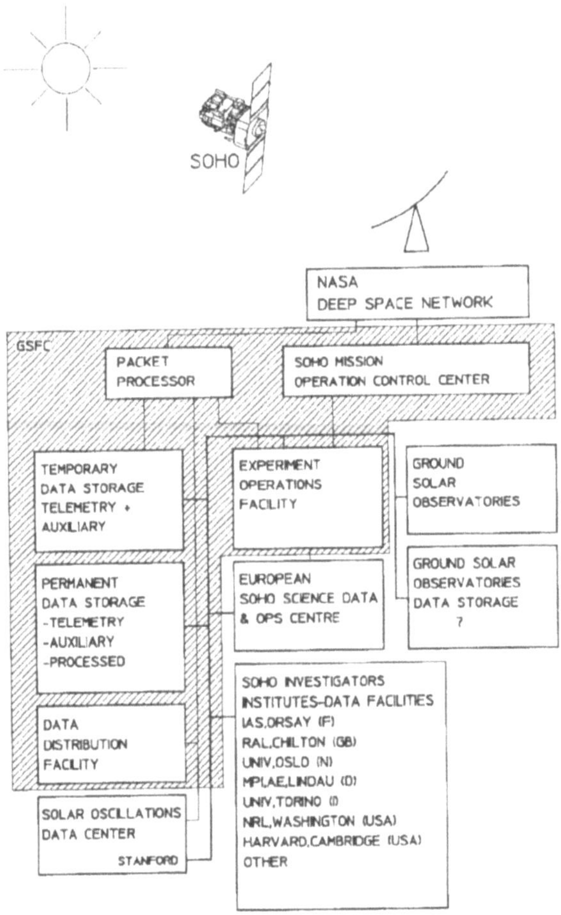

Fig. 1. SOHO ground system: basic functions for science operations

short (1.6 hrs) and one long (8 hrs) station pass per day. Science data acquisition during non-station pass periods will be stored on-board and played back during the short station passes. The MDI high data rate stream will be transmitted only during the long station pass. For 2 consecutive months per year continuous data transmission, including MDI high data rate, will be supported by DSN. Whenever there is data transmission, the basic science data (40 kbits/s) will be available in near real-time at the EOF. From the EOF the SOHO investigators will control the operation of the instruments via near-real time commands. These will be verified and sent immediately to the spacecraft.

Some SOHO instruments (CEPAC, CELIAS, VIRGO, GOLF) will generally operate automatically and will not need near real-time operational control except for surveillance of housekeeping data. Other instruments, those of the coronal imaging investigators, will be operated interactively every day in real or near real-time.

The EOF has the following functions:

- Provides the means with which the PI teams participating in the SOHO programme can monitor and control their instruments on-board the spacecraft.
- Is the center where the solar atmosphere investigators of SOHO will coordinate and plan the near real-time operation of their instruments, and will be the focal point on the one hand, for the overall SOHO science operations planning, and on the other, for coordinating science studies through the organisation of campaigns and workshops.
- Provides electronic interfaces with the appropriate data bases and networks to support the operation activities and to provide the necessary input from ground stations and other spacecraft data for the planning of the SOHO science operations.
- Provides data storage for science, engineering and housekeeping data; common data (attitude, orbit and spacecraft housekeeping) are also stored there. Cataloging capabilities are also available.

2. SOHO Operations Policy and General Plan

The Science Working Team (SWT) will set the overall science policy and direction for mission operations, set priorities, resolve conflicts and disputes, and consider Guest Investigator observing proposals. During SOHO science operations, the SWT will meet every three months to consider the quarter year starting in one month's time and form a general scientific plan. If any non-standard DSN contacts are required, the requests must be formulated at this quarterly meeting. The three-month plan will then be refined during the monthly planning meetings of the Science Operations Team (SOT), composed of those PI's or their team members at the EOF, which will allocate observing sessions to specific programs. At weekly meetings of the SOT, coordinated timelines will be produced for the instruments, together with detailed plans for spacecraft operations. Daily meetings of the SOT will optimize fine pointing targets in response to solar conditions and adjust operations if DSN anomalies occur.

While the Project Scientist will be responsible for the implementation of the scientific operations plan, execution of the plan will be carried out by the SOT. On a rotating basis, one of the PI's or their representatives at the EOF will serve as the Science Operations Leader (SOL), who will be responsible for the successful execution of the weekly plan. The SOL will serve for approximately 10 days, starting with the weekly planning meeting and continuing through the week of operations. The SOL will chair the weekly planning meeting and the daily meetings during the following week.

To provide operational continuity over the course of the SOHO mission, and from one SOL's tenure to the next, a Science Operations Coordinator (SOC), who is not a member of any of the PI teams, will support the SOL and SOT.

2.1 Monthly Planning Cycle

On a monthly time scale the instrument teams will meet to assess progress in achieving the scientific goals of their investigation and to discuss the objectives for operations starting in a month's time. This gives time for coordinated observations to be set up, arrangements for Guest Investigators to be made and any deficiencies in observing sequences to be identified.

Approximately 2 weeks later a SOHO wide meeting will discuss instrument health, solar activity and consider the operations for the month under consideration. SOL's will be appointed for each week and they will be responsible during the month for identifying any conflicts between the planned operations and the DSN schedule as they become available.

Inputs to the monthly meeting are made by each instrument team and common objectives are identified. The output of this meeting is a schedule showing when each instrument will be operating, whether joint or individual observations are being made, the types of solar features being observed, ground observatory support and a backup plan if these conditions are not met. Requirements for telemetry rate switching should be identified together with any spacecraft operations which may affect the observations, for example momentum dumping and station keeping. Conflicts between instruments for resources are resolved and disturbances identified.

2.2 Weekly Planning Cycle

A weekly meeting considers the week starting in approximately three days time and this is when the detailed plans for all the SOHO instruments are synchronised. It will be convened by the SOL designated to lead that week. The intention is to lay out a definitive plan with timings, flag status, disturbances etc. so that the daily meetings only consider deviations from the weekly plan. This meeting will have the conflict-free DSN schedule available. The weekly meeting will also be the forum for instrument teams to give advance notice of any special operations or changes to the plan for future weeks. The DSN forecast schedule will be available for the week commencing in 10 days time and the strawman proposal will be available for the week following that.

2.3 Daily Optimisation Meeting

The daily meeting convenes to hear about the state of the Sun, discusses fine pointing targets and whether any changes are necessary in view of yesterday's operations.

2.4 Coordinated Campaigns

Within SOHO: During agreed periods one or several experiment teams and, if agreed, teams from other spacecraft or at ground observatories will run, in collaboration, observation campaigns to address specific topics. The periods of two-months continuous near real-time observation will probably be the most convenient for campaigns that require continuous observation during more than 8 hours, or that require coordination with ground observations only feasible from particular observatories around the world.

With ground-based observatories: If the ground-based observatory is one that has electronic links that allow near real-time imaging transmission to and from the EOF, the coordination will be no different than if the ground based observatory were one of the SOHO experiments. If no real-time data transmission is needed or possible, the coordinated operation will be an agreed time of simultaneous observations. Generally speaking the coordinated observations with ground observatories will need a longer time lead in their planning to insure availability of the facility and coincidence of the SOHO real-time coverage.

3. Data

NASA will forward real-time science data to the EOF over the LAN to the PI workstations. Playback data will be sent to the EOF in the same manner with transmission delays from DSN (approximately 3 hours) and processing delays to turn the data around (approximately 2 hours).

3.1 Summary Data

The summary data will be used both to plan observations at the EOF and to provide an overview of the observations that have been obtained from SOHO. Table 1 provides a list of the proposed contributions from each PI team. The summary data will consist of three classes of data. The first two classes will consist of a representative image from each of the imaging experiments, and key parameters from the non-imaging experiments. The third class will be a list of observing programs and start/stop times of data sequences. Together these data will provide a synopsis of solar conditions and the science programs that have been carried out by the observatory.

The summary data will be the responsibility of the PI teams, and are to be generated as quickly as possible after receipt of data at the EOF. The data will be generated from quick-look science data and will have only preliminary calibrations performed on them. The summary data will therefore NOT be citable. An event log will be maintained in the summary data file . However, the contents of the log at the EOF will be allowed to change. The event log will provide a registry of events that may be of general interest. The log

TABLE I
Summary Data

MDI	Full disk magnetograms + full disk continuum images
EIT	Full disk images: Fe IX, Fe XII, Fe XV, and He II
UVCS	1.2-10 $R_\odot$ Ly α intensity and coronal temperature
LASCO	1.1-30 $R_\odot$ corona white light, 1.1-3 $R_\odot$ Fe XIV, Fe X, Ca XV
GOLF	Global magnetic field (6-hour mean)
VIRGO	Solar constant each day
SWAN	Heliospheric Ly α
CELIAS	Solar wind speed, heavy ion flux (5 min. averages)
CEPAC	Particle flux every 5 min
SUMER	Operation modes, time, area coverage
CDS	Operation modes, time, area coverage

may also include events identified by observatories other than SOHO, but which might be relevant to the SOHO observations. Each of the instruments will have an observation program summary data file even if they have only a few entries. The observation program should have both a planned and an executed data file. The EOF will hold an on-line copy of the Summary data and event time files.

4. Summary

The EOF is designed to allow for the operation of SOHO as a major observing facility.

The connections to the spacecraft instruments will be such that commands can be sent to the instruments for immediate execution. One will be able to decide where they want to point on the sun, what sequence of operations is desired and send those commands to the instruments.

The data, when received on the ground, will be immediately forwarded to the EOF where scientists will be able to view the current data from: their own experiment, other SOHO experiments, and certain ground based observatories. The ability to receive and view real time data will greatly facilitate our ability to optimize the scientific return from the experiments. The electronic connections will also greatly improve our ability to do joint research with ground based observatories.

The summary data to be provided by the SOHO PIs will greatly enhance the capability for joint research with scientists throughout the community. These data will be available electronically on a daily basis.

I Small Scale Structures

SOLAR MASS FLOW IN FINE-SCALE STRUCTURES

K. P. DERE
E. O. Hulburt Center for Space Research
Naval Research Laboratory
Washington DC 20375-5352, USA

Abstract. Observations of transient and steady velocities at chromospheric, transition region and coronal temperatures in the quiet Sun and coronal holes are reviewed. The relevance of fine-scale structures in governing the mass balance of the solar atmosphere is stressed. At present, a coherent picture of these mass flows does not exist. However, the current observational base of transition region and coronal velocity information is limited but should greatly improve with measurements from the SOHO satellite.

Key words: Sun – flows – velocities

1. Introduction

A major goal of the SOHO program will be to examine the transport of plasma through the solar atmosphere. HRTS spectra of Si IV formed in the transition region at 7 x 10^4 K, shown in Figure 1, graphically demonstrate the necessity of using high resolution data to understand the various flow patterns present on the sun. High velocity explosive events are denoted by the arrows but considerable variation on very small scales is also seen in the more typical profiles.

In the context of the SOHO mission, it is useful to try to understand the flows at various temperatures relative to the measured outflow in the solar wind. Table 1 presents estimates of the physical parameters needed to calculate the mass transfer throught typical structures into the solar wind. With these it is possible to predict the steady-state velocities required to provide a mass flux equal to that in the solar wind. The net mass flow F out of or into various structures is estimated by specifying the density n, the velocity v, the filling factor f, and the surface area A of the volume through which the material passes: F = nvfA. The filling factor can be important in estimating the area when the emitting plasma does not completely fill the volume of a resolution element of the observing system. The values included in the table have been used with little regard for the fine-scale nature of many of the flows and it is assumed that the solar wind originates in the polar regions. For the chromosphere we have used parameters characteristic of spicules which cover roughly 0.6% of the solar surface. At these temperatures, a spicular flow of 2 km s^{-1} is needed to supply the mass flux requirements of the solar wind. Consequently the 25 km s^{-1} flows in spicules are more than

Space Science Reviews **70**, 21–30.

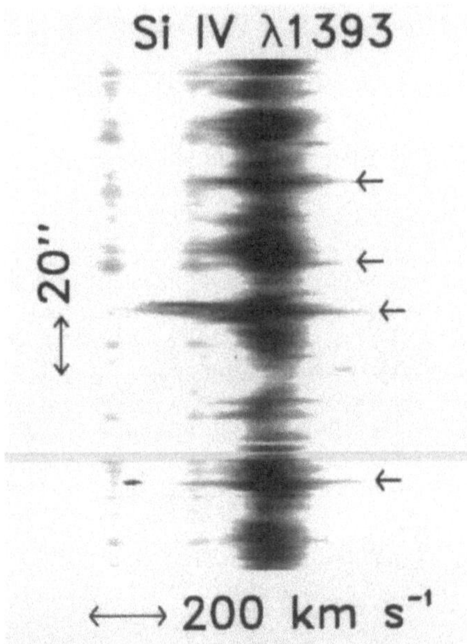

Fig. 1. HRTS profiles of Si IV in the quiet Sun

sufficient. The small fill factor in the transition region comes from the fact that features that appear to be resolved must consist of numerous subresolution structures that are sparsely distributed throughout the volume (Dere *et al.*, 1987). The mass flux requirements for the transition region lead to velocities of 100 km s^{-1} and this presents a real problem. Velocities of this order are observed but they are associated with transient events and do not seem to carry sufficient mass. Observations of doppler shifts of coronal lines in coronal holes (Rottman *et al.*, 1982) indicate velocities of the order required.

For transient eruptions and ejections, the net mass transfer from some solar region is evaluated with the expression $F = (\int_{\Delta V} n dV)$ B A. The integral is the total number of particles involved in each ejection, B is the birthrate (cm^{-2} s^{-1}) and A is the area (cm^2) of interest. Note that the velocity does not explicitly enter the equation.

1.1 Filling Factors

The filling factor can be a major unknown in determining the mass flux. It is the ratio of volume actually filled with emitting plasma compared to the total observed volume of the structure. Usually the filling factor is determined by comparing the intensities of lines produced with different sensitivities to the electron density in the source volume. For example, densities from a number of density sensitive line intensity pairs were compared with densities derived from emission measure-volume analyses

TABLE I
Solar Wind Mass Flux Parameters

	Chromosphere (10^4 K)	Transition Region (10^5 K)	Corona (10^6 K)	Solar Wind (1 AU)
$4\pi R^2$ (cm^2)	6.1 x 10^{22}	6.1 x 10^{22}	6.1 x 10^{22}	2.8 x 10^{27}
Area (A cm^2)	0.18 $A_\odot$ [a]	0.18 $A_\odot$	0.18$A_\odot$	"
Density (n, cm^{-3})	6 x 10^{10} [b]	6 x 10^9 [c]	2 x 10^8 [d]	7 [e,f]
Fill factor (f)	0.006 [b]	0.001 [g]	1. [h]	1.
nvfA (s^{-1})				9 x 10^{35}
Velocity (v, km s^{-1})	2	100	4	500 [e,f]

References: a, (Bohlin, 1977); b,(Beckers, 1972); c, (Doschek *et al.*, 1978); d, (van de Hulst, 1950); e, (Feldman *et al.*, 1977); f, (Schwenn, 1983); g, (Dere *et al.*, 1987); h, (Dere, 1982a).

using data from the NRL Skylab EUV spectroheliograph (Dere, 1982a; Dere, 1982b). These showed that the filling factor for active region coronal loop structures was unity. However, the filling factor rapidly decreased toward lower temperatures. At 10^5 K, filling factors are on the order of 0.001 (Dere *et al.*, 1987).

The filling factor is not an absolute quantity but is dependent on the resolution of the instrument. For example, the filling factor for M2 class flare loops in the 1-6 x 10^6 K regime was found to be unity for the NRL Skylab EUV spectroheliograms with 2" resolution (Dere *et al.*, 1979). An M class flare observed by with SMM was found to have a fill factor of between 0.003-0.03 (Wolfson *et al.*, 1983). A large part of the discrepancy between the two values can be attributed to the 14" spatial resolution of the SMM X-ray polychromator (XRP).

2. Mass Flow in the Quiet Sun

A summary of the average flow velocity from 10^4 to 10^6 K is shown in Figure 2 (Hassler *et al.*, 1991; Brekke, 1993). It should be noted that the flows are all downflows. At present, a coherent description of the flow pattern of the quiet Sun does not exist. For example, one might expect some of the average properties of coronal flows to conform to the equation of mass conservation for steady-flows along a flux tube: n v A = constant. The variation expected for constant cross-section A and constant pressure is shown in Figure 2. This clearly does not predict the observed dependence at temperatures above 10^5 K. Simple modifications, such as including a variation

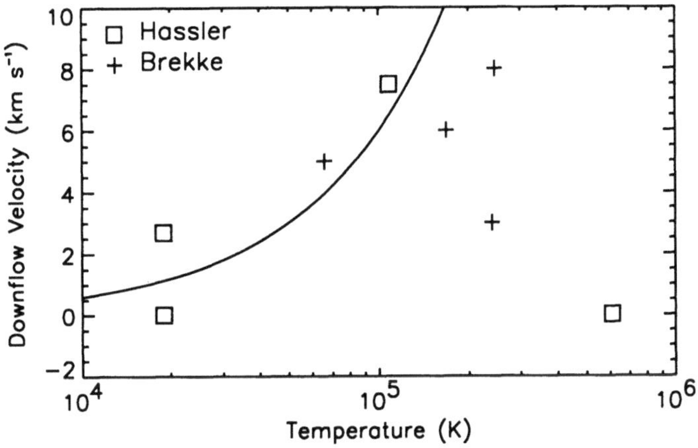

Fig. 2. Average Quiet Sun Downflow *versus* Temperature

in the loop cross-section, which are not observed (Klimchuk *et al.*, 1992), cannot reconcile the observed velocities with such a scaling. There is also the question of why the ultraviolet velocities at 10^4 K do not reflect the higher velocities observed in Hα spicules.

2.1 Chromospheric Flows: Spicules

The basic review article on the properties of spicules remains that of Beckers (1972). When viewed at the limb, spicules tend to move at a velocity of around 25 km s^{-1} with 2/3 of them moving outward and 1/3 falling back to the Sun. When Hα spectra of the quiet solar disk are analyzed in terms of a cloud model, velocities less than 6 km s^{-1} are typically found with perhaps a small asymmetry in favor of downflows (Grossman-Doerth and v. Uexkull 1971, 1973).

Suematsu *et al.* (1993) have recently studied the apparent motions of Hα spicules on the disk. They find that the great majority (76/96) go up and then back down. If interpreted as a ballistic motion, the initial velocity is of the order of 40 km s^{-1}. The extension velocity is about 18 km s^{-1}. Although measurements were not made such that a detailed profile could be reconstructed, the profile information was consistent with the apparent velocities observed. Suematsu *et al.* also note that spicules, which originate in the bright Hα and Ca network that lies at the base of coronal loops, tend to be strongly displaced from the vertical, indicating that spicules and coronal loops may lie on different field lines. Thus coronal loop and spicular flows are probably unrelated and spicular flows should be interpreted as a circulation of material.

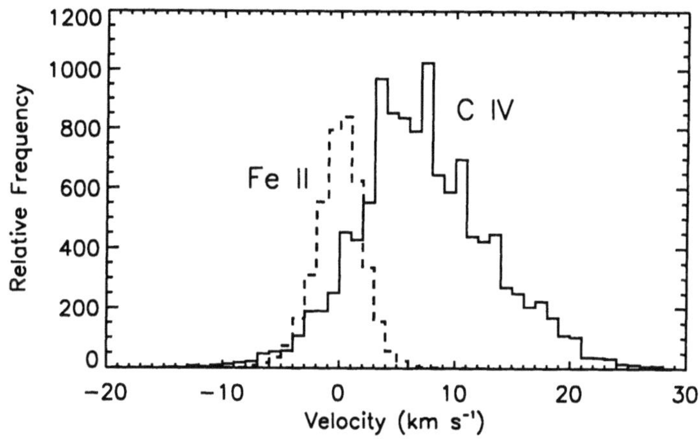

Fig. 3. Histograms of Quiet Sun Velocities

The HRTS instrument has recorded spectra of a number of chromospheric lines at a dispersion and spatial resolution that should reveal doppler signatures of spicular flows. For example, spectroheliograms of a narrow strip of the quiet sun were reconstructed from rastered spectra obtained during the third HRTS rocket flight (HRTS-3). These showed that ultraviolet lines of Fe II (λ1563, for example) were produced in structures quite similar to those seen in Hα-0.5Å. In network structures, there was a clear variation of temperature with height. C I λ1561 was clearly formed at lower temperatures than Fe II and C IV at considerably higher temperatures. However, there were no signatures of velocities near 25 km s^{-1} in the Fe II λ1563 profiles. Histograms of C IV and Fe II velocities in the quiet Sun from the third HRTS rocket flight (Dere *et al.*, 1984) are shown in Figure 3. When referenced to the narrow Si I lines in this region of the spectrum, the average C IV velocity is a downflow of 6 km s^{-1} with a standard deviation of 6 km s^{-1}. The average Fe II velocity is 0.0 km s^{-1} with a standard deviation of 2.1 km s^{-1}.

The only evidence for spicule-like 25 km s^{-1} flows in ultraviolet chromospheric lines appeared most strongly in lines of C I λ1560-1561. Very small regions of relatively high red and blue shifts (10-20 km s^{-1}) are seen in these lines and occasionally in other chromospheric lines at reduced contrast. The C I lines are emission lines with enough optical depth to cause a considerable self-reversal in the profiles. C I flow velocities can not be directly inferred from the profile. These features have been referred to as chromospheric jets. They occur mainly in the network cell-centers and have lifetimes on the order of 60s (Dere *et al.*, 1982). They do not appear to be related to the cell-center UV continuum bright points (Cook, 1991).

A study of the infrequent blue-shifted C IV profiles found two cases of spicule-like phenomena (Dere *et al.*, 1986b). C IV outflows of 10-20 km s^{-1} were found at the ends of two large Hα spicules with a commensurate apparent motion of the elongated C IV structure. However, many more of these should have been detected if 20 km s^{-1} motions are a common property of spicules and if they extend to transition region temperatures.

2.2 Explosive events

Explosive events, characterized by velocities of 100 km $s^{-1}$and sizes of 2", are dramatic examples of fine-scale eruptive structures. They are commonly observed with the HRTS in most regions of the Sun and are probably signatures of magnetic reconnection. From the third HRTS rocket flight (HRTS-3), we have a database of about 80 explosive events with which their characteristic physical parameters have been derived (Dere *et al.* 1986). The HRTS-3 observing sequence repeatedly rastered an area 10" wide with 6 rasters steps over a 200s interval. The birthrate B for the explosive events was determined to be 4 x 10^{-21} cm^{-2} s^{-1} in a coronal hole and 1 x 10^{-20} cm^{-2} s^{-1} in the quiet sun. Data from the HRTS-6 flight (Dere *et al.* 1991) found that the birthrates inside and outside a coronal hole were equal but the actual value could not be determined. In order to evaluate the mass or number content of an explosive event from observational material, the most direct approach would be to use density-sensitive line ratios to determine the density and images to measure the volume. The latter procedure is straightforward but measurements of density-sensitive line intensity ratios are extremely rare. So far, we have been able to determine the density from O IV λ1400 line ratios for only a single event. A major difficulty is the intrinsically low intensity of the O IV intercombination lines. An alternative is to use the emission measure ($\int n_e^2 dV$) which can be directly evaluated from a single line intensity and the observed volume. The total number content is then calculated from the expression $\int_{\Delta V} n\, dV = (\int n_e^2\, dV)^{1/2}(\mathrm{f}\ \mathrm{V}_{obs})^{1/2}$ where f is the volumetric fill factor and V_{obs} is the observed volume. A fill factor of unity produces the largest possible value for the mass or particle content of an explosive event. Dere *et al.* (1989) derived an average explosive event mass of 5 x 10^8 g or 3 x 10^{32} protons, using f=1.

A single explosive event observed during Spacelab-2 was bright enough, roughly 10 times brighter than the brightest events observed in the HRTS-3 data, to be detected in the wings of the O IV lines. The derived density was 7 x 10^{10} cm^{-3} (Dere *et al.* 1991). From a further analysis of these profiles, the blue wing of the C IV was found to have an emission measure of 3 x 10^{43} cm^{-3} and a volume of 2 x 10^{25} cm^3. The volumetric filling factor is then 3 x 10^{-4} in this single, very bright event. If we apply this value of the filling factor to the set of HRTS-3 explosive events, then we find that the explosive events that occur inside coronal holes contribute a net particle flux of 2 x 10^{32} particles s^{-1}, or 2.6 x 10^{-4} of the solar wind particle flux.

2.3 Coronal jets

Coronal jets were discovered in spectra taken during the first two HRTS rocket flights (Brueckner and Bartoe, 1983). They have sizes near 10" and velocities up to 400 km s^{-1}. In one case, 4 sequential accelerations at 5 km s^{-2} to 400 km s^{-1} were observed at the same site during a period of about 300 s. They occurred in a region of the quiet sun that was unremarkable in Hα. Since these observations were performed with a fixed slit pointing, only a small portion of the solar surface was seen and it appeared that these large velocity jets should be a rather common phenomenon when extrapolated to the full solar disk. Subsequent HRTS observation have included observing programs which raster larger fractions of the solar surface and indicate that the coronal jets are much rarer than first expected.

2.4 Small-scale ejections

Hermans and Martin (1986) have observed small-scale eruptive structures which appear to be the analog of large eruptive filaments. Their typical length is about 15" and ranges from 5 to 54". They are often associated with the cancellation of magnetic flux. Hermans and Martin estimate that about 600 occur across the solar surface every 24 hours which works out to a birthrate of 1.1 x 10^{-25} cm^{-2} s^{-1}, 4 orders of magnitude below the birthrate for explosive events.

2.5 The transition region intensity and velocity field

The most natural interpretation for transition regions flows would be in terms of mass conserving flows along magnetic flux tubes. Unfortunately, the general redshift of transition region flows makes this interpretation difficult, if not impossible. It is not even possible to very the proposition that *'what goes up must come down'*. Most of the profiles with a net blueshift consist of a normal gaussian at a typical 0-12 km s^{-1} redshift together with a highly blueshifted, nongaussian component. The outflows appear to be largely transient while the redshift patterns are often long lasting with accelerations of only a few km s^{-1} over 200s in discrete emission elements. Gebbie *et al.* (1981) have shown that the general pattern of network flows can be followed over a period of 3 hours. There is no statistical relationship between transition region velocity and intensity although a positive correlation is apparent in some areas (Dere *et al.*, 1984). When averaged over large areas, peaks and valleys in the intensity image correspond with similar features in the velocity image, but only in a local, relative way (Athay and Dere, 1989).

3. Coronal holes

3.1 Transition region flows

In the past, measurements of transition region line profiles in coronal holes on the disk have yielded contradictory results (Doschek *et al.*, 1976; Rottman *et al.*, 1981; Rottman *et al.*, 1982). The *Skylab* spectra analyzed by Doschek *et al.* (1976) show a general downflow in the coronal hole at all temperatures below 2 x 10^5 K in one case and essentially no flow in another case. Brekke (1993) has noted that the wavelength for the O V λ1218 line used by Doschek *et al.* (1976) is probably in error such that their O V velocities should be shifted to the red by 6 km s^{-1}. The rocket spectra of Rottman *et al.* (1981,1982) show relative outflow velocities of up to 7 km s^{-1} for O V λ625 in coronal holes on the disk. Brekke found a net downflow for O V of 8 km s^{-1} and 3 km s^{-1} for λ1371 and λ1218, respectively, in the quiet Sun with the 8 km s^{-1} value being the more reliable. This would suggest that the doppler shifts measured by Rottman *et al.* are indicative of near-zero velocities in the coronal holes, as in one case reported by Doschek *et al.*.

HRTS spectra of C IV of a disk coronal hole were taken as part of a large scale raster sequence. The spectral region included narrow chromospheric lines of Si I which were used to establish a nearly absolute wavelength scale, which is important in light of the typical redshift of transition region profiles. The C IV profiles showed a net redshift (downflow) in the coronal hole but with 26% of the profiles having a blueshift, as compared with 7% for the quiet Sun. Recently, the positions of the largest sources of the coronal hole outflows have been compared with the NSO Kitt Peak photospheric magnetogram of the area. These data are shown in Figure 4. The background grayscale pattern represents the magnetogram values and the large black line indicates the outline of the coronal hole determined from the He I λ10830 spectroheliogram. Positions of explosive events are shown by crosses. The thin black lines marks the location of the more coherent C IV outflows. It can be seen that most of these correspond to the regions of relative strong photospheric field that define the network. The C IV outflow velocities in these regions are on the order of 5-10 km s^{-1}. The intensity images do not outline any recognizable feature to associate with these flows.

3.2 Flows at Coronal Temperatures

There are only a few measurements of coronal line profiles in coronal holes and these tend to show a plasma outflow (Cushman and Rense, 1976; Rottman *et al.*, 1982). Cushman and Rense (1976) recorded line profiles of Si XI λ303, Mg IX λ368 and Mg X λ610 and determined line of sight velocities on the order of 13 km s^{-1} which translated to a radial outflow velocity of 16 km s^{-1}. Rottman *et al.* (1982) report an outflow of 12 km s^{-1} in Mg X λ625 in a disk coronal hole.

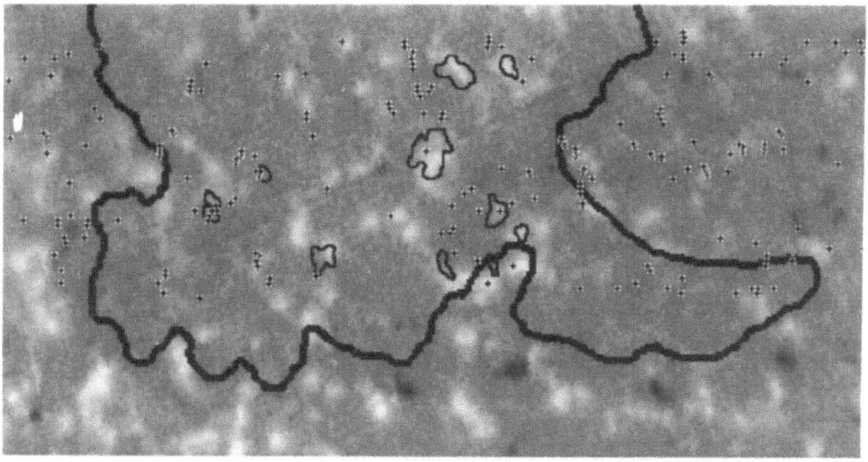

Fig. 4. C IV outflows in a coronal hole

3.3 MACROSPICULES

Macrospicules are candidates for a nonsteady source of the mass flux in coronal holes. They were described by Waldmeier in 1955 but interest in their properties was renewed by their rediscovery in the *Skylab* data (Bohlin *et al.*, 1975; Withbroe *et al.*, 1976). More recently, they have been observed during *Spacelab-2* in C IV (Dere *et al.*, 1989) and at 6 cm with the VLA (Habbal and Gonzalez, 1991). As their name implies, macrospicules are seen as spike-like structures that move outward into the corona at velocities of 10-150 km s^{-1}. Once they reach maximum altitude, they either continue to follow a ballistic trajectory, simply fade in place, or break up into knots or condensations.

4. Discussion

The observations required for building a comprehensive and detailed picture of the mass flow in the solar atmosphere are not yet complete. The SOHO instruments will help by observing relatively large fields of view for long periods of time. These must be coupled with further efforts to achieve a better theoretical understanding of the processes that drive steady flows and eruptive events.

References

Athay, R. G. and Dere, K. P.: 1981, *Astrophys. J.* **346**, 514.
Beckers, J. M., 1972, *Ann. Rev. Astr. Astrophys.*, **10**, 73.
Bohlin, J. D., 1977, in *Coronal Holes and High Speed Wind Streams*, J. B. Zirker (ed.), Colorado Associated University Press, 27.

Bohlin, J. D., Vogel, S. N., Purcell, J. D., Sheeley, N. R., Tousey, R. and VanHoosier, M. E.: 1975, *Astrophys. J.* **197**, L133.

Brekke, P.: 1993, *Astrophys. J.*, **408**, 735.

Brueckner, G.E.: 1981, in *Solar Active Regions*, F. Q. Orrall, ed., Colorado Associated University Press, Boulder.

Brueckner, G.E., Bartoe, J.-D.F.: 1983, *Astrophys. J.*, **272**, 947.

Cook, J. W.: 1991, preprint.

Cushman, G. W. and Rense, W. A.: 1976, *Astrophys. J.* **207**, L61.

Dere, K. P., 1982, *Solar Phys.*, **75**, 189.

Dere, K. P., 1982, *Solar Phys.*, **77**, 77.

Dere, K. P., Bartoe, J.-D. F. and Brueckner, G. E.: 1983, *Astrophys. J. Lett.*, **267**, L65.

Dere, K. P., Bartoe, J.-D. F. and Brueckner, G. E.: 1984, *Astrophys. J.* **281**, 870.

Dere, K. P., Bartoe, J.-D. F. and Brueckner, G. E.: 1986, *Astrophys. J.* **305**, 947.

Dere, K. P., Bartoe, J.-D. F. and Brueckner, G. E.: 1986, *Astrophys. J.* **310**, 456.

Dere, K. P., Bartoe, J.-D. F. and Brueckner, G. E.: 1986, *Solar Phys.* **123**, 41.

Dere, K. P., Bartoe, J.-D. F., Brueckner, G. E. and Recely, F.: 1986, *Astrophys. J. Lett.* **310**, L95.

Dere, K. P., Bartoe, J.-D. F., Brueckner, G. E., Cook, J. W. and Socker, D. G.: 1987, *Solar Phys.* **114**, 223.

Dere, K. P., Bartoe, J.-D. F., Brueckner, G. E., Cook, J. W. and Socker, D. G.: 1989, *Solar Phys.* **119**, 55.

Dere, K. P., Bartoe, J.-D. F., Brueckner, G. E., Ewing, J. and Lund, P.: 1991, *J. Geophys. Res.* **96**, 9399.

Dere, K. P., Mason, H. E., Widing, K. G. and Bhatia, A. K.: 1979, *Astrophys. J. Suppl. Ser.* **40**, 341.

Doschek, G. A, Feldman, U. and Bohlin, J. D.: 1976, *Astrophys. J.* **205**, L17.

Doschek, G. A, Feldman, U., Bhatia, A. K., and Mason, H. E.: 1978, *Astrophys. J.* **226**, 1129.

Feldman, W. C., Asbridge, J. R., Bame, S. J., and Gosling, J. T.: 1977, in *The Solar Output and It's Variations*, O. R. White (ed.), Colorado Associated University Press, Boulder, 351.

Gebbie, K. B., *et al.*: 1981, *Astrophys. J.* **251**, L115.

Grossman-Doerth, U. and v. Uexkull, M.: 1971, *Solar Phys.* **20**, 31.

Grossman-Doerth, U. and v. Uexkull, M.: 1973, *Solar Phys.* **28**, 319.

Habbal, S. R. and Gonzalez, R. D.: 1991, *Astrophys. J.* **326**, L25.

Hassler, D. M., Rottman, G. J. and Orrall, F. Q.: 1991, *Astrophys. J* **372**, 710.

Hermans, L. M., and Martin, S. F.: 1986, in *Coronal and Prominence Plasmas*, A. I. Poland (ed.), NASA Conf. Pub. 2442, 369.

van de Hulst, H. C.: 1950, *Bull. Astron. Inst. Neth.* **410**, 135.

Klimchuk, J. A., Lemen, J. R., Feldman, U., Tsuneta, S. and Uchida, Y.: 1992, *Pub. Astr. Soc. Japan*, **44**, L181.

Rottman, G. J., Orrall, F. Q. and Klimchuk, J. A.: 1981, *Astrophys. J.* **247**, L135.

Rottman, G. J., Orrall, F. Q. and Klimchuk, J. A.: 1982, *Astrophys. J.* **260**, 326.

Schwenn, R., 1983, in it Solar Wind Five, M. Neugebauer (ed.), NASA Conf. Pub. 2280, 489.

Shibata, K., *et al.*, 1992, *Pub. Astr. Soc. Japan*, **44**, L173.

Suematsu, Y., Wang, H. and Zirin, H.: 1993, *Astrophys. J.*, submitted.

Waldmeier, M.: 1955, *Ergebnisse und Probleme der Sonnenforschung*, Akademische Verlagsgellschaft, Leipzig.

Withbroe, G. L., Jaffee, D. T., Foukal, P. V., Huber, M. C. E., Noyes, R., W., Reeves, E. M., Schmahl, E., J., Timothy, J. G., and Vernazza, J. E.: 1976, *Astrophys. J.*, **203**, 528.

Wolfson, C. J., Doyle, J. G., Leibacher, J. W. and Phillips, K. H. J.: *Astrophys. J.* **269**, 319.

MAGNETIC FINE STRUCTURES IN CORONAL LOOPS

A. A. VAN BALLEGOOIJEN
Harvard-Smithsonian Center for Astrophysics,
60 Garden Street, Cambridge, MA 02138, U.S.A.

Abstract. The formation of magnetic fine structures and associated electric currents is considered in the context of the coronal heating problem. The penetration of field-aligned electric currents into the lower atmosphere is discussed. It is argued that currents strong enough to heat the corona can persist only for short periods of time. The formation of thin current sheets is discussed. It is argued that photospheric magnetic structures (flux tubes) play an important role in the generation of coronal currents.

Key words: Sun – Corona

1. Introduction

One of the main goals of SOHO is to understand the physical processes responsible for the heating of the solar corona. Magnetic fields are known to play a crucial role in this process, but the details of the energy transport and dissipation are still unclear. Many theories of coronal heating have been proposed (for recent reviews see Zirker 1993 and references therein). Here I focus on one possible heating mechanism: the dissipation of DC electric currents. This mechanism is relevant only to closed magnetic structures such as coronal loops.

2. Field-Aligned Currents

Electric current are generated in the corona as a result of fluid motions in the convection zone. Due to the high electrical conductivity of the solar plasma, convective flows interacting with the magnetic field at the two "ends" of a coronal loop cause twisting and braiding of the coronal field lines. If the time scale of the subsurface flows is long compared to the time for an Alfven wave to travel back and forth along the loop, the magnetic field will evolve quasi-statically through a series of equilibrium states. The magnetic pressure in coronal loops is generally much larger than the gas pressure, hence these equilibria are nearly *force free* with electric currents flowing nearly parallel (or anti-parallel) to the magnetic field lines.

Random motions at the loop footpoints naturally lead to a configuration in which some field lines have current flowing parallel to the magnetic field and other field lines have current flowing anti-parallel. If the transverse length scale of the motions is less than the width of the coronal loop, the net current integrated over the loop cross-section vanishes. One coronal loop may contain a large number of counter-flowing current channels. The spa-

Space Science Reviews **70**, 31–35.

tial distribution of these currents will depend on the nature of the footpoint motions and on the dissipation mechanism.

An estimate of the typical size of the current channels can be obtained as follows. The average heating rate for resistive dissipation is given by $f\eta J^2$, where η is the resistivity, f is the filling factor of the electric currents and J is the current density. The heating rate must balance the observed radiative and conductive losses; for active region loops this rate is $\sim 10^7$ erg cm^{-2} s^{-1}. The classical Spitzer resistivity of the coronal plasma is small, hence the required current density J is very large ($\sim 10^7$ [esu]). The change in the magnetic field across a coronal current layer must be less than the typical field strength ($\sim$ 100 G). This implies that the width of the current layers must be very small ($\sim$ 100 m).

One might suggest that this conclusion could be avoided if the resistivity is much larger than its classical value. However, most known processes for enhancement of the resistivity involve plasma microinstabilities (such as ion-acoustic and ion-cyclotron modes) which depend on the drift speed of the electrons relative to the protons. For typical coronal particle densities the threshold current density is $\sim 10^7$ [esu], similar to that required to heat the corona classically. Therefore, anomalous resistivity does not reduce the required electric current density, but rather reduces the required filling factor f.

What happens at the ends of the coronal loop where the field-aligned coronal currents flow into (or out of) the lower atmosphere? The answer to this question depends on the life time of the electric currents. I first consider the case that the currents persist for a time long compared to the time for an Alfven wave to travel from the corona down into the convection zone ($\sim$ 100 seconds). Then inertial effects are unimportant, so that the currents can cross the magnetic field lines only in the convection zone where the dynamical forces of convection are able to balance Lorentz forces ($\mathbf{j} \times \mathbf{B} \neq \mathbf{0}$). In the intermediate layers of the photosphere and chromosphere the currents must be field-aligned. The classical resistivity of the photosphere and chromosphere is much larger than that of the corona. Therefore, unless the resistivity of the corona is somehow enhanced relative to its classical value, most of the dissipation in the current circuit actually occurs in the lower atmosphere and does not contribute to coronal heating. Van Ballegooijen (1990b) showed that the coronal resistivity must be enhanced by a factor of at least 40, and to dissipate more than 50 % of the available energy in the corona requires an enhancement factor $\sim 10^5$. The dissipation time of such currents is at most a few seconds, which is inconsistent with the assumed long lifetime of the currents. Therefore, it does not seem possible to heat the corona by dissipation of strong, long-lasting currents.

A more plausible scenario is that strong (10^7 [esu]) currents persist in the corona only for a short periods of time. This scenario is consistent with the "nanoflare" model proposed by Parker (1988). According to Parker's model, the convective motions below the solar surface produce magnetic fine structures in the corona which are dissipated as a result of MHD instabilities ("nanoflares"). The direct effect of the footpoint motions is to produce electric currents which run all the way from the convection zone into the corona. I suggest that the footpoint motions produce current densities less than $\sim 10^4$ [esu], too small to cause significant heating in the corona or below. However, when an MHD instability occurs, the current is locally enhanced to about 10^7 [esu] for a short period of time. This causes rapid dissipation and release of the associated magnetic energy. In the "nanoflare" model the time scale for the instability is only a few seconds, so that the enhancement of the coronal current does not have time to propagate down into the lower atmosphere. Hence, the nanoflare model avoids the problems associated with resistive losses of in the lower atmosphere.

3. Formation of Current Sheets

A key question is how magnetic fine structures in coronal loops can be produced. Parker (1972) proposed that small-scale currents form naturally as a consequence of random footpoint motions. He considered an idealized coronal loop model consisting of an initially uniform field between two parallel plates. These plates represent the solar photosphere at the two ends of a coronal loop. The plasma in between the two plates is assumed to be perfectly conducting and the field is assumed to evolve through a series a equilibrium states. The question is: how does the field respond to slow, random motions of the plasma at the boundary plates? Parker (1972, 1983) argued that spatially continuous flow fields at the boundaries generally produce tangential discontinuities (TDs), i.e., current sheets with infinitesimal thickness. The argument is based on the idea that the field contains multiple magnetic flux systems; TDs form at the separatrix surfaces between those flux systems.

A different point of view was presented by Van Ballegooijen (1985, 1988a), who argued that multiple flux systems and TDs arise in Parker's model only if the imposed velocity fields at the boundary plates are themselves discontinuous. If the footpoint motions are incompressible and spatially continuous, the magnetic field remains continuous at later times but the field becomes increasingly tangled and there is a *cascade of magnetic free energy* to smaller and smaller spatial scales (van Ballegooijen 1985, 1986). The cascade is driven by the random intermixing of footpoints at the boundary plates and is not due to magnetic instability. The time scale of the cascade is determined by the time scale of the boundary motions. The theory predicts an exponential increase of the r.m.s. electric current density and a corresponding exponential decrease of the typical width of the current layers. Numerical simulations with simple, incompressible flow patterns have been performed by Mikic, Schnack and Van Hoven (1988) and Van Ballegooijen (1988b, 1990a). The results of the simulations agree with the theoretical predictions.

The cascade model can explain how the required high current densities are formed, but predicted values for the heating rates in active region loops are still about an order of magnitude too small (Van Ballegooijen 1986). In my opinion this disagreement should not be taken too seriously because the theoretical model includes a number of simplifying assumptions which make it difficult to accurately predict the coronal heating rate. For example, the curvature of the coronal loops is neglected in the present models. Perhaps more importantly, the model assumes incompressible boundary motions, whereas on the Sun the horizontal velocity in the photosphere has divergent flow in granulation cell centers and convergent flow in the intergranular lanes. This has the effect of concentrating the magnetic flux in the lanes and vertices between cells. The vorticity in these regions may be significantly enhanced (e.g., Karpen et al. 1993). Indeed, the photospheric magnetic field is highly intermittent and is made up of a collection of magnetic flux tubes. This may have a drastic effect on the coronal currents that are generated as a result of footpoint motions, and may increase the rate at which magnetic energy is transfered to smaller spatial scales. Therefore, more realistic models of the photospheric flow patterns and their effect on the coronal magnetic field are needed before a meaningful comparison between theory and observation can be made.

One approach is to assume that the regions in between the photospheric flux tubes are completely field free and that the flux tubes retain their identity for a long period of time (Parker 1983). In this case the flux tubes are topologically distinct from each other, and each coronal field line has two well-defined flux tubes associated with it (one at each end of the coronal loop). The set of field lines connecting the same pair of flux tubes constitute a well-defined magnetic flux system, and a coronal loop may contain a large number of such flux systems. Note that one photospheric flux tube may contain several flux systems. When the flux tubes are shuffled around in the photosphere, TDs form "spontaneously" at the separatrix surfaces between flux systems.

If the regions in between the photospheric flux tubes are not completely field free, the flux tubes are no longer topologically distinct from each other and it is no longer possible to sharply define different flux systems. A description of the coronal loop as a single magnetic flux system then seems more appropriate (Van Ballegooijen 1988a). Assuming that the velocity field at the photosphere is continuous, TDs in the magnetic field do not arise. However, very narrow current layers will develop on those field lines which are connected (on one end of the loop or the other) to a weak-field region in between photospheric flux tubes. The width of these current layers is proportional to the strength of the residual field in between the flux tubes. In the limit of zero residual field, these current layers become TDs.

The magnetic flux concentrations on the Sun continually slit up and merge with each other, so that a particular flux tube has only a finite life time. In fact, the life time of the flux tubes may be of the same order as the life time of granulation cells ($\sim$ 600 s). What effect does this have on the distribution of electric currents? If the regions in between the flux tubes are completely fields free, it is again possible to define different flux systems but the number of flux systems increases with time due to flux tube fragmentation. The typical distance between current sheets will therefore be less than in the case without fragmentation. More realistic models of photospheric field structures and flow patterns are required to determine what effect this may have on the structure of coronal currents.

References

Karpen, J.T., Antiochos, S.K., Dahlburg, R.B. and Spicer, D.: 1993, *Astrophys. J.* **403**, 769

Mikic, Z., Schnack, D.D. and Van Hoven, G.: 1989, *Astrophys. J.* **338**, 1148

Parker, E.N.: 1972, *Astrophys. J.* **174**, 499

Parker, E.N.: 1983, *Astrophys. J.* **264**, 642

Parker, E.N.: 1988, *Astrophys. J.* **330**, 474

Van Ballegooijen, A.A.: 1985, *Astrophys. J.* **298**, 421

Van Ballegooijen, A.A.: 1986, *Astrophys. J.* **311**, 1001

Van Ballegooijen, A.A.: 1988a, *Geophys. Astrophys. Fluid Dyn.* **41**, 181

Van Ballegooijen, A.A.: 1988b, 'Magnetic Fine Structure of Solar Coronal Loops' in R.C. Altrock, ed(s)., *Solar and Stellar Coronal Structure and Dynamics*, Sunspot, NM: National Solar Observatory, 115

Van Ballegooijen, A.A.: 1990a, 'Structure and equilibrium of coronal magnetic fields' in E.R. Priest and V. Krishan, ed(s)., *Basic Plasma Processes on the Sun*, IAU Symposium No. 142, 303

Van Ballegooijen, A.A.: 1990b, 'Magnetic heating of stellar chromospheres and coronae' in G. Wallerstein, ed(s)., *Cool Stars, Stellar Systems and the Sun*, Astron. Soc. Pacific Conf. Series, Vol. 9, 15

Zirker, J.B.: 1993, *Solar Phys.* **148**, 43

SMALL SCALE STRUCTURES IN THE SOLAR CORONA

S. R. HABBAL
Harvard-Smithsonian Center for Astrophysics, 60 Garden St, Cambridge, MA 02138, USA

Abstract. The observational characteristics of the small scale magnetic structures are summarized. The temperature structure and temporal variability of the emission from coronal bright points, that pervade the source region of the solar wind in coronal holes and the quiet sun, and from active regions are shown to be remarkably similar. Particular emphasis is given to observations, potentially feasible with SOHO, that could resolve some of the outstanding issues regarding the role of the small scale magnetic structures in the energy balance and properties of the solar wind.

Key words: Solar Corona -- Solar Wind -- Small Scale Structures

1. Introduction

The intermittent nature of the coronal heating process was first recognized from observations of the small scale magnetic structures, typically 10"-40" in spatial extent, also known as coronal bright points (e.g. Sheeley and Golub 1979, Habbal and Withbroe 1981, Porter et al 1987). This manifestation has recently led to the concept of heating the corona by microflares (e.g. Parker 1988, 1991). On the other hand, the filamentary and inhomogeneous nature of the coronal plasma, whether in active regions, quiet regions or coronal holes, (e.g. Cheng et al 1979, Koutchmy et al 1988, Golub et al 1990, Karovska et al 1994), as uncovered from observations with improved spatial resolution and image enhancement techniques, has been largely overlooked in explorations of coronal heating processes.

Through the analysis of multiwavelength observations of the solar corona, we discuss, in this review, the temperature structure of the coronal emission, its relationship to the underlying photospheric magnetic field, and the characteristics of its temporal variability. We show how some of the observed signatures of coronal heating in active regions, quiet regions and coronal holes are strikingly similar, and can be traced back to the small scale magnetic field. We conclude with recommendations for investigations that can be carried out with SOHO.

2. Temperature Structure of the Coronal Emission

In what follows, coronal emission will refer to emission at heights above the photosphere where plasmas at 10^6 K are observed. Shown in Figure 1 is a quiet Sun region observed simultaneously in Fe XIV 5303 Å and Fe X 6374 Å (Koutchmy et al 1988), which have a peak temperature of formation of

Space Science Reviews **70**, 37–46.

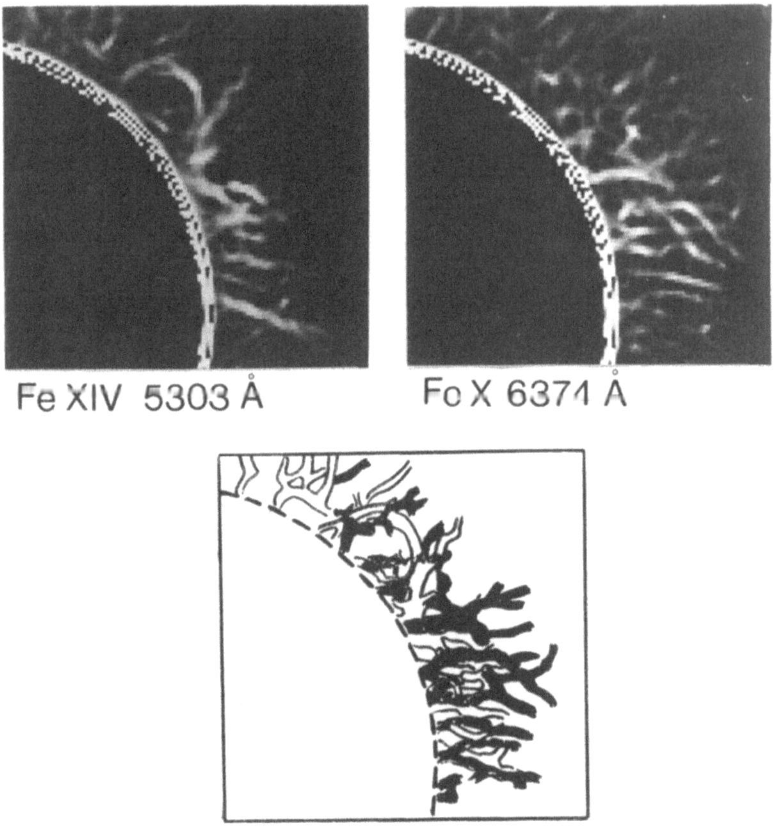

Fig. 1. Eclipse observations of a quiet Sun region made in the Fe XIV 5303 Å and Fe X 6374 Å lines (Courtesy S. Koutchmy). Image processing has been applied to these data (see Koutchmy et al 1988). An overlay of contours of the emission in these two spectral lines is shown in the lower panel.

1.8×10^6 and 10^6 K respectively. The image enhancement technique applied to these data exposes the filamentary nature of the emission. In addition, it shows very clearly how the emission from these different temperature lines is not cospatial, as evidenced by the overlay in the bottom panel of Figure 1. Other examples of the filamentary nature and coexistence of plasmas at different temperatures are given in Figures 2 and 3.

Figure 2 shows two consecutive scans, 5.5 mins apart, of active regions observed simultaneously in Si XII 521 Å and Ne VII 465 Å within the same 5'×5' field of view, with the Harvard EUV experiment on *Skylab* (see Reeves et al 1977). The spatial resolution is 5". These lines have a peak temperature of formation of 1.8×10^6 and 5×10^5 K respectively. Since these simultaneous observations were made through the same instrumental slit,

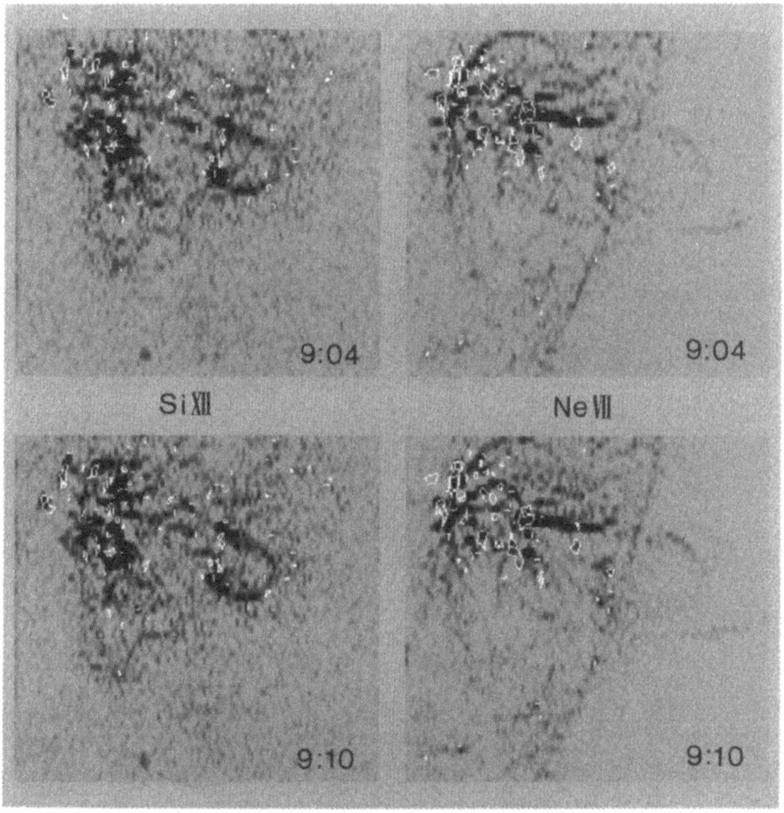

Fig. 2. Grey scale images of active regions observed simultaneously in Si XII 521 Å and Ne VII 465 Å on the disk and near the limb on 1973 November 23, in two consecutive scans, 5.5 min apart. The images have been enhanced using the technique proposed by Koutchmy et al (1988). The white contours indicate the areas of significant variability in the emission. The contour levels correspond to 5 to 20 % of the maximum value of δI (see eq. [1]) for Ne VII, and 5 to 37 % for Si XII. (From Arndt et al 1994)

there is no ambiguity about the spatial alignment at these two wavelengths, and the lack of cospatiality of the loops seen at these distinct temperatures is real.

Another example of the coexistence of multi-temperature plasmas is given in Figure 3 for a coronal hole observed in a south polar region with the Harvard EUV experiment on *Skylab*. Polar plumes detected in the Mg X 625 Å emission, with a peak temperature at 1.1×10^6 K, together with macrospicules, detected in C III 977 Å at 9×10^4 K, define the filamentary and inhomogeneous structure of the coronal hole at distances extending beyond 10" above the solar limb. Here too the distinct temperature structures are not cospatial.

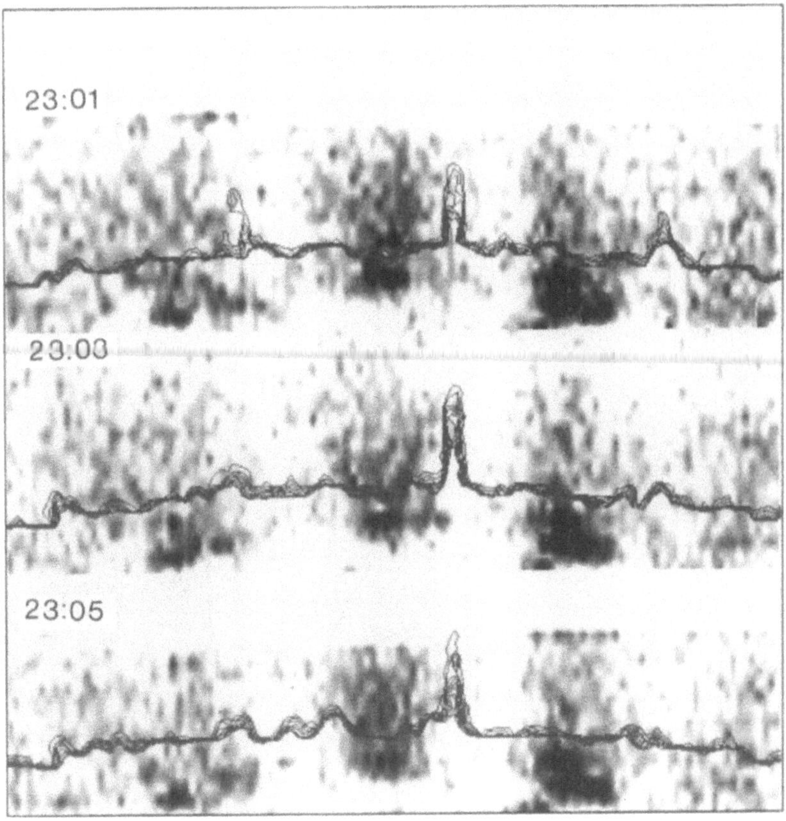

Fig. 3. Polar plumes (grey scale background) and macrospicules (contours) as observed simultaneously in Mg X 625 Å and C III 977 Å with the *Skylab*/EUV spectrometer, off the limb in a south polar coronal hole, on 1973 December 11. The time sequence corresponds to scans made 2 mins apart.

The last example is that of bright points observed with the Harvard EUV experiment on *Skylab* at disk center in a coronal hole (Figure 4). Readily identifiable against the reduced background coronal hole Mg X emission, they are indistinguishable from the bright network sources visible, for example, in C III. Bright points lie at the intersection of network cells as seen in the C III emission (see also Egamberdiev 1983). The coexistence of loop-like structures at different temperatures within the coronal bright points, in a fashion similar to that of active regions, can be seen from the time series shown in Figures 4b and 4c, of a coronal hole and a quiet region bright point. At present, coronal bright points are the best manifestation of the small scale structuring of the magnetic field that we can detect.

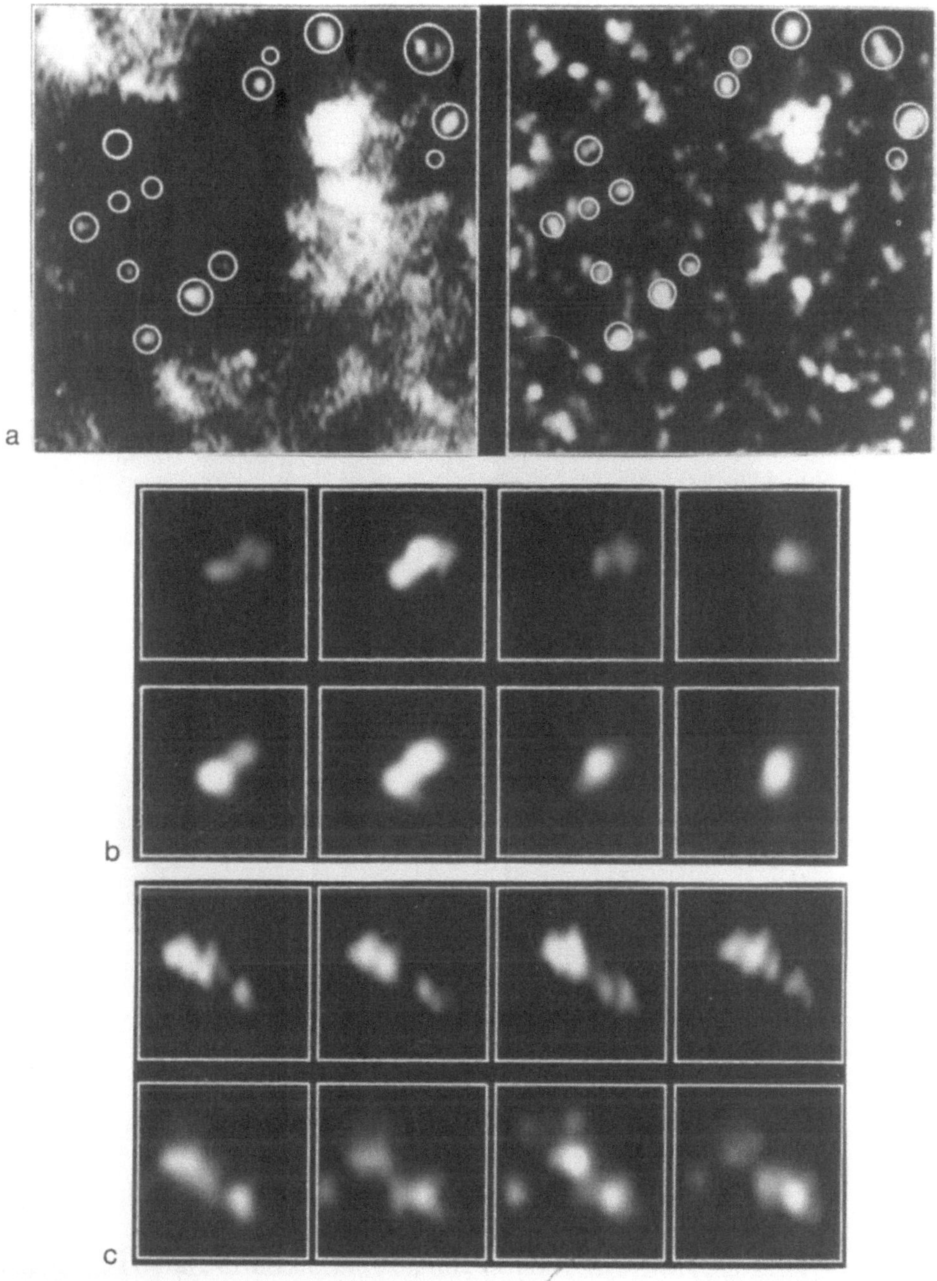

Fig. 4. (a) EUV spectroheliograms of a 5'x5' area of a coronal hole at sun center, taken simultaneously in Mg X 625 Å (left) and C III 977 Å (right) with the *Skylab*/EUV spectroheliometer on 1973 August 21. The bright points, as identified in Mg X, are circled. (From Habbal and Grace 1991) (b) Time evolution of a coronal hole bright point and (c) of a quiet region bright point. In both (b) and (c) the top row represents Mg X, and the bottom C III. The frames are 5.5 min apart, and 30" in width. (From Habbal et al 1990)

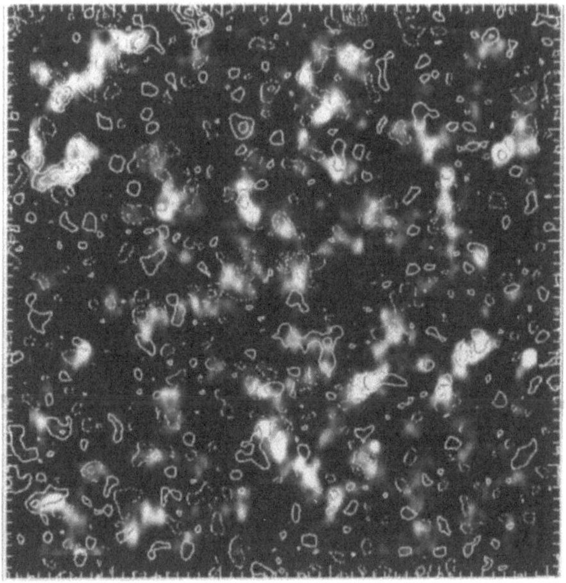

Fig. 5. Overlay of O VI 1032 Å emission in a quiet Sun region with contours of the line of sight photospheric magnetic field from a coregistered Kitt Peak magnetogram taken within 15-20 minutes of the O VI scan. Dashed and solid contours indicate opposite polarity magnetic field regions. (Courtesy J. Dowdy)

3. Relationship with the Photospheric Magnetic Field

To study the correspondence between the small scale structure and the underlying magnetic field we consider simultaneous observations of the line of sight photospheric magnetic field with coronal emission. Shown in Figure 5 is a quiet Sun region, observed in O VI 1032 Å on 1973 October 2 with the *Skylab*/EUV spectroheliometer, and a coregistered magnetogram taken within 15-20 minutes by the Kitt Peak magnetograph (see Dowdy 1993). This figure illustrates very clearly how the enhanced O VI emission is often straddling magnetic regions of opposite polarity. Although bipolar magnetic fields pervade the solar surface, not every bipole is associated with detectable emission. We also note that the emission is localized in the network boundaries as defined by the magnetic field.

We consider next the evolution of the magnetic field in time. The example of Figure 6 shows the temporal changes in the underlying line of sight photospheric magnetic field and the 20 cm radio emission characteristic of coronal emission which were measured simultaneously (see Habbal and Harvey 1988). The panels represent three selected times, a few hours apart. The magnetogram forms the background greyscale, and the radio emission, when present, is shown by contours of increasing brightness temperature.

Fig. 6. Time sequence of magnetograms and 20 cm radio emission shown by the white contours, taken simultaneously on 1985 September 8. The polarity of the magnetic field is defined by the outermost contour, being either white for positive, or black for negative. As the intensity level of the magnetic flux increases with each polarity, subsequent alternating white and black contours are used. (From Habbal and Harvey 1988)

The radio emission appears predominantly in regions of cancelling flux, i.e. where opposite polarity magnetic flux elements merge, and the flux decreases subsequently as a function of time. This can be deduced from a visual inspection of the magnetograms at the different times. In particular, we note the peak of the radio emission at 20:55 UT (upper right hand corner) which occurred when the distance between approaching magnetic elements of opposite polarity was reduced to less than 5". (Compare, for example, the location of the opposite polarities in this area at 17:40 and 21:00 UT).

4. Temporal Variability of the Coronal Emission

The time sequences shown in Figures 4b and 4c illustrate how bright points, whether in coronal holes or quiet regions, consist of small scale magnetic loops, each of which is intermittently heated more or less independently of the others. This example also shows how the temporal variability of the emission is not correlated at different temperatures, as evidenced by the often independent changes in C III and Mg X. Another example of the temporal variability of the emission in coronal holes is that shown in Figure 3. The time changes of the emission from the macrospicules is visually more dramatic than those of the polar plumes. Although anchored in bright points, which are varying in time, the Mg X emission in the plumes is too weak to show a significant variability on times scales of minutes. Yet, it is evident that these two distinct temperature structures evolve differently in time.

To study the characteristics of the variable emission in a more quantitative manner, we use a method developed by Habbal and Grace (1991) and apply it to the different *Skylab* observations presented earlier. The change in the intensity in any of the pixels making up a given scan is considered significantly variable if the absolute value of the difference in intensity in a given pixel, at two consecutive times, is greater than 2 σ, i.e. if

$$\delta I = |I_{t_1} - I_{t_2}| > 2\sigma \quad (1)$$

where $\sigma = \sqrt{\bar{I}}$, I_{t_1} and I_{t_2} are the intensities within each pixel of the images, taken at times t_1 and t_2, and $\bar{I} = \frac{I_{t_1}+I_{t_2}}{2}$. The spatial density of variability, $\frac{\delta n}{N}$, is then defined as the ratio of the number of variable pixels satisfying condition (1) to the total number of pixels N in a given area.

Applying this technique to a coronal hole and a quiet Sun region, we find that the variable emission is preferentially localized in coronal bright points (Habbal and Grace 1991), while in active regions it is predominantly localized around the footpoints of loops (see Figure 2). Whether in bright points, or in active regions, the variable emission occurs in different places at distinct temperatures, and does not always coincide with the regions of most intense emission.

Figure 7 summarizes these results in a more quantitative manner. The temperature dependence of the spatial density of the variability, $\frac{\delta n}{N}$, exhibits the same behavior for all three regions, with a prominent peak around 10^5 K. The spread of this measure as a function of time varies by only $\pm$ 0.05 to 0.15, as shown by the vertical bars. The shape of these curves persists when the criterion (1) is varied from 1 to 4 σ (see Arndt et al 1994).

On the other hand, we note that nonthermal motions in the corona are the same regardless of the large scale field, as shown by Cheng et al (1979) who found that, for solar coronal forbidden lines, the nonthermal broadening is essentially the same in quiet regions and active regions, and by Dere et al

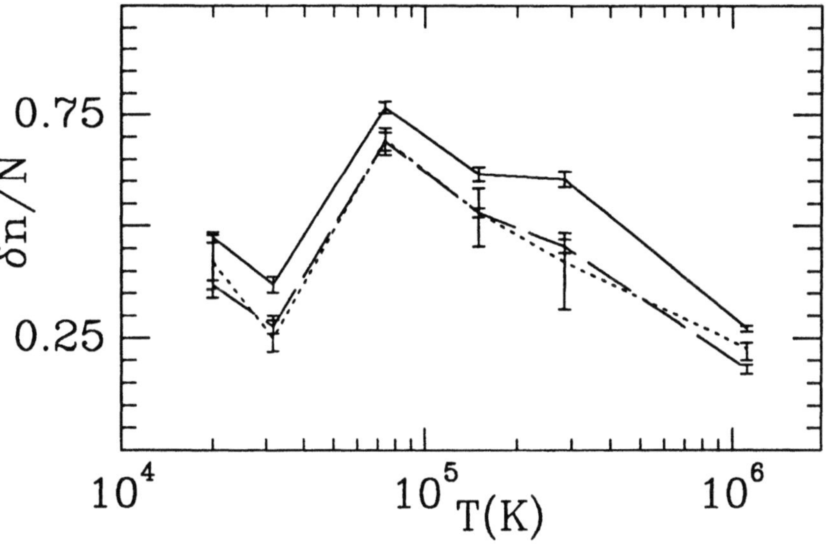

Fig. 7. $\frac{\delta n}{N}$ versus temperature for an active region (solid line), a quiet region (dashed line) and a coronal hole (dotted line).

(1989) who showed that, for lines formed around 10^5 K, they were essentially the same in coronal holes and quiet regions. These results together with those summarized in Figure 7 point to a common origin in the manifestations of coronal heating. Nevertheless, estimates from the analyses of these data show that the energy available from nonthermal motions (Dere et al 1989) or temporal changes in the emission (Habbal 1992) still falls short of supplying the energy needed to heat the corona as proposed by Parker (1988, 1991).

5. Summary

Multiwavelength observations acquired over the past two decades in the x-rays, EUV and visible light, have shown that the coronal emission is highly structured and inhomogeneous on a scale of a few arcseconds. Analyses of the temporal variability of the emission and the nonthermal broadening of spectral lines indicate that the characteristic signatures of coronal heating are independent of the large scale field. The small scale structures offer a promising tool for exploring the properties of coronal heating mechanisms.

It seems, therefore, worthwhile to dedicate some SOHO observations for the exploration of the properties of the small scale structures. Particularly suited for such studies are the CDS and EIT instruments which should yield simultaneous multiwavelength observations covering a wide temperature range. These should be acquired simultaneously with high resolution photospheric magnetic field measurements from ground-based observatories. Observations should be carried out over extended periods of time, such as hours and possibly days, within different large scale magnetic structures. Through careful planning, thorough data analysis and with improved image processing techniques, our understanding of the role of the small scale structure in the coronal heating process should be improved.

Acknowledgements

This work was supported by Air Force Grant AFOSR-91-0244 and NASA grant NAGW-249.

6. References

Arndt, M. B., Habbal, S. R., and Karovska, M.: 1994, *Solar Phys.*, , in press.

Cheng, C. C., Doschek, G. A., and Feldman, U.: 1979, *Astrophys. J.*, **227**, 1037.

Dere, K. P., Bartoe, J.-D. F., Brueckner, G. E., and Recely, F.: 1989, *Astrophys. J.*, **345**, L95.

Dowdy, J. F., Jr.: 1993, *Astrophys. J.*, **411**, 406.

Egamberdiev, Sh. A.: 1983, *Sov. Astron. Letts.*, **9**, 385.

Golub, L., Herant, M., Kalata, K., Lovas, I., Nystrom, G., Pardo, F., Spiller, E., and Wilczynski, J.: 1990, *Nature*, **344**, 842.

Habbal, S. R.: 1992, *Ann. Geophys.*, **10**, 34.

Habbal, S. R. and Grace, E.: 1991, *Astrophys. J.*, **382**, 667.

Habbal, S. R., and Harvey, K. L.: 1988, *Astrophys. J.*, **326**, 988.

Habbal, S. R., and Withbroe, G. L.: 1981, *Solar Phys.*, **69**, 77.

Habbal, S. R., Dowdy, J. F., Jr., and Withbroe, G.L.: 1990, *Astrophys. J.*, **352**, 333.

Karovska, M., Blundell, S. F. and Habbal, S. R.: 1994, *Astrophys. J.*, , in press.

Koutchmy, O., Koutchmy, S., Nitschelm, C., Sykora, J. and Smartt, R.N.: 1988, *in Proceedings of the 9th Sacramento Peak Summer Symposium*, ed. R.C. Altrock, 256.

Parker, E. N.: 1988, *Astrophys. J.*, **330**, 474.

Parker, E. N.: 1991, *Astrophys. J.*, **372**, 719.

Porter, J. G., Moore, R. L., Reichmann, E. J., Engvold, O., and Harvey, K. L.: 1987, *Astrophys. J.*, **323**, 380.

Reeves, E. M., Timothy, J. G., Huber, M. C. E. and Withbroe, G. L.: 1977, *Appl. Optics*, **16**, 849.

Sheeley, N. R., Jr., and Golub, L.: 1979, *Solar Phys.*, **63**, 119.

THE DYNAMICS OF CHROMOSPHERIC SPICULES

P. G. PAPUSHEV and R. T. SALAKHUTDINOV
Institute of Solar-Terrestial Physics(ISTP), Siberian Divison of Sciences, Russia

Abstract. We present the observational results on chromospheric spicules obtained at the Sayan observatory 50 cm coronograph. To investigate the evolution of chromospheric spicules, we analysed spicule spectra of strong chromospheric lines measured simultaneously at three altitudes above the solar limb during 5-60 min with a time resolution of 10 to 20 s. The spatial resolution was better than 1″, and the spectral resolution was 0.03Å in λ 6563Å. The appearance of a spicule at a given altitude is preceded by an sharp increase in line-of-sight velocity and/or in line half-width at a lower level. Generally, the evolution has a non-monotonous impulsive character. Changes of line-of-sight velocities and other parameters of the line profile can be represented as the superposition of slow, evolutionary changes and fluctuations with periods of about 80 to 120 s. The amplitude of line-of-sight velocity fluctuations is 2-3 km/sec and tends to increase with height. By studying the phase delays of the fluctuations at different heights, we found that the propagation velocity exceeds 300 km s^{-1}, and that the disturbances do not necessarily propagate upwards.

Key words: Sun - Chromosphere - Spicules

1. Introduction

The multitude of details observable in the chromosphere and transition region provides direct evidence of the inhomogeneous flow of energy in the higher layers of the solar atmosphere and its temporal fluctuations. One of the manifestations of this flow are the chromospheric spicules. The investigation of the origin and evolution of the spicules and their relationship to other fine-structure phenomena in the chromosphere and transition region (TR) is one of the main physical problems of the solar chromosphere.

Numerous observations have established the average characteristics of spicules. The luminosity of spicules is a little less than 1 erg s^{-1} cm^{-3}, and the internal energy at $N_h \sim 10^{12}$ and $T_e \sim 1.5 \times 10^4$ K is about 2 erg s$^{-1}cm^{-3}$. From the extremely broad line profiles of spicules one infers macroscopic velocities of 1.5-2 times the speed of sound.

Stygmatic spectra received in the EUV range in carbon and silicon lines show that the TR has a discrete fine-structure, too. Radiation of the TR reaches its maximum at a height of (1.5-2.5)$\times 10^3$ km and decreases exponentially with a scale height of about 1500 km (Withbroe, 1983). It is interesting to note that many spicule characteristics have about the same scale height.

The relationship between spicules and various other fine-scale phenomena observable on the disk in EUV lines is still not unequivocally established yet. Nevertheless, statistical studies of the properties of phenomena such as chromospheric jets observable in the carbon and silicon lines and dark fine Hα mottles in the chromospheric network indicate that all these phenomena

Space Science Reviews **70**, 47–51.

could have a similar origin. We therefore believe that the understanding of the physical nature of various other dynamical phenomena in the chromosphere and TR can be significantly improved by a better understanding of the evolution of chromospheric spicules.

2. Observations and Results

The observational material discussed below was obtained with the 530 mm Lyot coronograph (Nikolsky and Sazanov, 1967) of Sayan Observatory, which is located near the settlement of Mondy (Buryatya), 300 km to the south of Irkutsk-city and at a height of about 2000 m above sea level. Spicule spectra were photographically recorded in the lines Hα, Hβ, D_3 He I, and 10830 He I. In the case of the He I 10830 line, the photographs were taken from the screen of an Imaging Tube. The solar disk was always hidden behind an artificial "moon". The duration of our series of spectrograms varied from several minutes to one hour with a temporal resolution varying between 10 and 20 s. To simultaneously register spectra at different heights above the limb, a three-level image slicer was used. A multiplanimetric autoguide system was used to keep the slit at a given distance from the limb. The spectra were processed using standard computer algorithms which yielded the equivalent width, half-width and Doppler shifts of the various lines.

The spicule spectra, obtained with high spatial and spectral resolution, allowed us to observe several phenomena which are not marked in other works. One such phenomenon consists of Hα spicules being imposed on a weak diffuse background. The diffuse background was always observed in spicule spectra in the D_3 He I line. In spectra of Hα, however, the background is only observed when the scattered light level is very low. The intensity of the diffuse background measured on the spectrogram at a height (5.5-6)$\times 10^3$ km above the solar limb varies between (0.05-0.2)$\times 10^{-3}$ $I_\odot$. The average half-width of the Hα profile of the diffuse background varies between 0.3-0.4 Å, and the Doppler shifts do not exceed 0.2 Å. The diffuse background "cloud" surrounds a group of spicules of 8-15 arcsec size, separated from another group by a sharp dark gap. The distance between spicules in the group and the size of the dark interval are of the same order. Therefore, effects related to image scattering can be excluded and the described phenomenon is real. The reality of these phenomena is confirmed also by studying the evolution of the spicules. The spicules appear more often in locations connected with a diffuse background.

Limb observations made with an Hα birefringent filter show that chromospheric spicules appear as protrusions above a quasi-homogeneous strip of the chromosphere and, after that, the emission propagates upwards with a mean velocity of 25 km s^{-1}, reaching a height of (1-1.4)$\times 10^4$ km (Beckers, 1972). The observations made with an image slicer (Papusev, 1985) allow

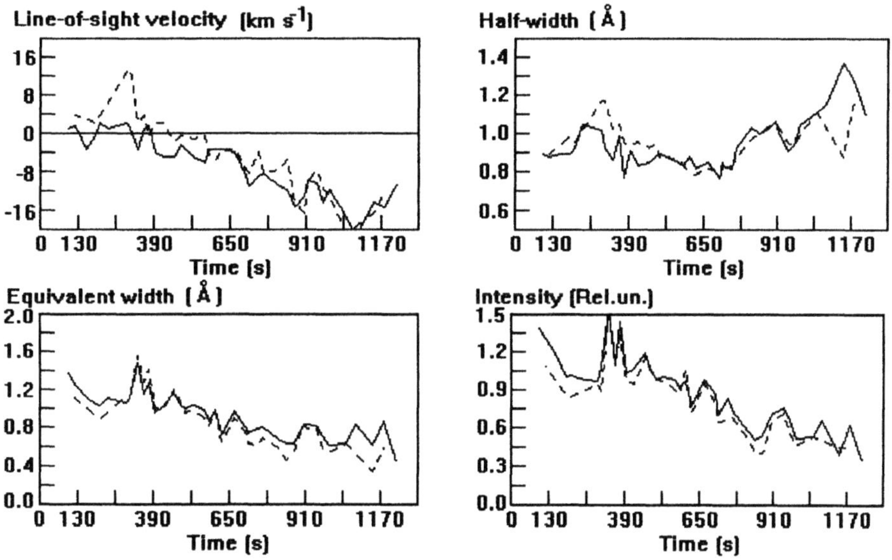

Fig. 1. Temporal variations of Hα line profile parameters and the line-of-sight velocity of spicules at two different heights. Solid line: 5000 km, dashed line: 8000 km

us to look also inside the quasi-homogeneous chromosphere, at least up to a height of (2-3)$\times 10^4$ km, where the arising spicules show spectral characteristics which differ from the "homogeneous" ones. From the temporal evolution one can conclude that the emission and, possibly, the material of spicules is propagating outwards. The upward propagation process is not a monotonous one and consists of several stages with a total duration of about 100-150 s. During this time interval there appear fairly intricate changes in the Hα profile in the lower chromosphere, which is just what leads to the appearance of emission in spicules at the height of (4.5-6.0)$\times 10^4$ km, where they are observed as isolated spectral features. Fig.1 shows the intensity, equivalent width, and half-width of Hα profiles as well as the line-of-sight velocity of spicules as a function of time.

A most typical feature of the evolution of chromospheric spicules is the non-monotonous variation of spectral line parameters. All line profile parameters reveal fluctuations with an amplitude of 30-40 %, and 30-200 s duration. These variations are accompanied by changes in the shape of the line profile itself. The disturbances show up as asymmetries of the profile and, in some cases, a clearly-defined secondary maximum is identifiable. However, the "middle" stage of the evolution is characterized by Gaussian profiles.

To investigate the time and amplitude characteristics of the Hα line profile variations, the initial plots were divided into time components (Papushev and Salakhutdinov, 1990). The results derived from studying the short-period (i.e., shorter than the spicules' lifetime) fluctuations, averaged over

five isolated spicules, are presented in Table I. One can see that:
- the oscillation amplitude of all parameters of the line profile (except the half-width) and of the line-of-sight velocity increases with duration;
- the amplitude of the line-of-sight velocity oscillations increases weakly with height, whereas the amplitude of the equivalent width and intensity oscillations decreases.

TABLE I

Characteristics of Short-period Pulsations of the Parameters of the Hα profile

Duration (s)	Line-of-sight velocity ($km\ s^{-1}$)	Equivalent Width (Å)	Halfwidth (Å)	Max Intensity (Rel.un.)
		Height 5000 km		
30- 60	1.4	0.054	0.054	0.043
60- 90	3.0	0.156	0.127	0.118
90-120	3.8	0.154	0.172	0.195
120-150	3.9	0.273	0.152	0.185
150-180	7.2	0.294	0.138	0.296
180-210	6.7	0.299	0.141	0.226
		Height 8000 km		
30- 60	2.0	0.044	0.053	0.032
60- 90	3.2	0.064	0.098	0.077
90-120	3.9	0.159	0.154	0.120
120-150	7.3	0.175	0.164	0.178
150-180	6.2	0.215	0.162	0.161
180-210	8.4	0.201	0.261	0.152

The velocity gradient along the axis of spicules is small. Therefore it can be used to reveal statistically the spatio-temporal characteristics of pulsating processes accompanying the evolution of spicules. The spectral analysis of the spatio-temporal distribution of the line-of-sight velocity indicates:
- the phase coherence of 50-160 s oscillation within single spicule groups and its significant changes from one group to another one with a spatial coherence length of about 22 arcsec;
- a phase delay of changes of the line-of-sight velocity measured at the maximum of the profile and the one measured at half intesity. The time delay is about 5 s for a height of 5×10^3 km and increases to 7 s at 8×10^3 above the limb of Sun;
- the power spectra of the velocity fluctuations vary with height; with increasing height, the amplitude of 50 s oscillations decreases and the amplitudes of 100 s (and longer) oscillations increases;
- velocity changes occur with small time delays at two heights separated by 3000 km, indicating propagation velocities of hundreds of $km\,s^{-1}$. Usually the time delay has a positive sign. The high velocity of these disturbances

has been noticed before (Pasachoff *et al.*, 1968; Hazan and Keil, 1984).

The most likely cause for such high velocities are pressure changes in the medium surrounding the spicule. An argument speaking in favour of this "external" cause is the high observed coherence of the velocity oscillations within a group of spicules. The idea that the physical processes inside spicules are controlled by external conditions appears to be quite promising, indeed: An earlier spicule model as a not fully expanded supersonic jet (Papushev, 1980) is able to explain not only the height to which spicules rise and the high nonthermal velocities but also the formation and origin of the fine structure in the transition region (Papushev and Salakhutdinov, 1989).

3. Future Programs of Observations

There are some outstanding questions concerning spicule data. One of the questions is concerned with the radial rise/fall velocity of spicules or, in other words, as to whether the observed line-of-sight velocity is the radial velocity projected onto the line of sight. For a direct determination of the radial velocity, one of us (Papushev, 1985) suggested that the frequency shift effect in the Thompson scattered light in spicules can be used in strong Fraunhofer lines, such as H and K CaII. The most favourable conditions for the observation of the weak depressions in the spectrum of spicule-scattered emission are given during total solar eclipses. It is also promising to use space-borne Lyot coronographs in conjunction with high-resolution spectrographs (0.03Å or better).

Another open questions relates to the thermodynamic properties of spicules and their immediate neighbourhood. To make progress in this area, we plan to measure time sequences of spicules and their surroundings in lines with excitation temperatures in the range $2 \times 10^4 \leq T \leq 10^5$ K. Further, it would be extremely useful to have simultaneous observations in the UV (e.g. Lyα) and in Hα. This will hopefully be possible in the not too distant future with the EUV instruments on board SOHO.

References

Athay, R.G., Holzer, T.E.: 1982, *Astrophys. J.*, **255**, 743
Beckers, J.M.: 1972, *Ann. Rev. of Astron. Astrophys.*, **10**, 73
Hazan, S.S., Keil, S.L.: 1984, *Astrophys. J.*, **283**, L.75
Nikolsky, G.M., Sazanov, A.A.: 1967, *Astron. zh.*, **44**, 426
Papushev, P.G.: 1980, *Solar Phys.*, **68**, 275
Papushev, P.G.: 1985, *Bulleten Abastumanskoi Astrofiz. Obzervat.*, **60**, 75
Papushev, P.G., Salakhutdinov, R.T.: 1989, *Proceedings of the XIII Consultation Meeting on Solar Physics*, **1**, 263
Papushev, P.G., Salakhutdinov, R.T.: 1990, *Issledovaniya po geomagnetizmu, aeronomii i fizike Solntsa*, **91**, 194
Pasachoff, J.M., Noyes, R.W. and Beckers J.M.: 1968, *Solar Phys.*, **5**, 131
Withbroe, G.L.: 1983, *Astrophys. J.*, **267**, 825

NUMERICAL SIMULATIONS OF SPICULE DRIVING MECHANISMS

A.S. ANDREEV
Crimean Astrophysical Observatory, Nauchny, Ukraine

and

A.G. KOSOVICHEV
Crimean Astrophysical Observatory, Nauchny, Ukraine
Institute of Astronomy, University of Cambridge, U.K.

Abstract. Spicules are known as one of the most prevalent small-scale dynamic phenomena on the sun, which are likely to give considerable contribution to coronal heating and mass supply. We discuss a model of the spicules driven by a train of slow MHD shock waves propagating along a vertical expanding magnetic flux tube. The shocks are initiated due to compression of the tube by the increasing external pressure in the lower chromosphere. Downflow of spicular material depends on radiative cooling and other dissipative processes.

Key words: Sun: atmosphere – Sun: corona

1. Introduction

Our models are based on the idea of spicule origin due to propagation of a shock-wave train, which lifts plasma in a vertical flux tube (Hollweg, 1982). These shocks are generated by perturbations in regions of relatively strong magnetic field in the low chromosphere. Because of the nonlinear wave-wake effect (Kosovichev and Popov, 1978) the perturbations give rise to quasi-periodical series of shocks which support spicule existence. The mechanism can take place not only in spicules but also in surges, macrospicules and other spicule-like features. In the model, flows in a vertical flux tube with the lower boundary located in the lower chromosphere and the upper one spread into the corona are described by a system of 1.5D MHD equations.

Previous simulations (Sterling and Mariska, 1990; Cheng, 1992) have shown that though some basic observed features can be reproduced in similar gas-dynamic models, there are still disagreements with observations, particularly, in maximum spicule heights and upward velocities. In these models the spicules were initiated by a piston or by a pressure perturbation localized in the lower atmosphere. However, the spicule driving mechanism, related to a complicated interaction between flows and the magnetic field, may be non-local. In order to understand it, we have considered the following types of initial perturbations: (i) a piston, moving up in the tube with a prescribed velocity; (ii) a modified Somov and Syrovatskii (1974) model of magnetic compression by convergence of magnetic force lines; (iii) local heating and compression of plasma in regions of relatively strong vertical magnetic field

Space Science Reviews **70**, 53–56.

in the lower chromosphere, presumably at boundaries between the chromospheric network cells; (iv) a rising magnetic loop acting as a piston. The results of these calculations (Andreev, 1992; 1993) have led to a conclusion that the mechanism (ii) of magnetic compression in the chromosphere may be most in line with the current observations. In this paper some new results of our study of the model of magnetic compression are presented. In this model, contrary to the piston model, magnetic effects are considered. Also, a more realistic driving mechanism is proposed, in which the possible non-local nature of the initial perturbations is taken into account.

2. Model

The equations describing vertical motions in a magnetic flux tube are

$$\frac{dz}{dt} = v, \qquad \frac{\partial v}{\partial t} = -A\frac{\partial p}{\partial s} - g, \qquad \frac{\partial}{\partial t}\left(\frac{1}{A\rho}\right) = \frac{\partial v}{\partial s},$$

$$\frac{\partial \epsilon}{\partial t} = -p\frac{\partial}{\partial s}\left(\frac{1}{\rho}\right) - Q_{\rm rad}, \qquad p + \frac{B^2}{8\pi} = p_{\rm ext}, \qquad BA = \Phi = \text{const},$$

where s is the Lagrange mass variable ($ds = \rho A dz$), z is the height above the photosphere, $A(t,z)$ is the area of a circular cross-section of the magnetic tube, v is the velocity, p and B are the gas pressure and the magnetic field strength inside the tube, respectively, Φ is the magnetic flux, $p_{\rm ext}$ is the total gas and magnetic pressure outside of the tube, ρ is the density inside the tube, ϵ is the internal energy density, g is the gravitational acceleration, $Q_{\rm rad}$ is the radiative cooling rate. Plasma ionization, which increases dissipation of the wave energy in the upper chromosphere due to bulk viscosity resulting from the relaxation of the state of the gas, is considered according to Hartman and MacGregor (1980). The radiative cooling is taken in the optically thin approximation, in the form suggested by Rosner et al.(1978) and MacNeice et al.(1984).

The equations have been solved numerically using the finite-difference technique of second-order accuracy, adopted by Kosovichev and Popov (1980) in which the discrete analogs of the basic laws of conservation of mass, momentum and energy are satisfied precisely, as are additional relations expressing balances of internal and kinetic plus gravitational energy.

For the initial flux-tube model we have chosen a model with linear increase of the flux-tube diameter with the height suggested by Ulmschneider et al. (1991). The temperatures inside and outside the tube are assumed to be equal at every given height. With the specification of the tube radius of 100 km and the magnetic field strength of 1500 G at $z = 0$, the structure parameters of the tube are determined uniquely from the hydrostatic equations.

We suggest that spicules are generated by an increase in external pressure resulting from the interaction of the magnetic tube with nearby magnetic structures or as a result of a strengthening external magnetic field. The perturbation is modeled as follows

$$p_{\text{ext}}(t,z) = p_{\text{ext}}(0,z)\left\{1 + \Delta a \exp\left[-\frac{(z-z_0)^2}{2\Delta z^2}\right]\sin\left(\frac{\pi t}{\Delta t}\right)\right\}, \text{if } t \leq \Delta t.$$

Here $\Delta a, \Delta z$ and Δt are the relative amplitude, the characteristic length and the duration of the perturbation respectively. The computations have been performed for $\Delta a = 0 - 5, z_0 = 0 - 500$ km, $\Delta z = 500 - 1000$ km, $\Delta t = 100 - 600$ s.

3. Results

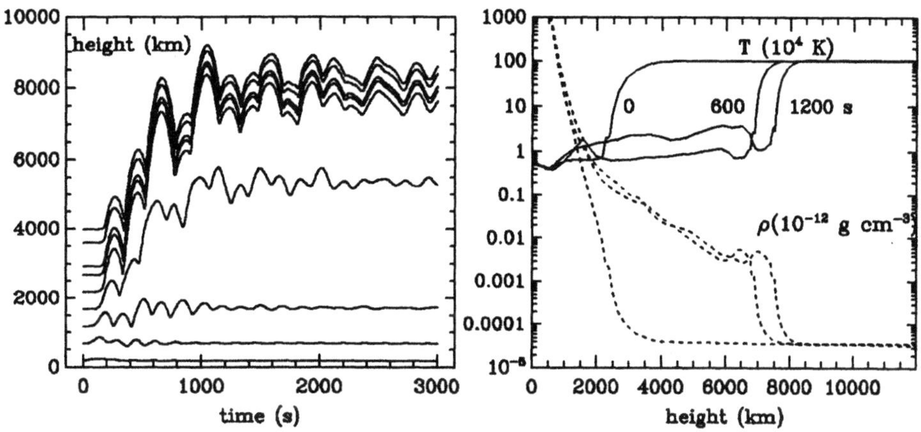

Fig. 1. The adiabatic model for $\Delta a = 1, z_0 = 0, \Delta z = 500$ km, $\Delta t = 300$ s. (a) Lagrangian motions of 9 mass elements which initially were at the heights of 200, 700, 1200, 1700, 2200, 2700 and 3600 km; (b) temperature and density in the tube at $t = 0$, 600 and 1200 s.

Figure 1 illustrates the typical results when the radiative cooling of the spicular material is neglected. The upward motion of the elements shows an oscillatory behaviour (Fig. 1a) because of the quasi-periodic slow MHD shocks generated in the tube with frequency of about the acoustic cut-off frequency of the chromosphere. The upper boundary of spicules can be identified in our model with the sharp region of transition between relatively cool ($T \approx 10^4 K$) and dense plasma to high-temperature coronal plasma (see Fig.1b). The boundary reaches the height of $\approx$ 8000 km and remains at this height during the relaxation process. The parameters of the rising column of plasma are similar to the observed spicular parameters. However, downward spicule motion cannot be explained in the model under the adiabatic conditions, just as it could not be explained in the piston model (Cheng, 1992).

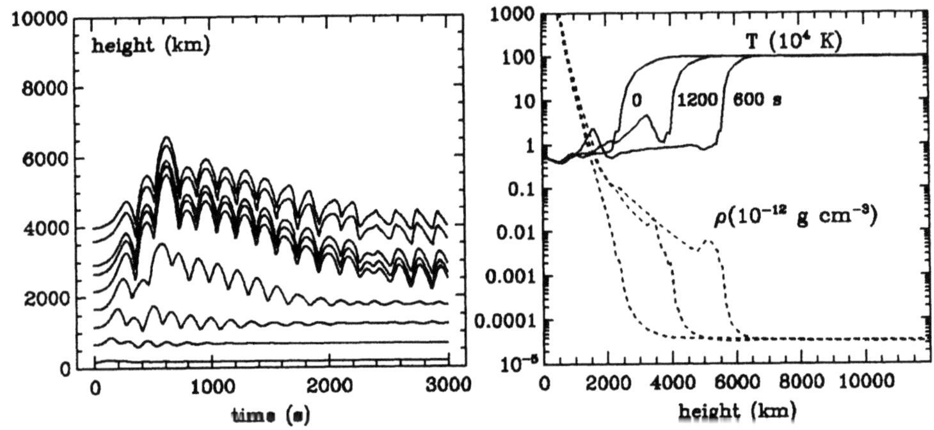

Fig. 2. The same as in Fig.1 for the nonadiabatic model.

In the model with the radiative losses turned on, the maximum spicule height is $\approx$ 5500 km (Fig.2a), which is significantly lower than in the adiabatic model. In this case there is slow downward motion of the spicular material after it reaches the maximum height. The material returns slowly because of continuing support by the quasi-periodic shocks in the wave wake for a long time after the initial impulse. The temperature and density variations (Fig.2b) are similar to those in the adiabatic model. In the case of a stronger perturbation of the external pressure ($\Delta a \simeq 3$) the maximum height is $\simeq$ 7000 km.

4. Conclusion

We conclude that the MHD model with compression of a vertical magnetic flux tube by enhanced external pressure in the lower chromosphere can reproduce some basic parameters of spicules. The model predicts highly inhomogeneous motions in the spicular material and its slow return to the solar surface. It has been demonstrated that radiative cooling plays an important role in spicule dynamics. This suggests that other non-adiabatic effects (e.g., energy exchange with the ambient plasma, turbulent viscosity and heat transfer) should be taken into account in the spicule models. This work is in progress.

References

Andreev, A.S. 1992, in: *Proc. XIV Consul.Sol.Phys.*, Karpacz, Poland.
Andreev, A.S. 1993, *Astron. Zh.*, in press
Cheng, Q-Q. 1992, *Astron. Astrophys.*, **266**, 537
Hartman, L. and MacGregor, R.B. 1980, *Astrophys.J.*, **242**, 260
Hollweg, G.V. 1982, *Astrophys. J.* , **254**, 806
Kosovichev, A.G. and Popov, Yu.P. 1978, *Keldysh Inst. Appl. Math. Prep.*, **No. 73**
Kosovichev, A.G. and Popov, Yu.P. 1980, *USSR Comp.Math.&Math.Phys.*, **19**, No.5, 168
MacNeice, P., McWhirter, R.W.P., Spicer, D.S. and Burgess, A. 1984, *Sol.Phys.*, **90**, 357
Rosner, R., Vaiana, G.S. and Tucker,, W.H. 1978, *Astrophys. J.*, **220**, 643
Somov, B.V. and Syrovatskii, S.I. 1974, *Proc. Phys. Inst. USSR Acad. of Sci.*, **74**, 35.
Sterling, A.C. and Mariska, J.T. 1990, *Astrophys.J.*, **349**, 647
Ulmschneider, P., Zähringer, K. and Musielak, Z.E. 1991, *Astron. & Astrophys.*, **241**, 625

THE CHROMOSPHERIC NETWORK DYNAMICS AS DERIVED FROM THE ANALYSIS OF CA II K AND HE I 1083 NM LINES

K. BOCCHIALINI and J.-C. VIAL
Institut d'Astrophysique Spatiale, Université Paris XI, Bat. 121, Orsay F-91405 (France)

and

S. KOUTCHMY
Institut d'Astrophysique de Paris - CNRS, 98 bd Arago, Paris F-75014 (France)

November 5, 1993

Abstract. We present results of line profile analysis of observations simultaneously performed around the Ca II K and He I (1083 nm) lines, using the Horizontal Spectrograph of the Vacuum Tower Telescope of NSO / SP. From the spectral analysis of a 83 min long sequence of CCD spectra, we derive some dynamical properties of the main components of the quiet chromosphere: i) the magnetic network, ii) the cell interior. We present a whole set of amplitude spectra near 5 and 3 min periods for the two lines; K_3 and He I velocity spectra extending up to 100 mHz are also considered, for the first time.

1. Introduction

Heating by MHD waves of the chromosphere and the corona is a viable mechanism which nevertheless needs detailed observations for confirmation (Zirker, 1993).

Observations of 5 min and shorter period waves in the corona have already been reported and discussed (Tsubaki, 1977; Koutchmy, Locans and Zhugzda, 1983). Moreover, the topic of chromospheric oscillations is continuously discussed, several groups being involved in both observations and analysis with improved methods; see e.g. Lites et al. (1993), Fleck et al. (1993).

Among the major issues, we note the relevance to the heating problem of the resonance period of the chromosphere near 3 min and the role of the magnetic field in magnetic elements, possibly generated by self-excited dynamos (see Lorrain and Koutchmy 1992) and the occurence and dissipation of high frequency (HF) waves, especially in the corona.

The observations we consider here were performed in order to analyse the low frequency (LF) period waves as well as the HF waves, from the chromosphere to the corona (Bocchialini et al. 1993a).

Two typical regions of the chromosphere were studied; we called them Active Network (AN) and Quiet Region (QR); the last one was selected as a region free of prominent and/or vertical magnetic field.

We use the sketch presented in Fig. 1 to illustrate the features we want

Space Science Reviews **70**, 57–63.

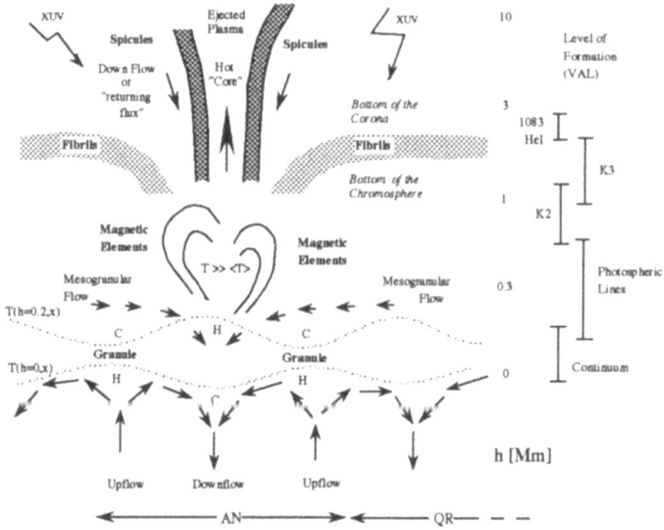

Fig. 1. Vertical cut across the chromospheric network: AN is defined as a magnetic region, QR as a non-magnetic region. The level of formation of lines is taken from Avrett and Loeser, 1992, and the height scale is logarithmic.

to discuss, considering the atmospheric properties from the photosphere up to a few thousand km. The cut across the chromosphere presented on Fig. 1 is similar to the ones proposed by Axford (1993) and Koutchmy and Loucif (1992). The heart of the AN is the bottom of the magnetic region, called by Axford the 'furnace', which is believed to be the ultimate source of the fast wind.

2. Observations

The observations have been achieved on March 22, 1993 at the VTT of NSO/SP; the new multi-channel Horizontal Spectrograph developed for the Advanced Stokes Polarimeter of HAO/NSO has been used. The data have been recorded on CCDs, simultaneously in the Ca II K and the He I lines. The final resolutions after all required corrections are: 1.2 arcsec/px and 5.8 pm/px for Ca II K and 16.2 pm/px for He I. The sequence of observations is 5000 sec long and the sampling interval is 5 sec.

We selected two regions: an active network element (AN), which is averaged over 6 arcsec along the 2.3 arcsec slitwidth, and a quiet region (QR), averaged over 3.6 arcsec; we also considered the average region (AV) over the whole 38 sec long slit. More details about the region seen with filtergrams and Hα slit-jaws images, and the data processing are given in Bocchialini et al. (1993a, b).

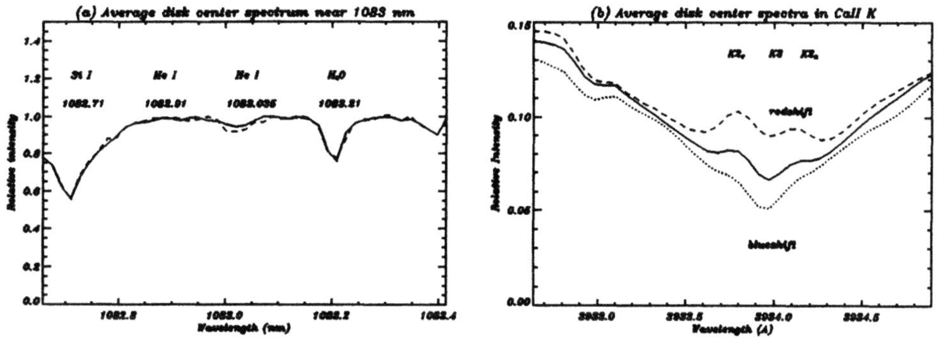

Fig. 2. Time average disk center spectrum (a) near 1083 nm, (b) in Ca II K; full line: AV, dotted line: QR, dashed line: AN; the spectra have been flat-field corrected but some irregularities remain, especially in (a), where the QR spectrum is not shown to avoid confusion.

3. Averaged line profiles in Ca II K and 1083 He I, and interpretation

AN and AV He I line profiles (average over the whole temporal sequence) are presented in Fig. 2a. We found that the average central line depression (1083.035 nm) is 8% of the continuum in AN, 2% in QR and 5% in AV, in good agreement with Giovanelli & Hall (1977). The FWHM is equal to 15±2 km s^{-1} in AN and AV. The most important result is the extended red wing, in both AN and AV, which suggests a downflow at least in AN; this red component could be due to the same downflow as the one systematically observed in transition region lines.

Ca II K line profiles in AN, QR and AV (obtained over the 5000 sec sequence) are shown in Fig. 2b. Intensities are scaled to White and Suemoto's profiles (1968). The measured FWHM of K_3 is equal to 9±1 km s^{-1} for the QR and to 8±2 km s^{-1} for the AV. The AN profile is too asymmetric for this quantity to be measured properly. Assuming that the velocity of AV at line center (K_3) is zero, we measured a velocity of 0.95±0.50 km s^{-1} (redshift) for AN and of −1.35±0.5 km s^{-1} (blueshift) for QR. Note that the red-shift shown in K_3 over AN is consistent with what we saw in the 1083 He I line. This suggests that the downflow well observed in transition region lines is also seen in the network, down to layers re-absorbing the Ca II K radiation (K_3).

4. Time series analysis: results and discussions

We performed a Fourier analysis of intensities at selected wavelengths measured on each spectrum along the whole time series.

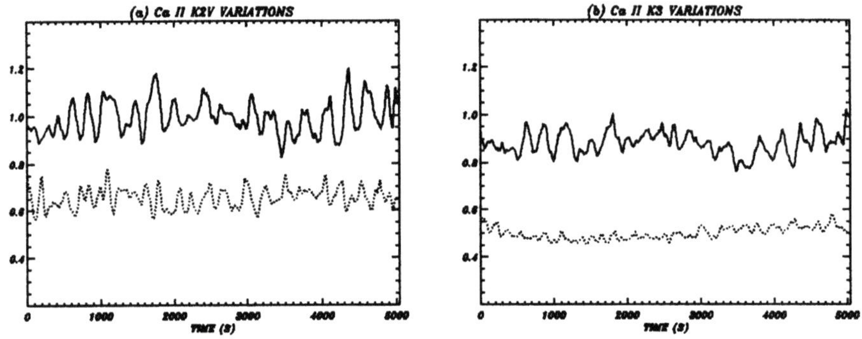

Fig. 3. (a) Ca II K_{2V}, (b) Ca II K_3 temporal variations over the 83 min sequence; full line: AN, dotted line: QR.

a) In Ca II K, we first present the temporal intensity variations of K_{2V} and K_3, for AN and QR, see Fig. 3. The amplitudes of K_2 variations are almost the same for AN and QR; in K_3, amplitudes are several times higher in AN than in QR. For AN, the amplitudes in K_{2V} and K_3 are quite comparable. In K_3, we see about 3 wave trains during the 5000 s sequence.

The temporal variations of K_3 Doppler velocity are also presented for both regions. To compute Doppler velocity, we used the fast method based on the subtraction of line wing intensity modulations and subsequent calibration using the *average* profiles of Fig. 2.

The amplitudes for AN and QR are quite comparable but it clearly appears that the periods of oscillations are different (Fig. 4a, b). This is confirmed by the *amplitude spectra* at LF (between 0 and 20 mHz), see Fig. 4c: around 3 min there is considerably more power in QR than in AN and the peak is centered around 7 mHz; at 5 min the spectra show similar amplitudes in QR and in AN; let us stress that the large ratio we measured in AN, between 5 min and 3 min oscillations, is in favor of the 5 min periods.

At HF (10 to 100 mHz), the amplitude spectrum in QR is higher than in AN (Fig. 4d); the significance of this result is considered below.

b) In He I, we computed the Doppler velocity near the line center, using the same straightforward method of line wing intensity differences (without assumption on the shape of the line profile). The visual comparison with Ca II K for AN shows an obvious correlation between the velocities observed in K_3 and in 1083 He I, the correlation being not obvious for QR. The amplitudes of the oscillations are typically three times higher in QR than in AN (Fig. 5a, b, c) near 5 min periods, and up to 8 times near 3 min periods!

We also present the amplitude spectra of the continuum, near 1083 nm, in AN and QR, to show the excellence of the signal/noise ratio (Fig. 5d) in both AN and QR: 5 min oscillations appear in intensity variations as reported by Koutchmy and Lebecq, 1986 and Title et al., 1990, as well as

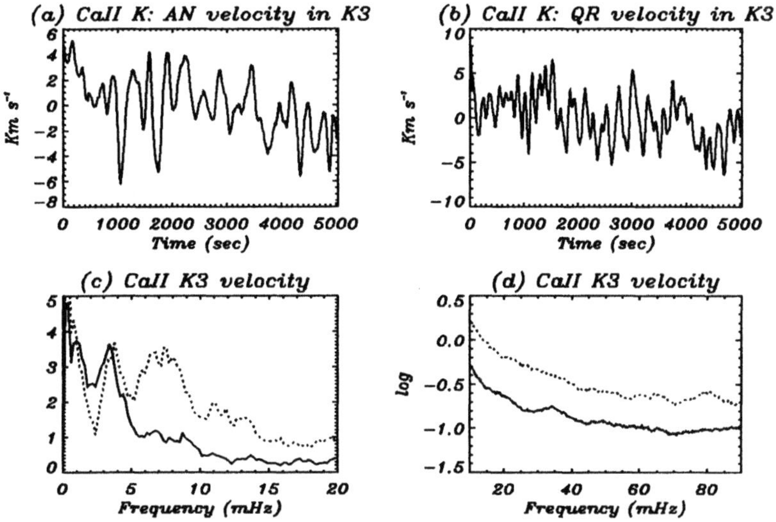

Fig. 4. Temporal variations of Ca II K_3 velocity (a) in AN, (b) in QR; amplitude spectra of K_3 velocity (c) at LF, (d) at HF; full line: AN, dotted line: QR

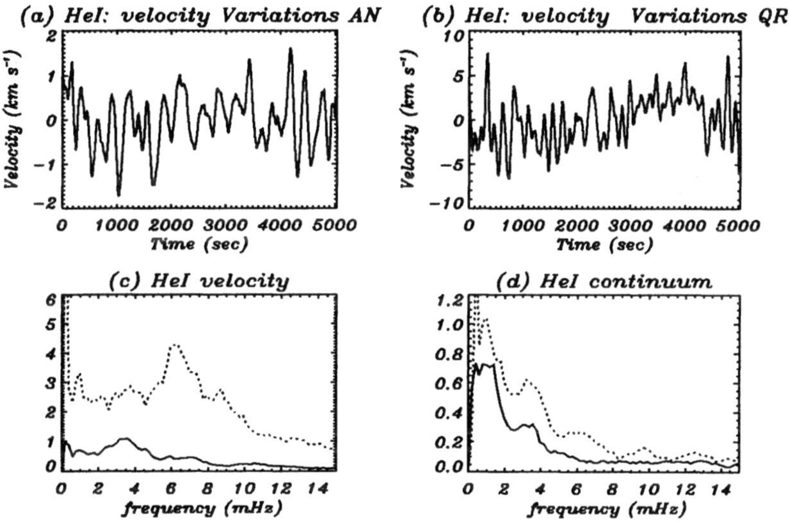

Fig. 5. Temporal variations of He I velocity (a) in AN, (b) in QR; (c) amplitude spectra of velocity at low frequency, (d) amplitude spectra of He I continuum; full line: AN, dotted line: QR.

very low frequency fluctuations.

In order to evaluate the results obtained at HF, we also performed an analysis of the spurious velocity signal due to random fluctuations present in our spectra (noise of different origins). We applied the same formula used

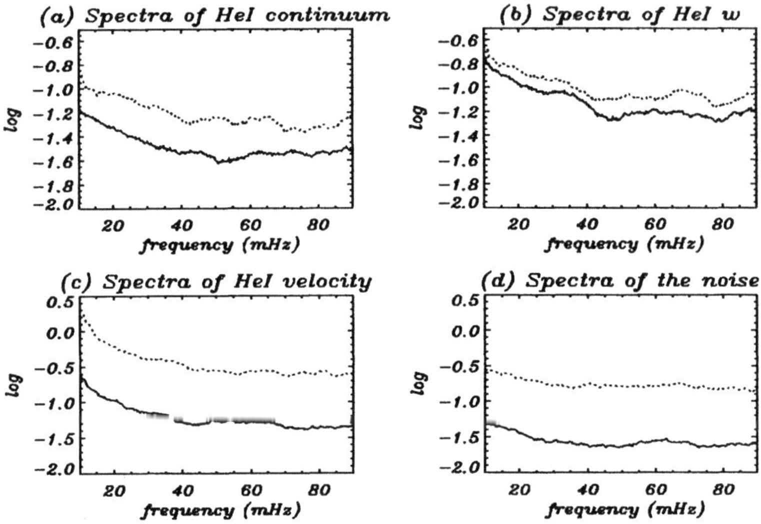

Fig. 6. He I amplitude spectra at high frequency (**a**) in the continuum, (**b**) in line intensity, (**c**) in Doppler shift or velocity; (**d**) amplitude spectra of the noise in simulated Doppler shifts of the pseudo continuum at HF; full line: AN, dotted line: QR. Note that velocity spectra (c) have not been corrected for the noise (d).

to compute the Doppler shifts in the 1083 He I line to the nearby pseudo continuum and obtained the results shown in Fig. 6d for HF.

At high frequency -10 to 100 mHz- the power measured in the continuum intensity, as well as in line intensity, and in velocity, is once again higher in QR, (Fig. 6a, b, c). In the He I line, both QR and AN exhibit HF power which is significantly above the noise level shown in Fig. 6d.

As for seeing effects, they are minimized by an excellent guiding, the compensation for the solar rotation, and essentially the spatial averaging (see §2) performed to improve the signal to noise ratio. We feel confident that seeing effects are negligible, a statement which certainly needs a thorough verification.

5. Conclusions

a) 3 min oscillations are prominent in QR: 3 min waves could be analysed, especially at the highest chromospheric level of He I 1083; in AN, 5 min oscillations are prominent. At HF, some power is observed above the identified noise level in both AN and QR (more in QR).

b) The definite occurence of 5 min oscillations at the He I level in AN, see Fig. 4c and 5c, as well as the absence of the chromospheric resonance near 3 min (especially if we assume that the small power observed there could be due to the influence of scattered light) point out to a mechanism of *wave propagation in the magnetic region* possibly connected with the heating. Further analysis is needed to conclude on that. We are now analysing our time series to look at phase and coherence spectra.

c) We also want to add a few recommendations concerning observations with SOHO: SUMER seems to be well-suited to the study of chromospheric regions up to the transition region. The best spatial and temporal resolution should be aimed at, even if during the data processing, addition of pixels in selected regions is performed to improve the signal to noise ratio, as done in the present work. The spectral window should have, as many lines formed at different temperatures as possible. In the 60 - 160 nm range, we suggest the 123.8 - 127.8 nm window where CI, Si I, Si II, C III, N V lines cover temperatures from below 10^4 K to 1.5 10^5 K. The suggested observing mode could use the 1"x240" slit, a spatial scan of 4 steps of 0.76" with a dwell time of two seconds, leading to a temporal resolution of 8 s. If possible, all lines pixels should be transmitted (or stored on board), which is the only way of measuring strong asymmetries in line wings. Temporal sequences should be significantly larger than one hour, in order that "events" (bright points, explosive events) be detected and statistically analysed.

6. Acknowledgements

We thank the staff of the VTT at NSO/SP for their efficient support during these observations; partial support was received from the "Action Spécifique Grands Télescopes" of INSU–CNRS and the GdR "Magnétisme dans les Etoiles" of CNRS. We thank B. Fleck and J. B. Zirker for their interest and comments.

References

Avrett, E. H. and Loeser, R.: 1992, *ASP conf. Series* **26**, 489
Axford, R.: 1993, *present conference* , 2nd SOHO Workshop
Bocchialini, K., Vial, J.-C., Koutchmy, S.: 1993a, *ApJ Letters* , in press
Bocchialini, K., et al.: 1993b, *PASP, NSO/SP Internat^{al} Summer Workshop* , submitted
Fleck, B., et al.: 1994, *in "Infra–red Solar Physics", D. Rabin et al., (eds)* **154**, p 65
Giovanelli, R. G. & Hall, D.: 1977, *Sol. Phys.* **52**, 211
Koutchmy, S. and Lebecq, C.: 1986, *A&A* **169**, 323
Koutchmy, S. and Loucif, M.: 1991, *p. 152 in "Mechanism of Chromospheric and Coronal Heating", Priest and Rosner eds* , Springer-Verlag
Koutchmy, S., Locans, V. and Zhugzda, Y. D.: 1983, *A&A* **120**, 185
Lites, B. W., Rutten, R. J., Kalkofen, W.: 1993, *Ap. J.* **414**, 345
Lorrain, P. and Koutchmy, S.: 1992, *A&A* **269**, 518
Title, A.M., et al.: 1990, *in "Solar Photosphere: Structure, Convection and Magnetic Fields", Stenflo Ed., IAU Symp.* **#138**, 49
Tsubaki, T.: 1975, *Sol. Phys.* **43**, 147
White, O. R., Suemoto, Z.: 1968, *Sol. Phys.* **3**, 523
Zirker, J. B.: 1993, *Sol. Phys* **148**, 43

DYNAMICAL FINE STRUCTURES OF THE CHROMOSPHERE

G. TSIROPOULA
Institute of Space Research, National Observatory of Athens, P.O. Box 20048, GR-11810 Athens, Greece

B. SCHMIEDER
Observatoire de Paris, DASOP, F-92195 Meudon Principal Cedex, France

and

C. E. ALISSANDRAKIS
Section of Astrophysics, Astronomy and Mechanics, Department of Physics, University of Athens, GR-15784 Athens, Greece

Abstract. We have studied the spatial structure and temporal evolution of a chromospheric region with chains of mottles at the junction of three supergranules. The observations were obtained with the Multichannel Subtractive Double Pass spectrograph operating in Hα at the Pic du Midi Observatory. From a statistical point of view the fine structures were stable in intensity over the observation period (15 min), while the line of sight velocity showed remarkable changes within a few minutes. A detailed analysis shows that the predominant pattern of bulk motion in the dark mottles is that of downflow at their footpoints and upflow at their tops, which is followed by downflows along the whole structure and that this process repeats itself several times during their lifetime.

Key words: Sun – chromosphere – fine structure

1. Introduction

High quality spectroheliograms indicate that the solar chromosphere possesses a rich diversity of highly dynamic forms distributed on a fine scale. The main chromospheric structural components are mottles organized into a complex geometric pattern over the solar disk. Usually they congregate into small groups which are called chains and larger groups which are called rosettes. As a rule they appear at the borders of the supergranule cells, where the magnetic field is concentrated in the form of isolated flux tubes due to the constriction of the lines of force by convective motions, which are directed from the centre of the cells towards the boundaries. Tsiropoula *et al.* (1993) analysing two dimensional Hα spectra of a chromospheric rosette obtained by the Multichannel Subtractive Double Pass (MSDP) spectrograph (Mein 1977) detected strong downflows at the roots of dark mottles and upflows in their upper part. The same authors studying the temporal variations of the velocity along the axes of some dark mottles of this same rosette found that this behaviour of downflows at their footpoints (towards the centre of the rosette) and upflows at their upper parts is followed by downflows along the whole structure and that this kind of motion is repeat-

Space Science Reviews **70**, 65–68.

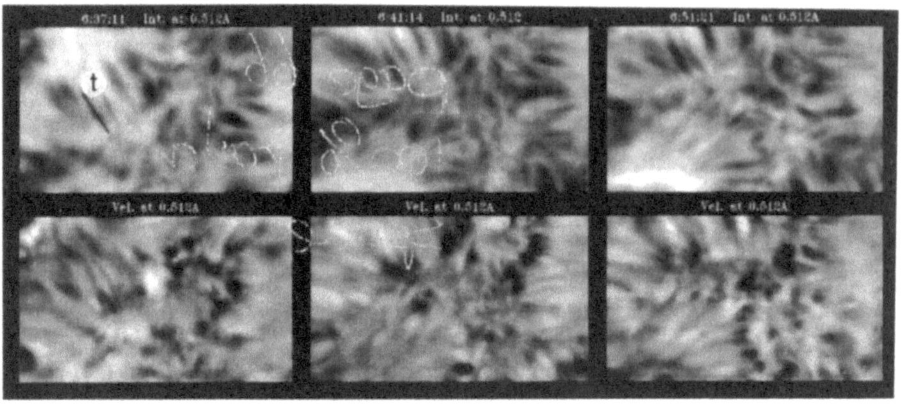

Fig. 1. The supergranule boundary region at 6:37:11 UT, 6:41:14 UT and 6:51:21 UT. The upper row shows the intensity and the lower row the velocity at ± 0.512 Å. Descending material is black in the velocity map.

ed several times during the lifetime of the mottle (Tsiropoula *et al.* 1994). A similar behaviour of ascending phases followed by descending ones is already reported by some authors for spicules, structures which are considered as the mottle counterpart seen at the limb. They also noticed that this behaviour is repeated several times before these structures diffuse away (Lippincott 1957, Nikolsky 1970).

2. Observations and data reduction

The present Hα observations were performed with the MSDP operating on the solar spectrograph installed at the focus of the 50 cm refractor of the Pic du Midi Observatory on June 17, 1986 at the disk centre (N5, W5) (Schmieder *et al.* 1991). The MSDP provided 11 simultaneous intensity images of the same field of view, 0.256 Å apart in the Hα line. From these data, maps of intensity fluctuations and Doppler shifts can be derived. By displacing the entrance slit a large field of view of $\sim 4' \times 5.5'$ is obtained with a repetition rate of 1 min and total duration of 15 min (from 6:37:11 UT to 6:52:21 UT). From the entire field of view a region was selected which was found at the boundaries of three supergranules ($27'' \times 40''$; Fig. 1).

3. Results

3.1 Statistical analysis

Cross-correlation (CC) is used for high precision detection of time changes. All time frames were carefully positioned. Each frame in the time series was considered as the reference frame and the CC function for both the intensity and the velocity was computed. Figure 2a shows the intensity CC function

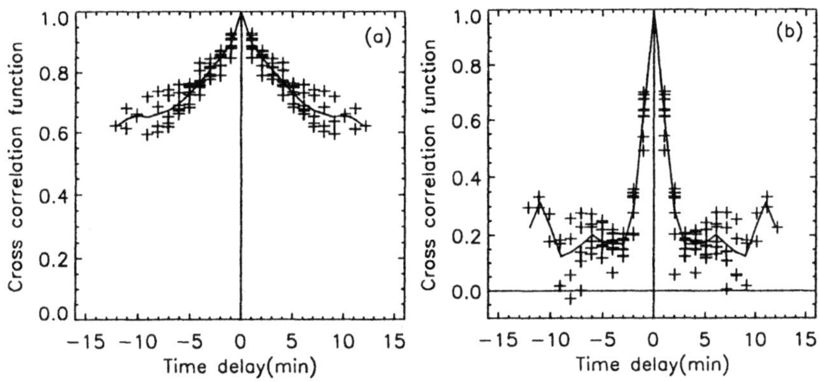

Fig. 2. Cross correlation function vs time of (a) Intensity and (b) Velocity for the whole region at ± 0.512 Å.

vs time at ± 0.512 Å. The solid line shows the average cross correlation function, while the crosses are the values of individual curves. The CC function at this wavelength decreases from 1 to 0.6 in 15 min. As the decay of the CC function is a measure of the lifetime of features we may conclude that the mean lifetime of the mottles is at least 15 min, in agreement with Bratsolis *et al.* (1993), who also pointed out that some methods used in the determination of the mean lifetime of solar features can lead to serious underestimations. Figure 2b shows the CC function of the velocity maps. It shows much more rapid changes than the intensity curve. It drops from 1 to 0.2 in about 3.5 min with an e-folding time of about 2.5 min.

3.2 Intensity and Doppler velocity variations along the axis of a dark mottle

Temporal and spatial variations of the intensity and the velocity were deduced along the axis of several mottles in order to derive their flow pattern as a function of time. The intensity and the velocity were averaged over a strip extending 0.5″ on either side of an axis traced on the best filtergram and then followed in each filtergram after the careful superposition of frames. An example is shown in Fig.3 (its axis and top are marked in Fig. 1). The intensity pattern remained almost constant during the observation period (Fig. 3a). The velocity pattern, however, changed appreciably (Fig 3b). The material moved downwards at the footpoint of the mottle and upwards at the top (the footpoint is at the left of the panels). This motion was followed by downflows in the whole structure and is repeated at least 3 times during the observation period.

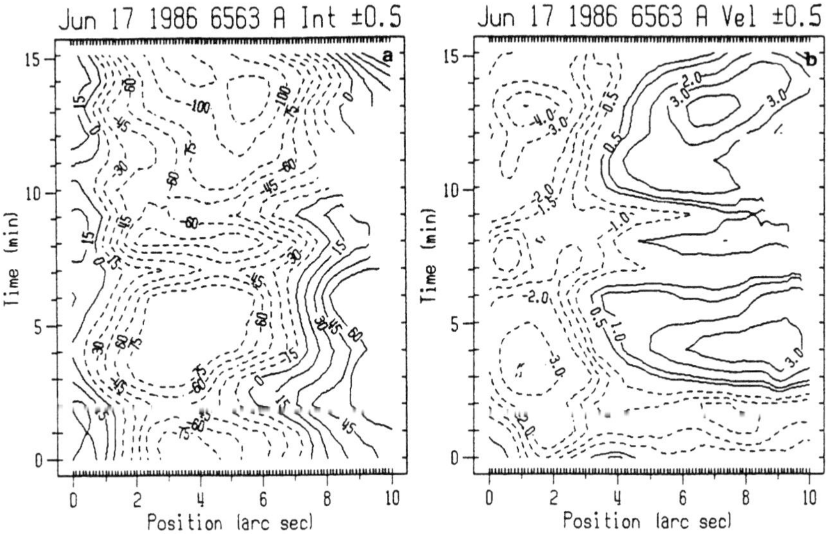

Fig. 3. (a) Intensity and (b) Velocity variations as a function of time along the axis of a dark mottle.

4. Discussion

An analysis of the behaviour of the line of sight velocity along the axis of a dark mottle shows that the predominant pattern of bulk flow is that of downflow at the footpoint and upflow at the top and that this motion has a reccurent character. This configuration is similar to the one proposed by Pikel'ner (1969) for the evolution of spicules. According to this model spicules are generated at the boundaries of supergranule cells by the interaction of regions with opposite magnetic polarities. As the field lines are forced together, the compressed gas trapped between them cools and flows down under gravity, while at the same time material is carried upwards as a result of the reclosure of the receding field lines. The upflow material may manifest itself in UV lines as an explosive event. SOHO will provide such possibility of observations.

References

Bratsolis E., Dialetis D.,Alissandrakis C.E.: 1993, *Astronomy and Astrophysics*, **274**, 940.

Lippincott, S. L.: 1957, *Smiths. Contr. Astrophys.*, **2**, 15.

Mein P.: 1977, *Solar Physics*, **54**, 55.

Nikolsky G. M.: 1970, *Solar Physics*, **12**, 379.

Pikel'ner S. B.: 1969, *Soviet Astronomy*, **13**, 259.

Schmieder B., Raadu M. A. & Wiik J.E.: 1991, *Astronomy and Astrophysics*, **252**, 353.

Tsiropoula G., Alissandrakis C. E. & Schmieder B.: 1993, *Astronomy and Astrophysics*, **271**, 574.

Tsiropoula G., Alissandrakis C. E. & Schmieder B.: 1994, *Mem. del' Osservatorio di Catania* (in press).

ENERGY INPUT FOR EXPLOSIVE EVENTS IN THE TRANSITION ZONE

G.M. SIMNETT
School of Physics and Space Research, University of Birmingham, Edgbaston, Birmingham B15 2TT, UK

Abstract. It is suggested that the energy input for explosive events in the transition zone comes from precipitating ions, typically of energies of a few $\times$ 10^2 keV/nucleon, accelerated in the high corona. The energetics of the process are discussed, together with implications for coronal heating.

Key words: Sun – Transition zone – Explosive events

1. Introduction

The purpose of this paper is to examine the hypothesis that the energy input for the explosive events seen in the transition zone comes from charged particles precipitating down from the solar corona, in a way somewhat analogous to the terrestrial aurora. One important difference is that the overall structure of the coronal magnetic field is not predominantly dipolar, as in the case of the Earth, but there is growing observational evidence for large magnetic loops extending high into the corona. Unfortunately, in the region of interest, which is at least up to a significant fraction of a solar radius above the photosphere, direct magnetic measurements are not yet possible. However, X-ray and EUV images show that there are loop structures which can link active regions on the Sun, and we can infer from radioheliograph data, sympathetic flares and coronal mass ejections that much larger structures are often present. Recently Yohkoh and SPARTAN images (presented at this workshop) have shown evidence of inhomogeneous X-ray emission in the 2 - 3 $R_\odot$ region, from which we can infer closed structures.

The most important consequence for our hypothesis is that the explosive events are quite secondary to an energy release that results from reconnection in the high corona. In such a situation the majority of the energy is distributed throughout the corona, as we shall see. This has a significant bearing on coronal heating. In models where the energy release is low down in the atmosphere some fraction of the observed output of the event is all that is available for coronal heating. If the explosive events are due to precipitating particles, then virtually all the energy in the primary reconnection, of which only a fraction gets down to the transition zone to produce an observable signature, will be available for coronal heating. This suggestion is an extension of the concepts advanced two decades ago by Tucker (1973) and Levine (1974) that coronal heating results from the reconnection of

Space Science Reviews **70**, 69–75.

magnetic fields, with dissipation of energy in current sheets.

2. Properties of the Explosive Events

Explosive events in transition zone lines were first discovered by the NRL High Resolution Telescope and Spectrograph (HRTS). Brueckner and Bartoe (1983) identified two types of event; relatively energetic blue-shifted jets and more frequent, but less energetic, events with more symmetrically-broadened line profiles. Further analysis (Dere et al, 1984, 1989) from the early HRTS rocket flights showed that the jets were not as common as first thought, but that the other, "explosive" events were extremely common. Brueckner and Bartoe (1983) had estimated a combined birth-rate of jets and explosive events over the whole Sun as 777 s^{-1}, or, following Parker (1988) $\sim 10^3$ s^{-1}. Dere et al (1989) showed that the birth-rate in polar coronal holes was only about 40 % of that over the rest of the quiet Sun. It was also demonstrated (their Figure 25) that the explosive events are most common between 40° and 50° latitude; there is a dearth of events near the equator, and also in the polar regions. Porter and Dere (1991) showed that the events occur in the solar magnetic network lanes at the boundaries of supergranular convective cells. It is significant that the events occur away from the strongest magnetic fields, because if they are caused by particle precipitation from the high corona, then they will occur predominantly where the high coronal magnetic field lines map down to the chromosphere. This will not normally be in the regions of the strongest fields which are closed in low-lying loops. We interpret the lack of events at the equator as due to this effect, while the shortage of events in the polar regions reflects the fact that over the polar coronal holes the field lines probably extend deep into interplanetary space.

Henoux and Dere (1994) quote spatial scales of 1500 km (2″) with some evidence for smaller scales. The more energetic microflares observed in the C IV (λ 1548) line by the UVSP on the Solar Maximum Mission (SMM) are located predominantly in one 3″ × 3″ pixel. The event lifetime is variable, but is typically 60 s, with few events above 150 s. Dere et al (1984) showed that the line-of-sight velocities are $\sim$100 km s^{-1} or so, and are independent of solar latitude. This is consistent with the events being highly turbulent. The line profiles show significant variation over timescales of 20 s. With HRTS the events are best observed in the C IV line, but they cover emission lines formed at temperatures from 2×10^4 K to at least 2×10^5 K. They are not seen in the lines of Si II, C I and O I (Brueckner and Bartoe, 1983, Dere et al 1984) which are produced in the upper chromosphere below 2×10^4K. Porter and Dere (1991) argue that the lower limit is real, but note that information on higher temperature lines is not available. Dere et al (1989) calculate the energy in an event as $\sim 10^{23}$ erg. However, they point out, correctly, that this is based on observations in one spectral line and could be an order of magnitude too low. Following Parker (1988) an energy of 7×10^{23} erg seems more realistic.

However, there is information on other transient events which may be related phenomena. One current observational problem is that instruments operate with varying degrees of temporal coverage, at different sensitivities, and over a fragmented energy range. Both the UVSP and HXIS instruments on SMM observed transient events of similar ($\sim 10^2$ s) durations to the explosive events. Fontenla et al (1989) reported transient brightenings in Lyα (λ1215), N V (λ1238) and C IV. (Porter and Dere (1991) argue from the similarity of lifetime and spectral distribution (i.e. blue excess) between explosive events and microflares seen by SMM in C IV that they are manifestations of the same type of energy release. Simnett and Dennis (1986) reported impulsive soft X-ray events (3.5 - 8.0 keV) lasting from 30 - 100 s, which occasionally had shorter-duration counterparts above $\sim$ 30 keV; some discrete hard X-ray spikes last only a few seconds (Lin et al, 1984). The soft X-ray data cannot measure blue-shifts, but the similarity in timescales is consistent with the X-ray events having a similar explanation, in rapid heating of the transition zone plasma from above, with an occasional short-duration high energy extension.

Porter et al (1993) studied the relationship between spikes in C IV and 3.5 - 5.5 keV X-rays and found a positive correlation, which suggests that plasma is generally heated above the 10^5 K attributable to transition zone lines. This confirms that the energy estimates made by Dere et al (1989) probably are too low, as suspected.

3. Energy Losses of Non-thermal Particles in the Corona

The lack of gamma ray and intense hard X-ray emission outside of flares precludes protons $\geq$10 MeV and electrons $\geq$10 keV from playing an important role in the energy transport. Therefore if non-thermal particles are important they are likely to be protons or helium nuclei below a few MeV/nucleon. If we adopt the premise that the ions are accelerated and trapped in large ($\geq$0.5 $R_\odot$) coronal loops, then to precipitate and heat transition region material they must penetrate (a) the mass column density in the coronal part of the loop and (b) the mass column density from the top of the transition region to the lowest depth at which the events are seen. The latter is around the 2×10^4 K limit. Within the transition region in the quiet Sun the column density appropriate to a bright network element is around 10^{-5} g cm^{-2}. However, coronal propagation, finite pitch angle, and scattering towards the end of the range will enhance this equivalent column to a value of $\geq 5 \times 10^{-5}$ g cm^{-2}, which is the range of a proton or helium nucleus somewhat above 200 keV/nucleon. Therefore the relevant part of the ion spectrum is in the few $\times$ 10^2 keV/nucleon range. Particles below $\sim$ 100 keV/nucleon will merely heat the corona.

The constraints on particle propagation in the corona are severe. At a mean density of 10^8 cm^{-3}, which from radio observations is a reasonable mean density for a coronal loop of height 0.5 $R_\odot$ above the photosphere, the column of gas along the loop (assuming it is semicircular) is equivalent to 3.7×10^{-5} g cm^{-2}, which is the range of 180 keV proton at 0° pitch angle. Therefore, ions in the MeV/nucleon range will rapidly develop a flattening of their spectrum, the energy losses serving to heat the gas in the loop. A 1 MeV proton "trapped" in such a loop will take typically two minutes to go between mirror points and will thermalize after a few "bounces". Therefore reconnection events in loops of this size will produce a pulse of particles in the loss cone which will propagate to the transition zone in times of the order of one minute. A velocity spread of accelerated particles of 6,000 km s^{-1} produces an event of 60 s duration for a path length of 0.5 $R_\odot$. Longer path lengths, from higher loops, will result in a higher minimum velocity for particles to reach the transition zone, so the energy deposition profile will tend to be similar for all practical loop sizes.

In summary, precipitating ions below the gamma ray production threshold will either heat the corona or lose their energy at the top of the transition zone. This is consistent with the observations (Brueckner and Bartoe, 1983, Dere et al 1984) that the explosive events have no detectable signatures in the cooler, chromospheric lines. If the energy is transported by ions, momentum transfer will produce the red wing of the line, while pressure-driven material will produce the blue wing; however, there will be mass motions in all directions, giving the appearance of turbulence.

4. Energy Considerations

The concept of energy flow in the corona is shown in Figure 1. The energy for transient solar events may be traced back to photospheric motions of high-density plasma. Such motions lead to magnetic fields which extend high into the corona. How high is a matter for speculation, but recent images from Yohkoh (K. T. Strong, this workshop) and SPARTAN (J. L. Kohl, this workshop) suggest some closed field lines out to at least 2 $R_\odot$ above the photosphere. Magnetic reconnection in these large loops will lead to particle acceleration.

There is no physical reason why reconnection and particle acceleration should not be occurring quasi-continuously. There are three situations to address:

- (1) where the accelerated particles act as a seed population for further acceleration, so that after some cycles there has been sufficient build-up to provide enough energetic particles to power a flare;
- (2) where the particles populate the large-scale closed regions of the

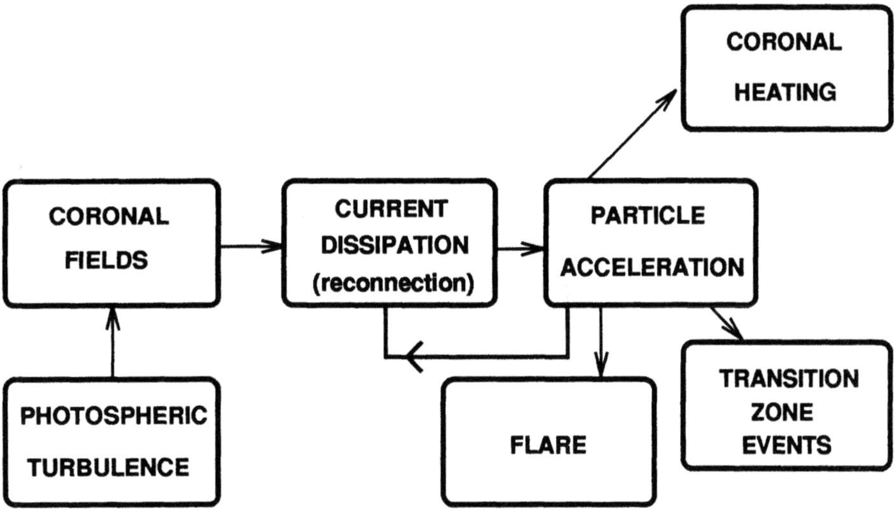

Fig. 1. The energy flow from the photosphere via coronal magnetic fields produces the particle acceleration which leads to transient events in the transition zone plus coronal heating. Major flares occur when accelerated particles serve as a seed population for reacceleration.

corona, losing their energy through collisions to the ambient gas, thereby heating it to EUV/soft X-ray emitting temperatures;

– (3) where there is the equivalent of a magnetospheric auroral event, whereby the sudden appearance of a substantial particle population in the loss cone will lead to explosive transition zone events.

Events occurring at chromospheric level, and possibly the very low corona, are normally considered to be the main sources of energy, both for flares, etc. and for coronal heating. Thus the fraction of energy escaping into the corona was all that was available for coronal heating. We are suggesting a rather different situation where the primary energy sources are in the high corona, and most of the energy is transferred to non-thermal ions. These ions are trapped in the closed coronal fields and via the chaotic nature of the reconnection events, bursts of particles are continually being dumped at sites where high coronal field lines map down to the transition zone. These sites are where the explosive events occur.

The energy in a typical explosive event is $\sim 7 \times 10^{23}$ erg (after Parker, 1988). If it has a duration of $\sim 10^2$ s and the typical dimension is 1,500 km (Henoux and Dere, 1994) this gives an average flux of protons of 0.5 MeV of 4×10^{11} cm^{-2} s^{-1}, which is a very modest number by flare standards. At the highest periods of solar activity the energy input into flares is $\sim 10^{28}$ erg s^{-1}. Parker (1988) has estimated the energy in explosive events and jets at 6×10^{27} erg s^{-1}, although he may have overestimated this due to reappraisal of the importance of jets (Henoux and Dere, 1994). Withbroe and

Noyes (1977) estimated the energy for coronal heating at 5×10^{29} erg s^{-1} . Therefore $\geq$1% of the energy dissipated in the corona is needed for transient transition zone events. Note that this partition of energy is determined (1) by the accelerated particle spectrum and (2) by the ambient gas density in the corona.

5. Other Considerations

It is fundamental to our argument that energetic particles are an important constituent of the corona. The energy source is high, some 0.5 to 1 $R_{\odot}$, or more, above the photosphere. Trapping will occur and the extent of this will depend on the height of the closed field lines; it is no accident that wherever we have sent spacecraft into large, closed magnetic fields we have found intense fluxes of non-thermal particles. We see no reason why high, closed coronal fields should be different.

Long-term particle trapping was advocated on the basis of observations over two decades ago (Simnett and Holt, 1971). Recent gamma ray observations (Kanbach et al, 1993), where >50 MeV photons were seen for over 8 hours after the 1991, June 11 flare, indicate trapping of protons > 50 MeV. This evidence, together with the Yohkoh and SPARTAN images referred to above, seems to put the existence of high, closed magnetic structures, which are central to our concepts, beyond doubt.

Additional evidence that low energy ions are very abundant during active times comes from Hα polarization observations. Recent measurements from the Mees Solar Observatory (Metcalf et al, 1993) have shown strong Hα polarization signals coming not only from the flare kernels but from locations well separated from them. The ion energy flux for the explosive transition zone events is some two orders of magnitude less that the energy in some of the Hα polarization events seen outside of the flare kernels. The higher flux means more energy in ions that can penetrate to sufficient depths to produce the Hα signal.

6. Conclusions

The explosive transition zone events give excellent cause to reconsider the way energy is transferred within the corona. The following points support our hypothesis that the events are caused by precipitating ions, mainly protons and helium nuclei, with energies typically in the few hundred keV/nucleon region:

- (1) the events avoid regions of strong magnetic field.
- (2) they occur at mid-latitudes.
- (3) they occur primarily at the transition zone at the top of the chromosphere; therefore energy input from above is favoured.

- (4) they contain both red- and blue-shifted material.
- (5) they are often seen at the boundaries of supergranualar convective cells; the weak field lines in these regions are are likely to extend into the high corona.
- (6) the energy demand is overall $\geq$1% compared to that needed to heat the corona; therefore they may be secondary effects of the coronal heating mechanism.
- (7) the durations of the events are consistent with precipitation of particles in the few hundred keV region out of a structure of typical dimension of 1 $R_\odot$.

Acknowledgements

The author thanks G. Poletto and G. Noci for their support for the SOHO workshop.

References

Brueckner, G.E. and Bartoe, J-D.F.: 1983, *Ap. J.* **272**, 329
Dere, K.P. Bartoe, J-D.F. and Brueckner, G.E.: 1984, *Ap. J.* **281**, 870
Dere, K.P. Bartoe, J-D.F. and Brueckner, G.E.: 1989, *Solar Physics* **123**, 41
Fontenla, J.M., Tandberg-Hanssen, E., Reichmann, E.J. and Filipowski, S.: 1989, *Ap. J.* **344**, 1034
Henoux, J.C. and Dere, K.P.: 1994, in *Advances in Solar Physics*, 7th European Meeting in Solar Physics, Catania, Italy
Kanbach, G. et al.: 1993, *Astron.Astrophys.Suppl.* **97**, 349
Levine, R.J.: 1974,*Ap. J.* **190**, 457
Lin, R.P., Schwartz, R.A., Kane, S.R., Pilling, R.M., and Hurley: 1984, *Ap. J.* **283**, 421
Metcalf, T., Mickey, D., Canfield, R. and Wulser, J-P.: 1993, Mees Observatory pre-print
Parker, E.N.: 1988, *Ap. J.* **330**, 474
Porter, J.G. and Dere, K.P.: 1991, *Ap. J.* **370**, 775
Porter, J.G., Fontenla, J.M. and Simnett, G.M.: 1993, *Ap. J.*, (submitted)
Porter, J.G., Moore, R.L., Reichmann, E.J., Engvold, O. and Harvey, K.L.: 1987, *Ap. J.* **323**, 380
Simnett, G.M. and Dennis, B.R.: 1986, in *Rapid Fluctuations in Solar Flares*, NASA CP-2449, ed. B.R. Dennis, L.E. Orwig and A.L. Kiplinger, p. 123
Simnett, G.M. and Holt, S.S.: 1971, *Solar Physics* **16**, 208
Tucker, W.H.: 1973, *Ap.J.* **186**, 285
Vernazza, J.E., Avrett, E.H. and Loeser, R.: 1981, *Ap.J.Suppl.* **45**, 625
Withbroe, G.L. and Noyes, R.W.: 1977, *Ann.Rev.Astron.Astrophys.* **15**, 363

JETS AND BRIGHTENINGS GENERATED BY ENERGY DEPOSITION IN THE MIDDLE AND UPPER SOLAR CHROMOSPHERE

ALPHONSE C. STERLING
*Computational Physics, Inc., Fairfax, VA 22031, USA**

KAZUNARI SHIBATA
National Astronomical Observatory, Mitaka, Tokyo, 181, JAPAN

and

JOHN T. MARISKA
E.O. Hulburt Center for Space Research, Naval Research Laboratory, Washington, D.C. 20375, USA

Abstract. Numerical simulations of energy depositions in the middle and upper solar chromosphere result in ejection of chromospheric material into the corona and heating of the chromospheric gas. These simulations may be capable of describing some of the features seen by the soft X-ray telescope on board the *Yohkoh* satellite.

Key words: Sun – Chromosphere - Heating - Spicules

1. Introduction

Sterling et al. (1991, hereafter Paper 1) and Sterling et al. (1993, hereafter Paper 2) investigated the consequences of depositing thermal energy in the middle and upper solar chromosphere using 1-D numerical simulations. Those calculations have implications for Hα, UV, and X-ray microflares, and have resulted in a new spicule model. Here we summarize the methods and results of those two papers, and discuss new implications of that previous work in terms of recently acquired data from the soft X-ray telescope (SXT) on board *Yohkoh.*

2. The Numerical Models

The models assume rigid magnetic flux tubes of length $\approx$ 20,000 km, and solve the hydrodynamic equations for conservation of mass, momentum, and energy. Approximate forms for radiation losses and heat conduction are included. A source term in the energy equation represents energy deposition into the chromosphere. Different simulations are executed by varying the duration, location, and magnitude of that energy source term. The two

* Current address: Institute of Space and Astronautical Science, Yoshinodai 3-1-1, Sagamihara, Kanagawa 229, JAPAN

Space Science Reviews **70**, 77–80.

papers explored an energy range for the input source ranging from 5 x 10^{24} to 5 x 10^{28} ergs.

Paper 1 limits the input energy source to a narrow region near the top of the chromosphere and assumes a semi-circular loop magnetic field geometry. Paper 2 allows for a broader range of source input regions and assumes a vertical open magnetic field. Other differences between the two models are of little consequence to the following discussion.

3. Results

Combining the results of Papers 1 and 2, we can make the following general statements. (We do not discuss less frequently occurring exceptions here.)

1. One of three different types of ejections of chromospheric matter into the corona results for a given set of input chromospheric energy source parameters:
   a) Pressure-Gradient Jets
   These are elongated, relatively iso-density and iso-thermal ejections resulting from the gradient in pressure between the ambient atmosphere and regions heated by the input energy. These types of jets generally form when the input energy source is distributed over the entire upper chromosphere.
   b) Two-Component Jets
   These jets are composed of a warm, denser pressure-gradient driven lower component and a cool, more rarefied shock-driven upper component. They generally form when only a localized region of the upper chromosphere is heated to sub-coronal temperatures.
   c) Gas Plugs
   These are regions of cool, dense (chromospheric) material surrounded by coronal gas, generally only a few thousand kms in extent. They form when a localized region of the chromosphere is heated to super-coronal temperatures. When the input heating is large enough, radiation and conduction at the boundary of the gas plug are not sufficient to maintain the cool gas plug for long, resulting in "evaporating gas plugs."
2. In some cases, pressure-gradient jets have properties similar to those of Hα spicules. This type of model spicule, which places the spicule driving force in the middle and upper chromosphere, differs fundamentally from model spicules resulting from earlier numerical studies by, e.g., Suematsu et al. (1982), Hollweg (1982), and Sterling & Mariska (1990), which placed the driving force near the base of the chromosphere. Several of the pressure-gradient and two-component jets have $\rho > 5 \times 10^{-13}$ g cm^{-3}, which is higher than typically quoted spicule densities. Thus we

do not include such features in our definition of model spicules. These jets may represent a category of surges (see Shibata et al. 1982).

3. Depositing energy in the chromosphere results in heating and expansion of the local chromospheric gas. When the input energy source heats the chromosphere to $> 10^5$ K, emissions in UV may result. These may, in fact, be the "UV microflares" observed by Porter et al. (1987). Similarly, heating the gas to $> 10^6$ K may produce UV and X-ray microflares. Assuming this to be the case, we find that some pressure-gradient jets, some two-component jets, and all gas plugs are associated with UV microflares. Only the gas plugs are associated with X-ray microflares. In addition, all three features would be associated with brightenings in Hα.
4. Solutions which produce jets with spicule-like features tend to *not* be associated with UV microflares, and are never associated with X-ray microflares.

4. Preliminary Comparisons with Yohkoh SXT Observations

Using data from the *Yohkoh* SXT, Shibata et al. (1992) have found that X-ray jets are a common feature of the solar atmosphere. The jets studied in that paper reached sizes of 5 x 10^3 - 4 x 10^5 km with velocities of 30 - 300 km s^{-1}. If the jets result from energy deposition in the chromosphere, then we would expect that they can be generated by the models discussed above. Shibata et al. (1992) suggest that a magnetic reconnection type of mechanism may be responsible for some jets. That view could be consistent with the models of Papers 1 and 2 if the reconnection is the source of the thermal energy deposition in the chromosphere.

In order for X-ray jets to be visible, the emission measure of the emitting gas in the regime observed by SXT would have to increase substantially over that of the background coronal plasma. This is what we find in the case of gas plug solutions: Paper 1 shows that the coronal gas below the plug is hotter and denser than the coronal gas above the plug. Thus the emission measure of the gas below is increased over the (nearly ambient) coronal gas above the plug. In a preliminary calculation for the gas plug in Fig. 6 of Paper 2, we have found that the emission measure of the gas from the region below the upward-moving gas plug is several orders of magnitude more intense than than that of the ambient corona. Here we define emission measure as $\int N_e^2 dV$, where N_e is the electron number density and dV is the volume of gas emitting in the temperature interval 1.2 – 2.5 MK. This more-intense region evolves upward with velocity $\approx 50 - 100$ km s^{-1}, and may appear as one type of X-ray jet seen by SXT. Higher velocities may result in cases of more energetic and evaporating gas plugs.

Some simulation results in Paper 1 show that in some cases a gas plug ejected from the chromosphere at one end of a coronal loop can traverse the loop and reflect off the chromosphere at the far end of the loop. We have yet to perform emission measure analyses for such a case, but we can speculate that such a feature may show up in SXT images as an X-ray loop filling from one end and extending to the other end. In some cases, the start and finish of the gas plug's trajectory would be particularly bright in X-rays because the pressure built-up is larger along the "lower legs" of the loop (see Figs. 3c and 4c of Paper 1). An example of an SXT feature that such simulations may describe is the 7 December 1991 X-ray bright point flare and associated plasma jet discussed in Strong et al. 1992. The simulations may also be able to explain some features of active region transient brightenings seen by SXT (Shimizu et al. 1992).

5. Conclusions

Numerical simulations of thermal energy depositions in the middle and upper chromosphere show, subject to the assumptions of the models, that:

1. The chromospheric gas becomes heated, and ejections of chromospheric material into the corona result. The heated chromospheric material should be visible as intensity enhancements in Hα, UV when $T_{chrom} > 10^5$ K (where T_{chrom} is the temperature the chromospheric gas is heated to at the location of energy input), and X-rays when $T_{chrom} > 10^6$ K.
2. Chromospheric jets which resemble spicules sometimes result, but jets with densities higher than those reported for spicules can also result. The spicule-like solutions are rarely associated with UV microflares, and never associated with X-ray microflares.
3. Coronal material below evolving gas plugs can have emission measure several orders of magnitude higher than that of the ambient corona. This material may appear as X-ray jets seen by SXT, and/or as other features observable by SXT.

References

Hollweg, J.V.: 1982, *ApJ* **257**, 345
Porter, J.G. et al.: 1987, *ApJ* **323**, 380
Shibata, K. et al.: 1982, *Solar Phys.* **77**, 121
Shibata, K. et al.: 1992, *PSAJ* **44**, L173
Shimizu, T. et al.: 1992, *PSAJ* **44**, L147
Sterling, A.C. & Mariska, J.T.: 1990, *ApJ* **349**, 647
Sterling, A.C. et al.: 1991, *ApJ* **381**, 313 (Paper 1)
Sterling, A.C. et al.: 1993, *ApJ* **407**, 778 (Paper 2)
Strong, K.T. et at.: 1992, *PSAJ* **44**, L161
Suematsu, Y. et al.: 1982, *Solar Phys.* **75**, 99

THE CORONAL CONTEXT OF TRANSITION REGION EXPLOSIVE EVENTS

D. MOSES and J.W. COOK
E.O. Hulburt Center for Space Research, Naval Research Laboratory

Abstract. Transition region explosive events are observed throughout the quiet Sun and represent an interesting local heating phenomenon. The coronal counterparts of these events, if they exist, were not observed in a sounding rocket campaign dedicated to this objective. The coronal instrument complement on the SOHO spacecraft provides an opportunity to extend this search for the coronal counterparts of the transition region explosive events, as well as to explore the correspondence of explosive events with large scale coronal structures, such as with coronal dark lanes.

Key words: Transition Region – Coronal Bright Points – Coronal Heating

1. Introduction

High velocity flows (> 50 km s^{-1}) are frequently found in small scale (~ 2 arc sec) quiet Sun features observed in the UV spectra of transition region material ($\sim 10^5 K$). These features are called explosive events and represent impulsive energy release in the transition region (see reviews in Cook 1991, Cook and Brueckner 1991). Explosive events are now believed to be the doppler signature of plasma jets, travelling at the local Alfven speed, which remove plasma and magnetic fields from regions undergoing magnetic reconnection. In the network, these reconnection events seem to be associated with the continual emergence of small magnetic bipoles at the supergranular cell centers and boundaries, their migration to the cell boundaries, and their eventual reconnection or coalescence with pre-existing flux. This recycling of flux takes place regardless of the epoch of the solar activity cycle. It is likely that the explosive event dynamics resemble the nanoflares Parker has proposed as the coronal heating mechanism (Parker 1988). So far explosive events are only observed at transition region temperatures ($\sim 10^5 K$). It is important to determine what, if any, coronal counterparts the explosive events may have, and to estimate the amount of plasma heating associated with an explosive event. Improved observations of the coronal phenomena associated with the reconnective explosive events observed in the transition region can be obtained with SOHO.

2. The Suborbital Campaign of 1987

An attempt at observing the coronal counterparts to explosive events was made in December 1987 with the near simultaneous sounding rocket flights of the NRL UV stigmatic spectrograph (HRTS) payload and the AS&E X-ray

Space Science Reviews **70**, 81–84.

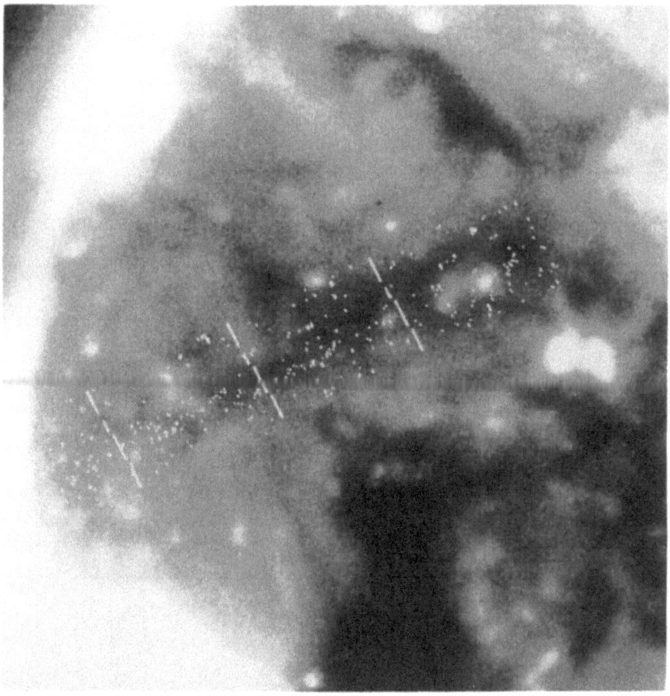

Fig. 1. X-ray image and superposed HRTS C IV explosive event map. Explosive events are more concentrated in the coronal dark lane region of the X-ray image.

imaging payload (Moses et al. 1994). In general, a direct coronal counterpart to the explosive events was not observed. The locations of explosive events compared to coronal structure are shown in Figure 1. Sites of small scale events in the HRTS C IV 1548 Å raster spectra with doppler signatures greater than 50 km s^{-1} are marked by points in Figure 1. This explosive event map is superposed on an approximately 900 arc sec x 900 arc sec coregistered portion of the soft X-ray image.

It can be seen that transition region explosive events did not correspond to coronal X-ray Bright Points (XBPs). However, a correlation was observed between the explosive events and the overlying large scale coronal structure. Throughout the X-ray image a pattern of dark lanes was evident in quiet Sun areas. One of these coronal dark lanes occurred in the region covered by the HRTS raster. The number density of C IV events was twice as large inside as outside this dark lane (4.6×10^{-3} vs. 2.3×10^{-3} explosive events per arc sec^2). The dark lane corresponded to an old magnetic neutral line. It seems likely that the magnetic configuration of the dark lane is conducive to small scale convergence and reconnection of opposite polarity magnetic features.

3. The SOHO Contribution

SOHO provides an excellent opportunity to extend this work on the coronal counterparts of transition region explosive events. SUMER has the UV spectral and spatial resolution needed to identify explosive events. The contrast between the C IV line center and the doppler broadened wing of a typical explosive event is such that efficient detection of explosive events can still be expected with the photon counting detectors being built for SUMER. EIT has the bandpass, spatial resolution, and temperature coverage to produce coronal images comparable to the AS&E sounding rocket instrument. Further insight will certainly be obtained from continual observation of the Sun compared to the "snapshot" nature of sounding rocket observations. However, the greatest advancement may come from the "cooler" EIT channels that span the gap between the highest temperature structures observed by the HRTS (O V at 2.5×10^5 K) and the lowest temperature plasma observable in the X-ray filtergrams ($\sim 1.5 \times 10^6$ K). Plots of the temperature coverage of the AS&E X-ray imager (Waljeski 1992) and the SOHO EIT (Dere 1993) are presented in Figure 2 for comparison. Since the basic result of the rocket study was that there were no X-ray coronal counterparts to the explosive events, perhaps the counterparts may be found at an intermediate temperature.

To further advance these studies, two sounding rocket SOHO "underflight" support missions were proposed and accepted. A flight of the HRTS instrument in essentially the same configuration as the 1987 flight (Brueckner 1993) will provide simultaneous coverage of the chromospheric and transition region signatures of explosive events. This should greatly enhance the SUMER observations, which are limited by dynamic range and wavelength coverage (i.e. the range of temperatures that can be simultaneously observed). A second sounding rocket flight, of the spare SOHO EIT instrument with a higher resolution focal plane detector (Moses 1993), will enhance the EIT observations by detection of explosive event counterparts that may be too small and faint to be detected by the SOHO EIT instrument (which must image the full disk). Such observations may finally test the importance of the explosive events to the energy balance of the upper solar atmosphere.

4. Acknowledgements

This work is funded by the NASA International Solar Terrestrial Physics Program and the Office of Naval Research. Support of the suborbital program efforts described in this paper is provided by NASA DPR W-14,541, DPR H-06877D, NAS5-25496, and NAS5-31619. Special thanks to Drs. K. Waljeski and K. Dere for assistance in the preparation of this paper.

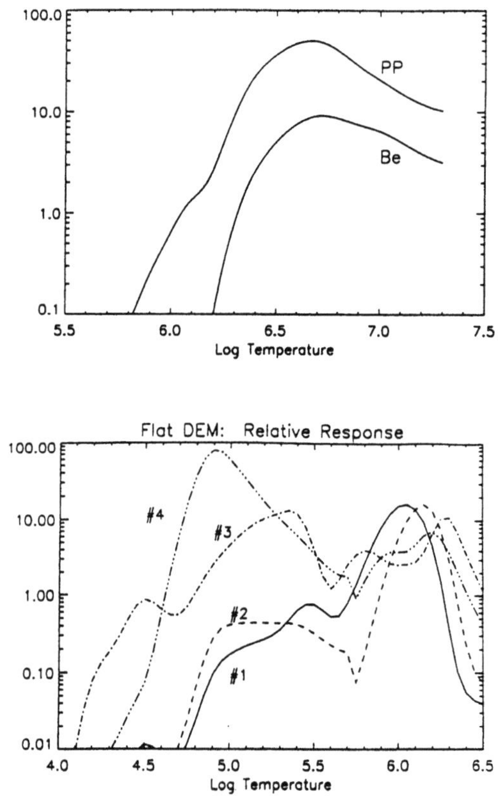

Fig. 2. The wavelength response of each instrumental observing channel convolved with a model solar spectrum for various temperatures. The relative temperature response of the AS&E X-ray imaging telescope (2 filter choices, upper plot) and the SOHO EIT (4 multilayer choices, lower plot) are shown. (The vertical axis is the relative instrumental response.)

References

Brueckner, G.E. (PI): 1993, *Very High Resolution UV Observations of the Sun*, NASA SR&T/Suborbital Program Proposal

Cook, J.W.: 1991, in *Mechanisms of Chromospheric and Coronal Heating*, P. Ulmschneider, E. Priest, and R. Rosner eds., Springer, p. 83

Cook, J.W. and Brueckner, G.E.: 1991, in *The Solar Interior and Atmosphere*, A.N. Cox, W.C. Livingston, and M.S. Matthews eds., Univ. of Arizona Press, p. 996

Dere, K.P.: 1993, Proceedings of the LASCO/EIT Consortium Meeting, Crystal City, VA

Moses, D. (PI): 1993, *Sounding Rocket Support for SOHO Investigations of the Plasma Properties of Solar Structures*, NASA SR&T/Suborbital Program Proposal

Moses, D., Cook, J.W., Bartoe, J.-D., Brueckner, G.E., Dere, K.P., Webb, D.F., Davis, J.M., Harvey, J.W., Receley, F., Martin, S.F., and Zirin, H.: 1994, *ApJ* , in press

Parker, E.N.: 1988, *ApJ* **330**, 474

Waljeski, K. M. H.: 1992, *Plasma Properties of Solar Coronal Active Regions*, Ph.D. Dissertation, Brandeis University:Waltham

THE EXTENSION OF EXPLOSIVE EVENTS FROM THE TRANSITION REGION TO THE CORONA

O. KJELDSETH–MOE
Institute of Theoretical Astrophysics, University of Oslo, Norway

and

C. C. CHENG
E.O. Hulburt Center for Space Research, US Naval Research Laboratory, Washington, D. C., USA

Abstract. We describe the properties of high velocity events in the corona and upper transition region and propose that they are the same phenomenon as the well studied explosive events seen in the lower transition region around T=10^5 K. Furthermore, we discuss how the SOHO spectrometers, CDS and SUMER, may be used to check this conjecture. Magnetic reconnection has been considered a strong candidate for the physical mechanism causing explosive events. We present a phenomenological model showing how some of the observed properties of explosive events may be explained by reconnection occurring in small magnetic loops.

1. Introduction

In this paper we will make two points:

- That energetic *high velocity events* observed in the *corona* and throughout the transition region constitute the same phenomenon as the *explosive events*, mainly observed in the C^{+3} lines at 155 nm with the High Resolution Telescope and Spectrograph, HRTS, and
- That important properties of explosive events may be explained assuming them to be flows from a reconnecting region in a small magnetic loop of length less than 3000 km.

2. Observations of Explosive and High Velocity Events

Transition region explosive events have been extensively studied and described by Brueckner (1980), Brueckner and Bartoe (1984), Cook *et al.* (1987), Porter *et al.* (1987), Dere, Bartoe and Brueckner (1989), and Dere *et al.* (1991). Most of the observations have been made using the C^{+3} lines at 155 nm, formed at T $\approx 10^5$ K. Detailed properties of explosive events are thus limited to their appearance in the lower transition region.

The coronal high velocity events, observed with the slitless spectrograph, SO82A, on Skylab, were only briefly mentioned by Brueckner, Patterson and Scherrer (1976). New quantitative descriptions based on a more extensive material from SO82A, have been given by Cheng and Kjeldseth–Moe

Space Science Reviews **70**, 85–88.

(1989, 1990), Kjeldseth–Moe and Cheng (1990,1991,1992) and Kjeldseth–Moe (1991, 1992). In the wavelength range of SO82A, 17–63 nm, one observes lines emitted at temperatures ranging from the low transition region (the He^{+0} and He^{+1} lines) to the corona (i.e. the Fe^{+15} lines emitted at $T \approx 2.5$ MK).

3. Are Explosive Events and Coronal High Velocity Events Identical?

Table 1 summarizes and compares the properties of explosive events and coronal high velocity events. There are clearly a number of similarities.

TABLE I

Explosive Events and Coronal High Velocity Events

Properties	Explosive events	Coronal HVE
Average extent	$\approx$ 1500 km	$\approx$ 2500 km
Range in extent	$\approx$ 1000–3000 km	$\approx$ 2000–5000 km
Velocity dispersion,	$\approx$ 200 km s^{-1}	$\approx$ 400 km s^{-1}
average and range	$\approx$ 100–400 km s^{-1}	$\approx$ 200–1350 km s^{-1}
Average lifetime	$\tau \approx 40$ s	Uncertain, $\tau \approx 1$ min
Range in lifetime	$\tau < 20$ s to minutes	Not known
Direction of flows	B/R/E: 35–25–45[(1)]	Probably isotropic
Lateral motion	$v_{Lat} < 20$ km s^{-1}	$v_{Lat} < 25$ km s^{-1}
Positional shift B/R wings	$\pm$ 500 km	700–800 km
Temperature range	$5{\cdot}10^4 K<T<2{\cdot}10^5 K$	$1{\cdot}10^5 K<T<2.5{\cdot}10^6 K$
Electron pressure[(2)]	P_e=1.65 dyn cm^{-2}	$P_e \geq 1.4$ dyn cm^{-2}
Location in latitude	Extended, mid latitudes	Up to 60
Location rel. to features	Edges of magnetic regions	Both Q.R. and A.R
Global frequency[(3)]	4000 or $5.5{\times}10^4$	Not known
Full disc birthrates[(3)]	44 s^{-1} or 600 s^{-1}	Not known

1) Isotropic in appearance. Respectively emission in the blue, red or both wings of the C^{+3} lines prominent in the given ratio.

2) From Dere et al. (1991) for EE and from Vernazza and Mason (1978) for CHVE.

3) Values according to Cook *et al.* (1987) and Dere, Bartoe and Brueckner (1989), respectively.

Spatial extents are very similar. Likewise the position asymmetry, the shift with wavelength in the feature position across the direction of dispersion, is the same for the two phenomena.

Life times are not known for the coronal high velocity events since there are few time series in the SO82A material. Furthermore, the exposure time used, 60 seconds, is long compared with the expected life time.

Locations and velocity dispersions are also very similar. The values for the coronal high velocity events may be caused by Alfvén velocities being generally higher at high temperatures or higher for the perhaps intrinsically stronger events observed with the Skylab instrument.

The most striking discrepancy between the types of events is the difference in frequency and birth rate. These are very uncertain even for the well studied explosive events. Thus, the two values quoted in Table 1 derive from different series of observations with the same instrument, HRTS.

The SO82A spectroheliograms show at most a few coronal events on the disc at any time. However, in the heliograms only the strongest events, i.e. those having highest contrast, may be distinguished from the background of other solar features. Since explosive events are not particularly bright, it may be impossible to detect most of their upper transition region and coronal counterparts in the heliographic images.

Thus, in spite of the difference in apparent frequency, the explosive events and the coronal high velocity events may be the same phenomenon. This conjecture can be checked, using the UV spectrometers on SOHO. CDS will have sufficient spectral resolution to measure the maximum velocity, and may, with the advantage of a slit spectrograph, observe the true frequency of these in the upper transition region and corona. SUMER will improve on the detailed velocity studies of HRTS and extend the observations to higher temperatures, $T=3\cdot10^5$ K.

4. High Velocity Events as Supersonic Flows in Small Loops

It has been argued that the high velocity or explosive events are caused by magnetic reconnection. More properly one should speak of a reconnection–like instability, since in this case one has an event and not an ongoing and "steady" process.

A simple, conceptual model for this has been tried, where gas flows at supersonic speeds (i.e. the Alfvén speed) out of a small reconnecting region along the legs of a magnetic loop. This flow may give the observed, strongly shifted emission in the transition region and coronal lines, both in the blue and red directions.

The calculations have been done for a number of loop and flow parameters of which the most significant are: the size of the loop relative to the width of the spectrograph slit or angular resolution of the instrument, the angle between the axis of the loop and the direction of the slit, and the position of the reconnection or instability point along the loop. Several simplifying assumptions have been made:

- The loop is an half–circular arch
- Both the velocity and the line emission are unchanged along the full

length of the loop

- The loop is located at heliocentric angle 45°, since explosive events appear to cluster at this latitude

The properties of explosive events that we particularly have tried to simulate, are:

- The shift in position between the blue and red wings of the emission, and
- The 35:25:45 distribution between the blue, red and "explosive" (i.e. having simultaneously blue- and redshifted emission) events.

Comparing the model results with the observations, one may conclude that these properties may be simulated only if the loop is fairly small, not much exceeding 4″, or 3,000 km, in length. Furthermore, the observed average redshift of the C^{+3} 155 nm lines of 5 km s^{-1} has to be included to obtain the distribution among the categories of blue, red, and "explosive" events.

For a "long" loop, i.e. L > 5″, there is little agreement between the simulated and observed explosive events. The simulated features are too extended angularly, reflecting the large size of the loop, with too few "explosive" cases, compared with the purely red or blue shifted explosive events.

References

Brueckner, G.E., Patterson, N.P. and Scherrer, V.E.: 1976, *Solar Physics* **47**, 127

Brueckner, G.E.: 1980, *Highlights of Astronomy* **5**, 557

Brueckner, G.E. and Bartoe, J.-D.F.: 1984, *Astrophysical Journal* **272**, 329

Cheng, C.-C. and Kjeldseth-Moe, O.: 1989, *Bulletin of the AAS* **21**, 835

Cheng, C.-C. and Kjeldseth-Moe, O.: 1990, 'XUV Streaks Observed from Skylab' in B. Schmieder and Eric Priest, ed(s)., *Flares 22 Workshop: "Dynamics of Solar Flares"*, Observatoire de Paris, DASOP, 101

Cook, J.W., Lund, P.A., Bartoe, J.-D.F., Brueckner, G.E., Dere, K.P. and Socker, D.G.: 1987, 'Statistical Properties of Small High-Velocity Events in the Solar Transition Region' in J.L. Linsky and R.E. Stencel, ed(s)., *Cool Stars, Stellar Systems and the Sun, 5th Cambridge Workshop, Boulder, Colorado July 7-11, 1987*, Springer, 150

Dere, K.P., Bartoe, J.-D.F., and Brueckner, G.E.: 1989, *Solar Physics* **123**, 41

Dere, K.P., Bartoe, J.-D.F., Brueckner, G.E., Ewing, J., and Lund, P.: 1991, *Journal of Geophysical Research* **96**, 9399

Kjeldseth-Moe, O.: 1991, *Ann. Geophys. Suppl.* **9**, C297

Kjeldseth-Moe, O.: 1992, *Ann. Geophys. Suppl.* **10**, C441

Kjeldseth-Moe, O. and Cheng, C.-C.: 1990, 'Explosive Events Observed in the Solar Corona' in E. Leer and P. Maltby, ed(s)., *Mini-Workshop on Physical Processes in the Solar Transition Region and Corona*, ISBN 82-7121-004-1, 37

Kjeldseth-Moe, O. and Cheng, C.-C.: 1991, 'Energetic High Velocity Events in the Solar Transition Region and Corona' in E. Leer and P. Maltby, ed(s)., *Mini-Workshop on Coronae and Winds in Late-Type Stars*, ISBN 82-990556-3-6, 55

Kjeldseth-Moe, O. and Cheng, C.-C.: 1992, 'Reconnection and Energetic High Velocity Events in the Solar Transition Region and Corona' in P. Maltby, ed(s)., *Mini-Workshop on Magnetic Reconnection in Physics and Astrophysics*, ISBN 82-7121-010-6, 147

Porter, J.G., Moore, R.L., Reichmann, E.J., Engvold, O. and Harvey, K.L.: 1987, *Astrophysical Journal* **323**, 380

FINE STRUCTURE AND THE EMISSION FILLING FACTOR

O. KJELDSETH–MOE, N. BRYNILDSEN, P. BREKKE and P. MALTBY
Institute of Theoretical Astrophysics, University of Oslo, Norway

Abstract. There is observational evidence for an extreme fine structure in the solar transition region, much smaller than 1″ in size (Dere *et al.*, 1987, 1988) Corresponding to this extreme fine structure there appear to be an equally complex dynamical structure. We review the evidence for such dynamical extreme fine structure as demonstrated by the frequent appearance of multiple velocities, i.e. distinctly different velocities in the transition region occurring within the angular resolution element. Multiple velocities are prominent in active regions and particularly near sunspots, where velocity components may be supersonic. However, multiple velocities are frequent also in quiet regions. The consequences of such fine structure for modeling the transition region will be outlined. Finally the appropriate CDS and SUMER observations needed to extend our knowledge of a finely structured transition region and corona, spatially or in time, are discussed.

1. Introduction

We wish to emphasize the observed multiple velocities in the transition region as clear evidence of its fine scale structure on a scale $\phi \ll 1''$. Multiple velocities are the simultaneous existence of distinctly different velocities seen within an angular resolution element of $1''\times 1''$.

A filamented transition region may consist of twisted strands or fibrils inside complex loops. Each fibril has a velocity associated with it. These may range from near zero in fibrils where the gas is not flowing, to supersonic velocities in fibrils where one has a standing shock. The flows may go in several directions, since both the fibrils and the enveloping structure may be twisted. Since a number of fibrils, with individual sizes of 4–40 km, are likely to occur within a resolution element, the resulting line profiles may show the multiple velocities.

Multiple velocities thus constitute new evidence for a transition region with an extremely fine spatial scale, and give an additional opportunity for studying the properties of this fine structure. Furthermore, the complex dynamics of the transition region is in itself of great significance for the understanding and modelling of its structures and properties.

2. Evidence for a Filamentary Fine Structure

The likely presence of a very fine scale structure has been pointed out on several occasions, based on energy considerations (e.g. Nicolas and Kjeldseth–Moe, 1979; Kjeldseth–Moe *et al.*, 1984), and on the presence of transition region emission at high altitudes above the limb (Feldman, 1983).

Lately, and most thoroughly, the sub–resolution fine structure has been discussed by Dere *et al.* (1987, 1988). Their analysis derives filling factors

Space Science Reviews **70**, 89–92.

by comparing observed emission measures, measured or estimated electron densities, and spatial extent of emitting regions of the C^{+3} lines at 155 nm, using spatially resolved UV spectrograms from the High Resolution Telescope and Spectrograph – HRTS.

The observed extent of emission features on the disc in the C^{+3} lines is typically L ≈ 2500 km. From limb observations the vertical extent is of the same order. This "macro" structure may correspond to the size of an individual, small magnetic loop. The electron density was found from the density sensitive O^{+3} intercombination lines near 140 nm. Where these lines were too weak, an electron pressure of 0.14 dyn cm^{-2}, appropriate for the quiet Sun, was assumed. Filling factors were derived for each "macro" structure, obtaining values of ≈ 0.0001–0.01. Depending on the number of fibrils inside a "macro" structure the corresponding fibril diameters are 4 to 40 km.

3. Multiple Velocities and a Dynamical Transition Region

The HRTS spectrograms shows that areas with a complex velocity structure, i.e. multiple velocities, occur frequently. In these regions the line profiles cannot by any close approximation be represented with a single Gaussian. Examples of such line profiles are given in Figure 1.

Multiple velocities were first observed in or near sunspots, where they are commonly occurring in the sunspot downflow regions (Kjeldseth–Moe *et al.* 1988, 1993) At these locations one or more of the components show supersonic downflow velocities typically in the range 50–80 km s^{-1}, but speeds as high as 200 km s^{-1} have been observed (Brekke, Kjeldseth-Moe and Brueckner, 1990). The high flow velocity is the reason why multiple flows were first recognized in these observations.

It is more interesting that multiple velocities are frequently seen away from sunspots, in both quiet and active locations. In quiet regions multiple velocities are observed at approximately 15% of all locations (Brekke *et al.*, 1992). In active regions they are even more prevalent. This estimate of the frequency of multiple velocities is conservative. Only distinct multiple components, with intensities well above the noise level, have been included.

Line profiles with multiple velocity show varying degrees of complexity, having two or three distinguishable components, in addition to emission from gas nearly at rest, which is always present. Typically the higher speed for a pair of multiple velocity components is in the range 30–50 km s^{-1}, i.e. near the sound speed. Very high velocities of 100 km s^{-1} are present in about 1% of the quiet regions (Brekke *et al.*, 1992).

Both red– and blue–shifted components are present at a number of locations between Sun center and limb. Thus, we have flows occurring in all

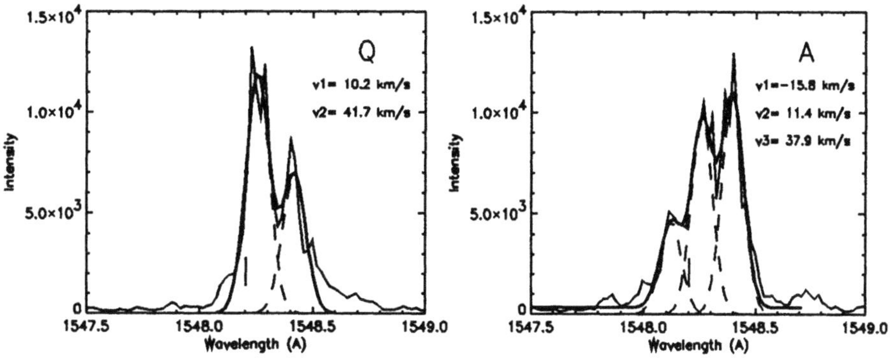

Fig. 1. Sample profiles of the C^{+3} at 154.8 nm, in active and quiet regions, showing multiple velocities. Figure shows the observed profiles and the results of a 2 and a 3 component fit.

directions. Flow velocities are both sub– and supersonic, and as many as three distinct multiple components can be distinguished in some cases.

Furthermore, there may be more multiple velocities than is possible to distinguish in the data. Multiple components have only been fitted to line profiles that show an obvious complex velocity structure or asymmetry.

4. A Dynamical Model of the Transition Region

Our proposed model depicts a transition region consisting of fine strands or fibrils inside larger "macro" structures, which are also of a limited size, L $\leq$ 3000 km. Each fibril would then be nearly isothermal, but individual fibrils might exist with a range of temperatures. The fibrils would be thermally isolated from each other by the magnetic field. This might tie in with the model of Antiochos and Noci (1986) for the low transition region.

Flows in the fibrils might be caused by the siphon mechanism. With the high electrical conductivity in the solar atmosphere, it is likely that the flow is directed along the magnetic field lines. The velocities will be either subsonic throughout the fibril loop or the flow could become supersonic at a critical point, which, for a twisted fibril, may not be near its summit, and remain so until a shock is reached.

Line profiles consisting of one or more subsonic Gaussian line components, may be compatible with one or several siphons with subsonic velocities.

Observed line profiles with one subsonic and one supersonic component may be interpreted as caused by a shock. Since the transition region lines are formed within a restricted range in temperature, the temperature change through the shock will have to be small.

In some cases, particularly near sunspots, the observed line profiles have 3 components, of which one or two may be supersonic. These types of line profiles suggest that gas in more than one flux tube contributes to the observed line emission. The fraction of observed line profiles that require 3 components amounts 30%, which is too large to be neglected (Kjeldseth–Moe *et al.*, 1993).

It is obvious that the flows might become very complex in a fibril siphon flow. With simple siphon flows along typical loops we might expect regular patterns of upward and downward directed flows. This is not found in the HRTS–rasters from Spacelab 2 which we have studied.

5. Observing Multiple Velocities with CDS and SUMER

SUMER has sufficient angular and spectral resolution to register transition region fine structure and to study multiple velocities. It will extend the available range in temperature in the lower transition region to f.ex. include the O^{+5} resonance lines at 103.2 nm and 103.7 nm.

With CDS a temperature range from 10^5 K to 10^7 K is available, i.e. the instrument covers both the transition region and the corona. The spectral resolution of CDS is sufficient to observe the supersonic components of multiple velocities in the transition region.

Both instruments have sufficient S/N to allow the detection of multiple velocity components, and their time resolution is a few seconds in the stronger lines allowing them to study a fine structured, non–static transition region and corona.

References

Antiochos, S.K. and Noci, G.: 1986, *Astrophysical Journal* **130**, 440

Brekke, P., Kjeldseth–Moe, O., and Brueckner, G.E.: 1990, *Astrophysics and Space Science* **170**, 135

Brekke, P., Brynildsen, N., Kjeldseth–Moe, O., Maltby, P., and Brueckner, G.E.: 1992, *26th ESLAB Symposium ESA SP–346* , 211

Dere, K.P., Bartoe, J.-D.F., Brueckner, G.E., Cook, J.W. and Socker, D.G.: 1987, *Science* **238**, 1237

Dere, K.P., Bartoe, J.-D.F., Brueckner, G.E., Cook, J.W. and Socker, D.G.: 1988, *Solar Physics* **114**, 223

Feldman, U: 1983, *Astrophysical Journal* **275**, 367

Kjeldseth–Moe, O., Andreassen, T., Andreassen, Ø, Maltby, P., Bartoe, J–D.F., Brueckner, G.E. and Nicolas, K.R.: 1984, *Adv. Space Research* **4**, 63

Kjeldseth–Moe, O., Brynildsen, N., Brekke, P., Engvold, O., Maltby, P., Brueckner, G.E., Cook, J.W., Dere. K/P. and Socker, D.G.: 1988, *Astrophysical Journal* **334**, 1066

Kjeldseth–Moe, O., Brynildsen, N., Brekke, P., Maltby, P. and Brueckner, G.E.: 1993, *Solar Physics* **145**, 257

Nicolas, K.R. and Kjeldseth Moe, O.: 1979, *Bulletin of the AAS* **11**, 420

TURBULENT TRANSITION IN SOLAR SURGES

R. B. DAHLBURG and J. T. KARPEN
Naval Research Laboratory, Washington, D.C., USA

Abstract. It has been suggested that a surge can be modelled as a jet travelling in a sheared magnetic field, and that the transition to turbulence of this "MHD tearing jet" can explain several key observed features. In this paper we present our preliminary results of the transition to turbulence *via* secondary instabilities of the MHD tearing jet. Our results confirm that turbulent transition can decelerate the surge, with decay times which compare well with surge data. Furthermore, we find that the turbulent MHD tearing jet forms magnetic field-aligned velocity filaments similar to those often observed in the surge flow field.

Key words: Solar surges – magnetohydrodynamics – turbulence

1. Introduction

Solar surges are collimated mass eruptions often associated with sites of magnetic field parasitic polarity. Carbone *et al.* (1987) have suggested that the surge can be modelled as a jet travelling in a sheared magnetic field, which we call a "MHD tearing jet". Using an analytic turbulence model, their research indicated that transition to turbulence could explain several features of the surge, including deceleration and the early time correlation with Type III radio bursts. In this paper we present some preliminary results of our extended studies of this simplified surge model. We show elsewhere that the MHD tearing jet can transition to turbulence *via* a secondary instability process (Dahlburg and Karpen 1993). There are three steps in this process: [1] the 1d primary quasi-equilibrium is first destabilized by a linear 2D resistive primary disturbance (*e.g.*, Einaudi and Rubini, 1989); [2] the primary disturbance grows until it saturates to form a 2D secondary equilibrium state (Ofman *et al.*, 1993); and then [3] the secondary equilibrium state is destabilized by a 3D secondary instability which grows on the Alfvénic timescale (Dahlburg and Karpen, 1993). In this paper we consider what happens as the secondary mode attains a finite amplitude, and relate this process to the evolution of the surge.

2. Methodology

Following Carbone *et al.*, we model the surge as an incompressible jet travelling in a sheared magnetic field. The mean velocity and magnetic fields are specified, respectively, to be: $U_0(z)\hat{\mathbf{e}}_x = \mathrm{sech} z\hat{\mathbf{e}}_x$, and $B_0(z)\hat{\mathbf{e}}_x = \tanh z\hat{\mathbf{e}}_x$. Primary perturbations of this system are determined by an equation of the Orr-Sommerfeld type (Einaudi and Rubini, 1989), of which the fastest growing modes are 2D (Michael, 1955). We initialize our 3D calculations with a

Space Science Reviews **70**, 93–96.

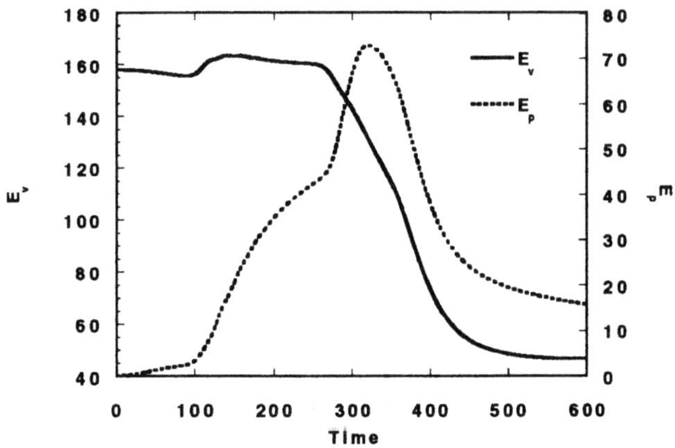

Fig. 1. Total kinetic energy (E_v) and the perturbed kinetic energy (E_p) as functions of time. Times are in units of the Alfvén transit time for the initial magnetic shear width. Velocities are measured in units of the far field Alfvén speed.

dose of the unstable 2D primary mode large enough that the system begins close to a secondary equilibrium. Then we add a smaller amount of a 3D primary mode to initiate activity in the third direction. The numerical simulations are performed with a semi-implicit Fourier collocation – Fourier pseudospectral algorithm which we have generalized from its previous application to neutral sheets (Dahlburg *et al.*, 1992).

3. Results

Figure 1 shows the time evolution of the total kinetic energy, E_v, for a run with $R = R_m = 200$, and streamwise (x) and spanwise (y) wavenumbers equal to 0.5. There is a slight increase in E_v around $t = 100$, followed by a very rapid decay phase initiated at about $t = 260$. The kinetic energy subsequently decays to about 30% of its initial value in roughly 200 Alfvén times, reflecting the rapid deceleration of the jet. Figure 1 also shows the perturbed kinetic energy, E_p, of the jet. The secondary mode is evident in E_p around $t = 90$. The perturbed kinetic energy increases its value by a factor of about 20 within 220 Alfvén times, followed by a rapid decrease in value of 80%. A comparison of E_v with E_p indicates that the perturbed velocity field is a significant fraction of the mean velocity field, with $\Delta v/v \approx 1/2$ or larger for the later portions of the simulation.

To illustrate better the morphological evolution of the surge model, we next show two plots of the velocity field magnitude ($|\mathbf{v}|$), with isosurfaces chosen at 90% of the peak velocity magnitude. Figure 2 shows a velocity magnitude isosurface at $t = 90$. The cross-stream direction (z) is here vertical, so the contours vary strongly in this direction. At this time the velocity

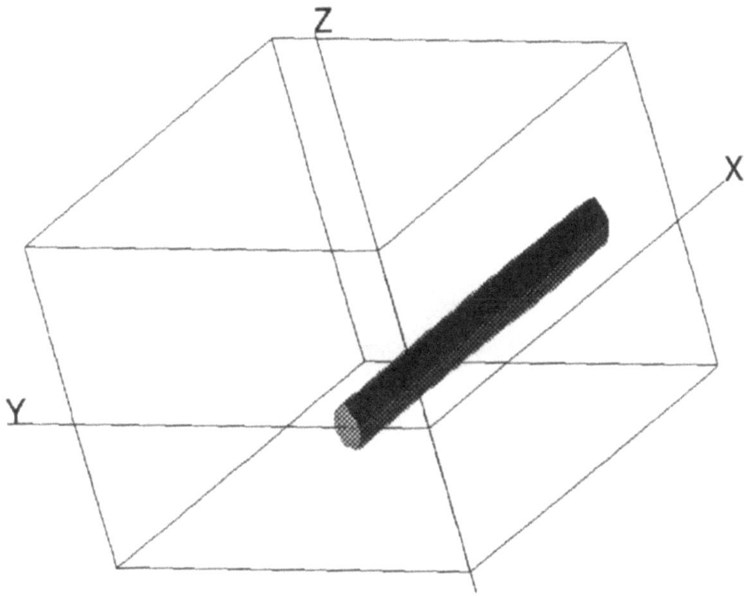

Fig. 2. Contours of velocity field magnitude at $t = 90$. The isosurface has a value of .93.

field is dominated by the saturated 2D mode, hence most of the additional structure is in the streamwise (x) direction. The 3D disturbance is visible from the wavelike irregularity in the spanwise (y) direction. Figure 3 shows an isosurface of $|\mathbf{v}|$ at $t = 360$, when the perturbed kinetic energy is near its maximum. At this time the secondary mode has developed to the point that the structural variation is very large in the spanwise direction. A tube-like feature aligned with the streamwise direction is evident at this time. This velocity configuration resembles the fine structure often observed in surge flow fields (*e.g.*, Foukal 1990).

4. Discussion

Our simulations indicate that the MHD tearing jet is a promising model for the solar surge. Transition to turbulence of this configuration decelerates the jet, and also leads to the development of magnetic field-aligned velocity filaments. Our simulation has a duration of 600 Alfvén times. For a physical Alfvén time of $0.2\ s - 2\ s$ this corresponds to an event with a duration of about 2 to 20 minutes. Hence our event times are consistent with typical observed surge lifetimes, *viz.*, 10 - 30 minutes.

Note, however, that because of our initialization, we did not reproduce the growth of correlation between the fluctuating velocity and magnetic field obtained by Carbone *et al.*. This growth of correlation is important because it is a way to eliminate the perturbed electric field responsible for

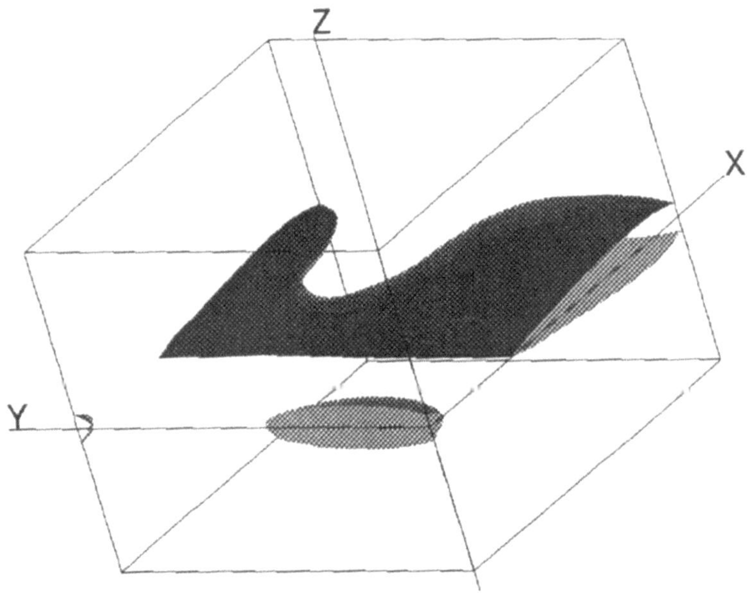

Fig. 3. Isosurface of velocity field magnitude at $t = 360$. The isosurface has a value of .79.

accelerating the electrons which produce Type III radio bursts. Our initial fluctuations, the unstable primary eigenmodes, are uncorrelated, and hence they remain uncorrelated during the simulation (*e.g.* Dahlburg *et al.*, 1988 and references therein). To obtain correlation growth for the MHD tearing jet, a seed correlation is required in the initial configuration (Einaudi, 1993 private communication). We are currently working on this aspect of the surge problem.

Acknowledgements

The calculations reported here were performed on the NASA Ames Research Center Cray Y-MP under a grant from the NAS program. This work was supported by the NASA Space Physics Theory Program and the NASA SR&T Program.

References

Carbone, V., Einaudi, G., and Veltri, P.: 1987, *Solar Phys.* **111**, 31
Dahlburg, R. B., Picone, J. M., and Karpen, J. T.: 1988, *J. Geophys. Res.* **93**, 2527
Dahlburg, R. B., Antiochos, S. K., and Zang, T. A.: 1992, *Phys. Fluids B* **4**, 3902
Dahlburg, R. B., and Karpen, J. T.: 1993, *Astrophys. J.* , submitted
Einaudi, G., and Rubini, F.: 1989, *Phys. Fluids B* **1**, 2224
Foukal, P.: 1990, *Solar Astrophysics*, Wiley: New York, 354
Michael, D. H.: 1955, *Proc. Cambridge Phil. Soc.* **51**, 528
Ofman, L., Morrison, P. J., and Steinolfson, R. S.: 1993, *Phys. Fluids B* **5**, 376

OBSERVED REDSHIFTS OF TRANSITION REGION/CORONA LINES

P. BREKKE
Institute of Theoretical Astrophysics, University of Oslo

Abstract. Solar UV observations reveal a redshifted emission at transition region temperatures, commonly interpreted as a net downflow of plasma. In earlier investigations the magnitude of the redshift has been found to increase with temperature, reaching a maximum at T= 10^5 K, and then to decrease towards higher temperatures. These observations, mostly from Skylab, suggested no significant shift of the O V line at 1218 Å formed at 2.4 $\times 10^5$ K. The variation of the downflow velocity with temperature is, however, uncertain since there are few reliable observations of lines formed at higher temperatures.

Using spectrograms from the High Resolution Telescope and Spectrograph - HRTS we find an average net redshift of the O V lines at 1218 Å and 1371 Å at all locations extending from disk center to solar limb. A discrepancy between the observed flow velocity in the two lines is probably caused by uncertainty in the available laboratory wavelength of the intercombination line at 1218 Å ($2s^2\ ^1S_0$—$2s2p\ ^3P_1$).

The observed shift in O V is compared with corresponding measurements of lines formed at other temperatures (Si IV, C IV, N IV, O IV, and Fe XII). Large variations in the shift are found along the instrument slit. Thus, blueshifts are also observed with the sites of the largest upflow located in the sunspot umbrae and in a quiet region close to an active region.

Key words: Transition Region – Corona – Mass flow – UV Radiation

1. INTRODUCTION

Ultraviolet observations of the Sun show a net redshift near disk center in the emission lines from the transition region. During the last decades, this phenomenon has been observed with several UV instruments with different spatial resolution (e.g. Doschek, Feldman, & Bohlin 1976; Hassler, Rottman, & Orral 1991). Systematic redshifts have also been observed in stellar spectra of late type stars (e.g. Ayres, Jensen, & Engvold 1988). Both solar and stellar flows in the transition region have been most extensively studied at temperatures around T = 10^5 K. Measurements of the variation of the flow with temperature are somewhat ambiguous. However, the net downflow velocity has been observed to increase with temperature, reaching a maximum at T= 10^5 K and then decreasing again at higher temperatures.

In this paper we present measurements of Doppler shifts in the O V lines at 1218 Å and 1371 Å obtained during two rocket experiments with the High Resolution Telescope and Spectrograph – HRTS (Bartoe & Brueckner 1975). We find an average net redshift of the O V lines at 1218 Å and 1371 Å at all locations extending from disk center to solar limb. Several workers have

Space Science Reviews **70**, 97–102.

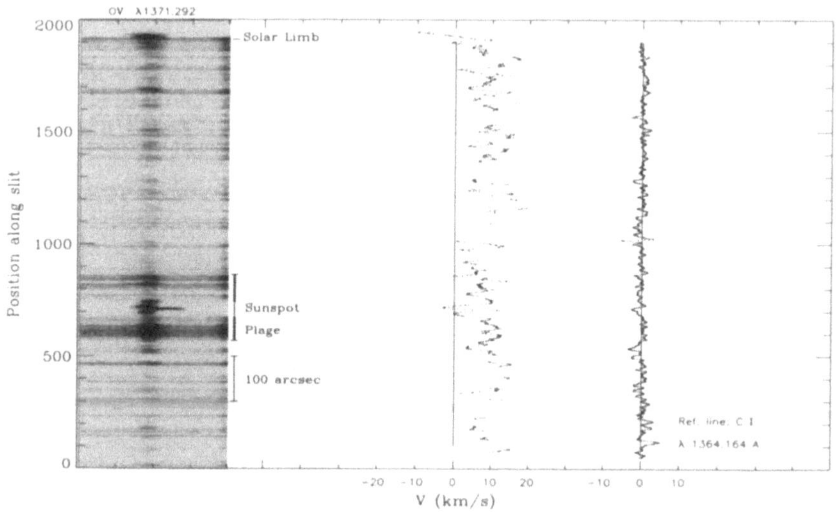

Fig. 1. Spatial variation of line-of-sight velocities in the O V line at 1371 Å and a chromospheric reference line (C I λ1364.164 Å) observed with HRTS 1. Positive values are redshifts. To the left is a section of the HRTS 1 spectrogram showing the O V profile extending from approximately disc center to the solar limb. The active region McMath 13766 has been marked. The scale along the spectrograph slit is given in pixel units where 1 pixel corresponds to approximately 0."5.

compared their numerical models of flows in the solar transition region with the earlier observations suggesting no significant shift in the O V at 1218 Å. Thus, our result could put new constraints on such models and will also be of importance for modeling stellar atmospheres.

Similar observations have been suggested using the Solar Ultraviolet Measurements of Emitted Radiation (SUMER) on SOHO. With SUMER's high spatial and spectral resolution, combined with the extensive wavelength coverage, it will be possible to study flows at temperatures ranging from 10^4 K to 2×10^6 K.

We discuss variation of velocities in the O V lines over the entire field of view of the HRTS instrument. A description of the photometry and of the various steps in the reduction process, including corrections for geometrical distortion and wavelength calibration is given by Brekke (1993).

2. OBSERVED REDSHIFT OF TRANSITION REGION LINES

In Figure 1 the spatial variation of the observed line-of-sight velocities in the O V line at 1371 Å is displayed together with a section of the HRTS 1 spectrogram showing the O V profile. The slit covered approximately one solar radius. The O V line is clearly redshifted at most positions along the

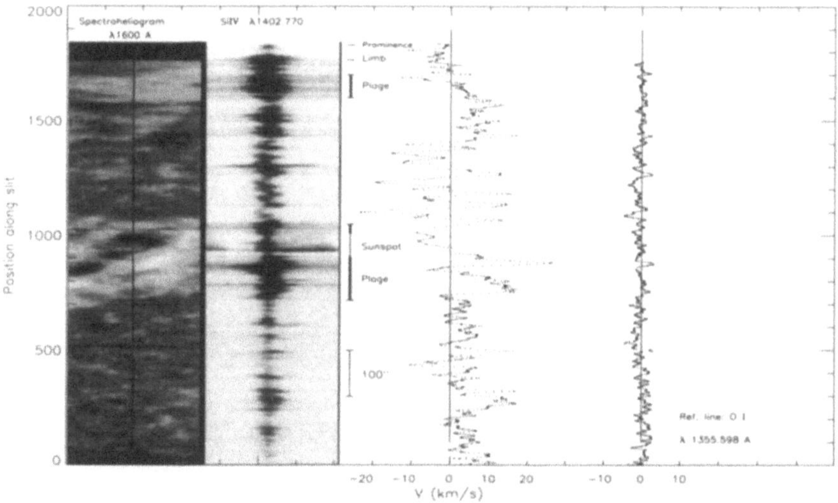

Fig. 2. Spatial variation of line-of-sight velocities in the Si IV line at 1402 Å observed with HRTS 2. To the left is a section of a spectroheliograph image showing the network structure in the chromosphere and the image of the Si IV profile. The chromospheric O I reference line at 1355 Å is plotted on the same velocity scale. The slit crossed the relative large active region McMath 15139.

slit corresponding to an average line-of-sight velocity of 8 km s^{-1} in the quiet regions. There is considerable spatial variation along the slit, where the flow velocities are varying between –3 km s^{-1} to almost 20 km s^{-1} at the solar disk. The blueshift is located above the sunspot umbra and outside the limb. For comparison, the observed shift of a nearby chromospheric reference line (the C I line at 1364.164 Å) has been plotted on the same velocity scale. The measured redshift shows no strong center-to-limb variation. Figure 2 shows the spatial variation of the line-of-sight velocities in the Si IV observed with HRTS 2 and demonstrates that much more complex variations along the slit are also present. In fact the largest velocity variation is found in a quiet region where the velocities are changing from −25 to +25 km s^{-1} over a distance of 20" . Thus, significant blueshifts are also present.

When measuring the 1218 Å line we subtracted the strong background continuum from the Lyλ and adopted the laboratory wavelength from Brown (1980) of 1218.344 Å. Using this value we find an average net redshift of approximately 3 ± km s^{-1} in the quiet regions. This is approximately 5 km s^{-1} less than the value obtained from the line at 1371 Å. The difference is probably due to an error in the adopted laboratory wavelength of the 1218 Å line, which is an intersystem line and thus, more difficult to measure in laboratory. The result suggests a laboratory wavelength close to 1218.325 Å if the two lines are to have the same redshift. In addition to O V we have

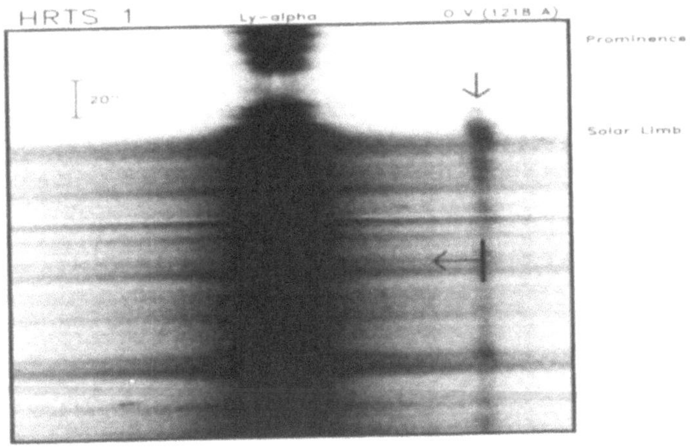

Fig. 3. The effect of the Lyα wing on the position of the O V line is to shift the O V line towards blue relative to the off limb position. Without correcting for the Lyα wing little or no shift will be found.

also measured Si IV, C IV, N IV, O IV, and Fe XII. The observed Doppler shifts in transition region and coronal lines has been summarized in Table 1. and our results from HRTS has been compared with previous observations of Doppler shifts in Figure 4.

The results from the observed wavelength the OV 1218 Å line, as presented by Doschek et al. (1976) should be regarded with some caution for the following reasons:

- The wavelength determination of the O V line is influenced by the Lyα line from hydrogen at 1215.67 Å.
- The solar wavelength calibration is uncertain owing to few chromospheric reference lines in this wavelength region.
- The laboratory wavelength of the oxygen line at 1218 Å is uncertain.

The observed position of the O V 1218 Å line is shifted up to 4 km s^{-1} towards blue because of the slanted Lyα wing as illustrated in Figure 3. This was not corrected for in the Skylab data. In addition Doschek et al. (1976) used the off-limb position as a zero reference which could also introduce systematic errors since horizontal flows are observed above the limb in the HRTS material. Also, the effect of the Lyα wing is neglectable above the limb.

3. SUMMARY

In earlier investigations the magnitude of the redshift has been found to increase with temperature, reaching a maximum at T= 10^5 K, and then to decrease towards higher temperatures as shown in Figure 4. Some ob-

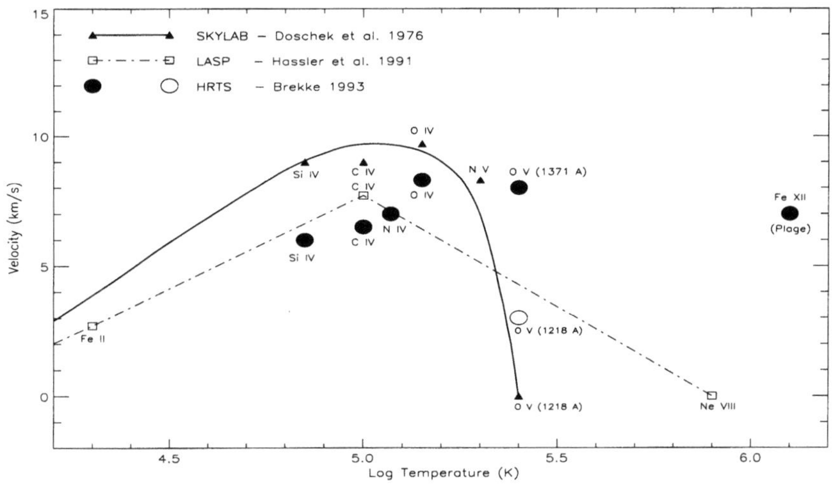

Fig. 4. A comparison of observed redshift as a function of temperature of formation.

servers find no significant shift in the O V line at 1218 Å at disk center (e.g. Doschek et al. 1976). It should be noted that the commonly quoted average velocity variation with temperature above 10^5 K to a large extent depends on this particular observation of the 1218 Å line. A more recent observation by Hassler *et al.* 1991 supports the evidence that the net redshift is reduced or disappear at higher temperatures. The result of Doschek et al. (1976) has been extensively used in comparison with numerical simulations of flows in the transition region (e.g. Spadaro, Antiochos, & Mariska 1991, Hansteen, 1993). Some authors find the redshifted emission to be replaced by blueshifted emission at temperatures around 2.5×10^5 K where O V is formed.

The present results from HRTS demonstrate that systematic net redshifts can exist at higher temperatures in quiet regions, network structures and active regions. The O V lines are clearly redshifted in both quiet Sun and plage areas. Thus, the existing theoretical models do not reproduce the observed shifts at higher temperatures. We suggest that the discrepancy between the two lines are due to an error in the given laboratory wavelength for the 1218 Å line. Also the coronal line Fe XII (1349 Å) is redshifted above the plage area suggesting that the downflow may persist into the corona. No strong center-to-limb variation of the redshift is found in our material. This suggests that other mechanisms should be considered to explain the observed shifts (e.g. Brekke 1993).

One should note that the the Si IV and O IV lines both show larger flow velocities in the plage relative to the average quiet region. In fact the plage flow at T $\sim 10^5$ K has approximately twice the speed found in quiet regions while the O V lines have nearly the same velocities in both quiet regions and the plage.

Our result demonstrates that extended observations with existing and future instrumentation are needed. Great improvements can be expected in the determination of average flows in all temperature regimes in the solar atmosphere utilizing the SOHO scheduled to be launched in 1995. Improved laboratory wavelengths for the most important species are also of great importance. In addition there is a need for improved theoretical models of the dynamics of the solar transition region and corona.

TABLE I
Observed average redshifts in transition region; HRTS 1

Ion	Laboratory Wavelength (Å)	Temperature of Formation (K)	Line of Sight Velocity Quiet Region (km s^{-1})	Line of sight Velocity Plage (km s^{-1})
Si IV	1402.770[1]	7.0×10^4	6 ± 2	15 ± 2
C IV	1548.202[2]	1.0×10^5	6 ± 2	13 ± 2
C IV	1550.774[2]	1.0×10^5	6 ± 2	13 ± 2
N IV	1486.496[3]	1.2×10^5	7 ± 2	16 ± 2
O IV	1401.156[4]	1.4×10^5	8 ± 2	16 ± 2
O V	1371.292[5]	2.4×10^5	8 ± 2	8 ± 2
O V	1218.344[6]	2.3×10^5	3 ± 2	4 ± 2
Fe XII	1349.40[7]	1.3×10^6	—	7 ± 4

1 — Moore 1965, 2 — Bockasten, Hallin, and Hughes 1963, 3 — Hallin, 1966, 4 — Bromander 1969, 5 — Bockasten & Johansson 1968, 6 — Brown 1980, 7 — Sandlin et al. 1977, (Solar observation).

References

Ayres, T. R., Jensen, E., & Engvold, O. 1988, ApJS 66, 51
Bartoe, J.-D. F., & Brueckner, G. E. 1975, J. Opt. Soc. Am., 65, 13
Bockasten, K., Hallin, R., & Hughes, T. P. 1963, Proc. Phys. Soc. London, **81**, 522
Bockasten, K., & Johansson, K. B. 1968, Arkiv f. Fys. 38, No 31. 563
Brekke, P. 1993, Dr. Scient Thesis, University of Oslo.
Bromander, J. 1969, Ark. Fys., 40, 257
Brown, C. M. 1980, A&A. 88, 273
Doschek, G. A., Feldman, U., & Bohlin, J. D. 1976a, ApJ, 205, L177
Hallin, R. 1966, Ark. Fys., **32**, 201
Hansteen, V. H. 1993, ApJ, 402, 741
Hassler, D. M., Rottman, G. S., & Orral, F. Q. 1991, ApJ, 372, 710
Moore, C. E. 1965, NSRDS-NBS, 3, Section 1
Sandlin, G. D, Brueckner, G. E., & Tousey, R. 1977, ApJ, 214, 898
Spadaro, D., Antiochos, S. K., & Mariska, J. T. 1991, ApJ, 382, 338

TRANSITION REGION LINE-SHIFTS IN THE REBOUND SHOCK SPICULE MODEL

V. H. HANSTEEN and Ø. WIKSTØL
Institute of Theoretical Astrophysics, University of Oslo, Norway

Abstract. Spectral emission lines created in the solar chromosphere - corona transition region show net red-shifts. It has been proposed that this may be the result of the return of spicular material. We simulate a spicule numerically using the rebound shock model and find that the resulting hydrodynamic evolution leads to a perceived up-flow in transition region spectral lines even though the average velocity in the line forming region is directed downward. The explanation for this apparent paradox is found in the correlation between density and velocity in the waves generated by the rebound shock 'spicule'.

Key words: Solar chromosphere – Spicules – Transition Region – Down-flows

1. Introduction

Though the spicule phenomena certainly is important for the dynamics of the outer solar atmosphere no credible spicule model has yet been constructed. Among the models that have gained attention is the "rebound shock" model of spicule formation (Hollweg 1982, Sterling & Hollweg 1988). In this model, a shock train propagating upwards from the photosphere lifts and heats the chromospheric plasma. Numerical models of the rebound shock scenario have been presented by Sterling & Hollweg (1988), Sterling & Mariska (1990), and recently by Cheng (1992 a,b,c).

Spicules should perturb the solar transition region and corona and one expects to find evidence of their presence in the velocities inferred from spectral lines formed in the transition region. Analysis of spectral line profiles show that transition region lines display a net red-shift (*e.g.* Doschek et al. 1976, Dere et al. 1984, Hassler et al. 1991, Brekke 1993). For example the observed mean red-shift for the C IV $\lambda 1548$ line formed at 10^5K, while varying considerably, is typically 5km/s. The possible connection between the observed red-shifts and the return of spicular material was pointed out by Pneuman & Kopp (1977) and later by Athay & Holzer (1982). In Cheng's (1992c) rebound shock simulations the average velocity at the $T = 10^5$K level is found to be directed downward and of the same order as the line shift observed in the C IV $\lambda 1548$ line.

The goal of this paper is to point out that knowledge of an average gas velocity is not sufficient to insure that the shift in wavelength has the corresponding amplitude or even sign as the gas velocity. We find, in contradiction to Cheng (1992c), that though the rebound shock train may give rise to an average downward directed velocity in the transition region, the line

Space Science Reviews **70**, 103–106.

shifts resulting from these motions are on average *blue-shifted* from which, if observed, a net up-flow would be inferred.

2. Line emission

We reconsider the emission of optically thin resonance lines from multiple ionized metals formed in the transition region in the presence of atmospheric perturbations. To a good approximation (*e.g.* Jordan & Wilson 1971) collisional excitation is balanced by spontaneous de-excitation. Since we are considering resonance lines we may confidently identify the population of the lower level with the total population of the radiating ion n_{ij}. Employing these assumptions the emergent intensity from an optically thin transition of energy $h\nu_{ul}$ and an emission profile equal to the absorption profile ϕ_ν, may be written

$$I_\nu = \frac{h\nu_{ul}}{4\pi}\int_0^{s0} \phi_\nu n_e n_i g(T_e) ds,$$

with

$$g(T) \equiv \frac{n_{ij}}{n_i}\exp(-h\nu_{ul}/kT_e)T_e^{-\frac{1}{2}}\Gamma(T_e).$$

The intensity integration is carried out along the line-of-sight, n_e is the electron density, n_i is the total density of the atom to which the radiating ion belongs, and $\Gamma(T_e)$ is a slowly varying function of the electron kinetic temperature T_e. The total frequency integrated intensity of the emission line is thus seen to vary proportionally with the squared density and the ionic concentration n_{ij}/n_i, where the variables are evaluated at some appropriate near the maximum of ionic concentration.

The response of an optically thin transition region line to a compressive perturbation depends on how the variation in the intensity integrand correlates with the line shift generated by the velocity u in the line forming region. For a linear, propagating acoustic wave the basic effects may be outlined by considering a sinusoidal sound wave with frequency ω propagating along the atmosphere in the z direction. The particle velocity u_z varies in phase with the pressure perturbation Δp and the density perturbation $\Delta\rho$;

$$\frac{u_z}{c_s} = \frac{\Delta\rho}{\rho} = \frac{\Delta p_m}{\gamma p}\sin[\omega(t - z/c_s)],$$

where Δp_m is the amplitude of the pressure perturbation and the sound speed c_s is given by $c_s^2 = \gamma p/\rho$ where γ is the usual ratio of the specific heats (c_p/c_v). The phase relation between the density and velocity perturbation will cause the line intensity to vary with the line-of-sight velocity, scaling roughly with the square of the density perturbation in a fully ionized plasma, leading to a net shift of the spectral line if we integrate over a wave period $P_w = 2\pi/\omega$ (Eriksen & Maltby 1967). Thus, *blue-shifted* spectral lines are expected when perturbations propagate upwards through the transition region.

The above discussion was made without reference to departures from ionization equilibrium and variations in the temperature, which will effect both the intensity integrand and the extent of the region of line formation. In addition, the shock train that constitutes the rebound shock model represents highly non-linear waves propagating in a time-varying environment as the atmosphere is heated and cools. The consequences of these non-linear effects are taken into account in the numerical calculations presented below.

3. Results and Discussion

In constructing a numerical model of the rebound shock scenario an initially static chromosphere – transition region – coronal model is subjected to an impulsive perturbation at the lower boundary. The evolution of this perturbation is followed by solving the equations of mass, momentum, and energy conservation along with the ionization rate equation to obtain the variation of density (ρ), velocity (u), temperature (T), and population densities (n_{ij}) of elements i in ionization states j throughout the atmosphere as a function of time. We include thermal conduction and optically thin radiative losses in the energy equation. We have also included a heating term in the chromospheric region of the atmosphere in order to maintain chromospheric temperatures against radiative losses. Details of the numerical method used in integrating the equations may be found in Hansteen (1993) and Hansteen & Wikstøl (1993).

At time $t = 0$s a half cycle of a sinusoidal wave with a period $P = 180$s is generated at the bottom of the atmosphere. As the pulse propagates upwards through the atmosphere it steepens, forming a shock before reaching the chromosphere-corona transition region. The shock front lifts the transition region to a height several thousand kilometers above its equilibrium position before propagating into the corona. In the wake of the pulse, material falls back to the underlying chromosphere eventually generating another 'rebound' shock. This sequence of events repeats with steadily decreasing amplitudes.

In the table below we show the average temperature of line formation, the average gas velocity, and the average resonance line line-shifts for transition region lines formed at various temperatures. The most striking result is that though the average gas velocities are all positive and thus directed downward the average line shifts are all negative (blue-shifted) indicating a net up-flow. This result indicates that a line formed in an atmosphere where the density varies will show enhanced emission when the density is high, scaling roughly as the density squared.

TABLE I

Average material and line shift velocities (km/s) for various lines formed in the transition region (positive sign corresponds to downward velocities)

$T_e(K)$	$u_{gas}(km/s)$	$u_{sh}(km/s)$	Ion	λ(Å)
6.0×10^4	1.0	-6.8	C III	977
7.5×10^4	1.1	-3.5	O III	834
1.0×10^5	1.0	-7.2	C IV	1548
1.3×10^5	1.4	-4.0	O IV	789
1.8×10^5	1.6	-6.3	O V	629
2.5×10^5	2.3	-6.0	O VI	1034

We may separate the effects of ionization fraction and density by comparing the net line shifts of C IV 1548Å and O IV 789Å. These lines originate from ions that are formed at approximately the same temperature and accordingly experience the same velocity and density perturbations. The difference between the ions is that while the ionization state of C IV is highly dependent of the velocity, density and previous history of the plasma, O IV is not and is close to ionization equilibrium at all times.

On the basis of the results presented here we conclude that the upwardly propagating shock fronts with subsequent down-flow of material characteristic of the rebound shock spicule model *cannot* explain the average red-shift observed in transition region spectral lines formed in the temperature range 5×10^4K to 3×10^5K.

References

Athay, R.G., Holzer, T.E. 1982, ApJ 255, 743
Brekke, P. 1993, this issue.
Cheng, Q.Q 1992, A&A 262, 581
Cheng, Q.Q 1992, A&A 266, 537
Cheng, Q.Q 1992, A&A 266, 549
Dere, K.P., Bartoe, J.-D.F., and Brueckner, G.E. 1984, ApJ 281, 870
Doschek, G.A., Feldman U., and Bohlin, J.D. 1976, ApJ 205, L177
Eriksen, G., Maltby, P. 1967, ApJ 148, 833
Hansteen, V.H. 1993, ApJ 402, 741
Hansteen, V.H., Wikstøl, Ø. 1993, A&A (submitted)
Hassler, D.M, Rottmann, G.J., and Orrall, G.Q. 1991, ApJ 372, 710
Hollweg, J.V. 1982, ApJ 257, 345
Jordan, C., Wilson,R. 1971, in Physics of the Solar Corona, ed. C.J.Macris, Dordreicht : Reidel, 219
Pneumann, G.W., Kopp, R.A. 1977, A&A 55, 305
Sterling, A.C., Hollweg, J.V. 1988, ApJ 327, 950
Sterling, A.C., Mariska, J.T. 1990, ApJ 349, 647

ANALYSIS OF A REDSHIFTED PLASMA FLOW OVER A SUNSPOT

J. MARTIN LAMING*
SFA Inc., Landover MD 20785, U.S.A.

Abstract. The ultraviolet spectrum of a redshifted plasma flow appearing over a sunspot observed during the first flight of the High Resolution Telescope Spectrograph (HRTS I) is analysed, and interpreted as a radiatively cooling plasma. For most of the lines emitted from this plasma, the assumption of ionization equilibrium during the cooling is good. However for He II (and other ions with a single electron outside of closed shells), this is not the case. Integrating differential equations for the various ionization fractions of helium and the temperature allows an approximate determination of the abundance of helium relative to other elements whose lines appear in the spectrum of the plasma flow.

Key words: Sun – UV – Transition Region – Sunspots – He abundance

1. Introduction

Several redshifted plasma blobs have been observed by the HRTS instrument on various of its flights. Our interest in the HRTS I redshifted flow is due to the relative weakness of the He II 1640 Å multiplet compared to the blob in other lines (e.g. the C IV doublet at 1550 Å) emitted from the same plasma temperature. Indeed the general behaviour of the stationary component of these lines over the sunspot is not too different from the quiet sun, but the redshifted plasma is qualitatively different. Some of the redshifted blobs appearing in others of the HRTS flights, e.g. the HRTS II data, do not show such a large difference. They have the line intensities and background similar to other solar regions around the sunspot. The redshifted blobs of emission in the HRTS I data are most intense in lines formed at around $1 - 2 \times 10^5$ K falling off markedly for lines formed towards 10^4 K, whereas the opposite trend is apparent in the HRTS II observations. The redshifted velocity of the HRTS I plasma blobs remains constant over this temperature range (Nicolas et al. 1982).

The HRTS (Brueckner, Bartoe & Van Hoosier 1975) instrument is a stigmatic spectrograph. Along the slit direction, the spatial resolution is 0.5-1.0" (0.8" was achieved on the HRTS I rocket flight of 1975 July 21 (Brekke et al. 1991)) and the spectral dispersion is 4.774 Å mm^{-1}. Spectra are recorded on Kodak 101-01 film, which was then digitized with a PDS microdensitometer. The stigmatic properties of the instrument mean that each spectral line appears as a patchy image, corresponding to different spatial regions of the sun. This property is valuable in allowing us to isolate spectra solely

* also Naval Research Laboratory, Washington DC, 20375, U.S.A.

Space Science Reviews **70**, 107–110.

from the redshifted plasma blob (spatial extent $\sim$ 1") over the sunspot without contamination from nearby but unrelated regions. The data reduction is discussed in detail elsewhere (Laming & Feldman 1994 (paper I)). Here, we will concentrate on more details of the interpretation.

2. Interpretation: The He Abundance

It has been argued previously (Laming & Feldman 1992) that the intensities of lines in He II are very sensitive to the effects of non-equilibrium ionization, and that in fact one needs to invoke such a model to explain these intensities relative to other lines from different ions formed at the same temperatures. Such effects also hamper determinations of the He abundance due to inadequacies in the modelling. The weakness of the He II emission in these data mentioned above is possibly indicative of a plasma in true ionization equilibrium, the modelling of which is substantially simpler than that of non-equilibrium ionzation, making a He abundance determination in principle feasible. In addition, the redshift largely removes the problem of absorption of Ly β 256 Å radiation from the solar disk which limited the effectiveness of a study of He II in solar prominences (Laming & Feldman 1993a, 1993b).

A differential emission curve was constructed (see paper I for details). It was found that this curve could be interpreted as due to a radiatively cooling plasma blob, in which the number of particles is conserved. In ignoring thermal conduction along the flow, we are implicitly assuming that the plasma flow is fragmented into blobs, each of which is isothermal. This is discussed in more detail in the next section. Furthermore, we find that radiative cooling at constant pressure gives a better fit to the data than at constant density. In the constant pressure case, the product of the differential emission measure D, the radiative cooling rate Λ, and the electron density n_e is constant. Thus an expression for the He abundance $A_{\rm He}$ relative to that of Hydrogen $A_{\rm H}$ may be written, where $I_{\rm obs}\left(1640\text{Å}\right)$ is the observed intensity in the 1640 Å multiplet

$$\left(q_{\rm H}+q_{\rm He}\frac{A_{\rm He}}{A_{\rm H}}\right)^2\frac{A_{\rm He}}{A_{\rm H}}=\frac{I_{\rm obs}\left(1640\text{Å}\right)n_e\Lambda}{\left(I_{1640}/\text{unit}\,DEM/\text{unit}\left(A_{\rm He}/A_{\rm H}\right)\right)\langle D\Lambda n_{\rm e}\rangle}$$

where $q_{\rm H}$ and $q_{\rm He}$ are the ionizations fractions of H and He in the temperature region over which He II 1640 Å is emitted, and (I_{1640}/unit...) is the expected intensity in the 1640 Å multiplet in this temperature region per unit differential emission measure and per unit He abundance relative to H. These quantities are determined by integrating simultaneous differential equations for the temperature and various charge states of He. In this calculation, the radiative cooling curve of Gaetz & Salpeter (1983) is used for

TABLE I
He Abundance Results.

parameter	units	values	
T_f	K	10^4	4×10^3
I_{1640}/atom	ergs sec^{-1} $atom^{-1}$	2.22×10^{-12}	5.44×10^{-12}
$I_{1640}/\mathrm{unit}DEM/\mathrm{unit}A_{He}/A_H$	ergs sec $^{-1}$cm $^{-3}$	2.5×10^{-24}	4.0×10^{-24}
$n_e^2\Lambda$	ergs $sec^{-1}cm^{-5}$	12.9	98.4
q_H		0.81	0.70
q_{He}		1.29	1.07
n_e	cm^{-3}	9×10^{11}	1.4×10^{12}
A_{He}/A_H		0.078	0.22

$T > 10^5$ K. For $T < 10^5$ K a form with a term which extrapolates their curve for elements C and heavier to lower temperatures, together with other terms denoting the cooling by photons from radiative recombinations to form neutral H, He and ionized He^+. In this aspect, the He abundance was assumed to be 0.1 relative to H. The results of the integration are largely insensitive to the exact value chosen here, but ideally one would iterate until the abundance derived at the end of the calculations agrees with that put in here. Table 1. shows the results determined from two separate calculations, one with an assumed final temperature for the cooling flow of 10^4 K, the other with 4×10^3 K, chosen to represent typical chromospheric and sunspot temperatures respectively. With an observed intensity in the 1640 Å multiplet of 400 ergs $sec^{-1}cm^{-2}$ we determine a solar abundance for helium relative to hydrogen in the range 0.078-0.22, this range being determined by the range of plausible temperatures at which the cooling stops. While not being as precise as determinations from solar evolution models, (though these still have as yet to be finalised uncertainties arising from questions of helium diffusion and other related thermodynamics), we believe that the problems confronting other works (Laming & Feldman 1992, 1993) have been obviated to an extent that this work qualifies as an "abundance determination", albeit somewhat imprecise, rather than a "model consistent with a helium abundance of 0.1 relative to hydrogen". The main problem we face here is that most of the lines we observe are formed at significantly higher temperatures in the flow than He II, and so some extrapolation is required. The observation of more lines with temperatures of formation around 10^4 K, i.e. O I, Si II, etc would improve the accuracy, both by determining the final temperature and giving a more precise reference differential emission measure.

3. Interpretation: Fragmentation of the Flow

In the previous section we stated that the data were best fit by a plasma that is isothermal and cooling radiatively at constant pressure. Here we discuss some of the implications of this. First of all, for thermal conduction to be unimportant in such a flow, to allow our treatment, the flow must be fragmented. In order that constant pressure be maintained, the size of these fragments must be smaller than the product of the sound speed and the cooling time. Taking a typical cooling time at 10^5 K and 10^{11} cm^{-3} electron density of 0.2 second, and the sound speed as 2.5×10^7 cm sec^{-1} gives a size of less than 50 km. The pressure is presumably maintained by trapped magnetic field internal to the blob and an ambient magnetic field. In this case, the ambient magnetic field must be about 5 Gauss stronger than the trapped field. We are unable to say much more, since the time over which the blob is emitting radiation observable by the HRTS instrument (about 10 seconds) is presumably substantially shorter than its total lifetime, so questions involving its origin or formation involve considerable extrapolations.

Acknowledgements

I am happy to acknowledge helpful discussions with Drs Uri Feldman and Spiro Antiochos.

References

Bruekner, G.E., Bartoe, J.-D.F., Van Hoosier, M.E.: 1977, in: *Proc. OSO 8 Workshop*, Univ. of Colorado Press, Boulder, p. 380.
Brekke, P., Kjeldseth-Moe, O., Bartoe, J.-D.F., Brueckner, G.E.: 1991, *ApJS* **75**, 1337
Gaetz, T.E., Salpeter, E.E.: 1983, *ApJS* **52**, 155
Laming, J.M., Feldman, U.: 1992, *ApJ* **386**, 364
Laming, J.M., Feldman, U.: 1993a, *ApJ* **403**, 434
Laming, J.M., Feldman, U.: 1993b, *ApJ* **409**, 869
Laming, J.M., Feldman, U.: 1994, *ApJ* , in press
Nicolas, K.R., Kjeldseth-Moe, O., Bartoe. J.-D.F., Brueckner, G.E.: 1982, *Solar Physics* **81**, 253

SPECTROSCOPIC DIAGNOSTICS FOR CDS AND SUMER

HELEN E. MASON
Department of Applied Mathematics and Theoretical Physics
Silver Street, CAMBRIDGE CB3 9EW

Abstract. The CDS and SUMER instruments will make extensive use of spectroscopic diagnostics to determine the physical parameters of the solar transition region and corona. A great deal of effort is going into the development of these techniques to ensure that they are accurate. This requires interfacing solar analysis programs with the best available atomic data calculations. Recent developments in this work will be presented, with specific examples of useful spectral line ratios.

1. Introduction

An UV spectroscopic diagnostics working group has been formed from CDS and SUMER co-investigators and associated scientists. The aim is to focus on the spectroscopic diagnostic software requirements for these two instruments. The group at the Institute of Theoretical Astrophysics, University of Oslo, Norway is already responsible for a large section of the CDS analysis software. The spectroscopic diagnostics working group met for the first time in June 1993 at Cosners' House, Abingdon, UK (convenor Dr H.E. Mason). The discussions were wide ranging, covering different aspects of electron density and temperature determination, differential emission measures, elemental abundance variations and basic atomic data, including both electron excitation, ionisation and recombination. Table I shows some of the CDS spectral line pairs which will be used for electron density diagnostics. Further details and examples of observing sequences can be found in the CDS blue book (edited by R.A. Harrison) and the SUMER red book (edited by K. Wilhelm).

2. The Atomic Data and Analysis Structure (ADAS)

The Atomic Data and Analysis Structure (ADAS) package has been developed for fusion research by Prof H.P. Summers at JET (Joint European Torus). ADAS provides an accurate method of atomic data analysis, storage and processing. It makes use of a database of the best available atomic rates and prepares derived atomic data required for plasma modelling and spectral interpretation. It is a very powerful suite of subroutines, which solves the generalised collisional-radiative model for equilibrium or non-equilibrium conditions. It ties together the ionisation/recombination rates with the low lying ground and metastable level populations. This provides an accurate solution to the time dependent equations for the ionisation ratio and level populations, including electron density effects. Theoretical line intensity

Space Science Reviews **70**, 111–114.

ratios can be used for spectroscopic diagnostics. ADAS has been adopted as the core of further software development for CDS. The atomic data input needs to be optimised for application to the solar UV spectrum. The output products will be available for CDS/SUMER solar analysis routines. Theoretical models can be linked with ADAS to study the important plasma processes in the solar atmosphere.

TABLE I

Some CDS Spectral Line Pairs for Electron Density Diagnostics

Ion	$T_e(K)$	Wavelengths (Å)		
Ne V	3.2 x 10^5			416.20/482.99
Mg VII	6.3 x 10^5			319.03/367.66
Mg VIII	8.0 x 10^5			430.46/436.73
Si IX	1.0 x 10^6			345.13/341.95
Si X	1.2 x 10^6			356.04/347.40
Fe X	1.0 x 10^6	174.53/175.27		
Fe XI	1.3 x 10^6	178.06/179.76	180.41/182.17	
Fe XII	1.5 x 10^6	186.88/193.51		338.26/364.47
Fe XIII	1.7 x 10^6		202.04/203.79	320.80/359.64
Fe XIV	1.9 x 10^6	211.32/219.12	264.79/274.20	334.17/353.83
Fe XV	2.0 x 10^6		284.15/321.79	

3. Atomic Data Requirements

The electron excitation rates required to analyse the CDS and SUMER spectra were the subject of an Atomic Data Assessment Workshop held in March 1992. The proceedings from this workshop, including individual reports on all of the iso-electronic sequences of solar interest, are in press as a single issue of Atomic Data and Nuclear Data Tables (Volume 57, May 1994, edited by J. Lang). The iron ions Fe IX - Fe XIV were assessed by H.E. Mason. New calculations are required for many ions. The recommended data are being compiled in a machine readable form, input into ADAS, and made accessible for other CDS/SUMER software.

The Iron Project co-ordinated by D. Hummer (Munich), P.J. Storey (University College London, UCL) and K. Berrington (Queens' University of Belfast, QUB) aims to calculate new electron excitation rates using the R-matrix close coupling codes, with very accurate targets, including relativistic effects. The first stage is to compute collisional excitation rates for the fine structure transitions in the ground configuration of many ions needed for the interpretation of the infra-red lines from astrophysical objects. The second stage is to calculate new electron excitation rates for all the ions of iron, including several excited configurations. For example the target which P.J. Storey and H.E. Mason are using for Fe XIV includes the configurations $3s^23p$, $3s3p^2$, $3s^23d$, $3p^3$, $3s3p3d$, $3p^23d$, $3s3d^2$, $3p3d^2$, $3s^24s$, $3s^24p$, $3s^24d$, $3s^24f$, $3s3p4s$, $3s3p4p$, $3d^3$. This work is being carried out specifically to address the needs of CDS and SUMER. Other electron excitation calculations of astrophysical interest are in progress at QUB (Prof A. E. Kingston).

4. Calculation of synthetic spectra

We require the ability to calculate synthetic spectra in the wavelength range 150-1600$\AA$ for comparison with the CDS and SUMER observations. "Quick look" synthetic spectra, compiled from previous solar observations, are already available. These will be used for planning purposes but not in analyses of the actual observations. Theoretical, or semi-theoretical G(T) functions (see CDS blue book, edited by R.A. Harrison) have been produced by several groups. The function G(T) is proportional to the fraction of ions in the relevant stage of ionisation (predominantly a function of temperature), the collisional excitation processes, the radiative cascade rate, and the number density ratio of Hydrogen atoms to electrons. G(T) is determined from atomic parameters, usually assuming simple atomic models and near equilibrium conditions. A computer code to produce synthetic spectra in the UV wavelength region has been developed by B. Monsignori-Fossi and M. Landini at Arcetri Observatory. This is being updated with new atomic rates. Similar work is in progress by K.P. Dere and K. Waljeki at the Naval Research Laboratory. This effort is being co-ordinated and linked to ADAS to produce very accurate G(T) functions and hence theoretical spectra for general solar models (quiet Sun, active region, coronal hole etc) in equilibrium.

5. The derivation of differential emission from line intensities.

R.A. Harrison and A.M. Thompson have reported (RAL-91-092) on the performance of a number of integral inversion codes used in the calculation of differential emission measure (DEM) from line intensities. The codes from Glasgow University (A. Thompson) and Arcetri Observatory (B. Monsignori-Fossi) will be developed for CDS/SUMER software.

6. Basic interpretation software.

Electron density and temperature diagnostic line ratios will form an important part of the CDS and SUMER observations. It is necessary to specify sets of optimum spectral lines for different solar features. The best atomic data for these lines should be used to compile diagnostic plots and to enable "quick look" electron density contours to be produced from the solar observations.

More accurate analyses of the observations would require interfacing the atomic models (ADAS) with the solar models to provide physical parameters as a function of position and time. The dynamic nature of the solar atmosphere requires an iterative approach. It is likely that various groups will want to develop their own theoretical models, so the emphasis should be on how best to interface the different components of the software.

7. Developing and testing software on existing solar spectra

The SERTS (Solar EUV Rocket Telescope and Spectrograph) spectra overlap in wavelength region with CDS. They provide an excellent dataset for studying potential spectroscopic diagnostics. The SERTS team (Goddard Space Flight Center) have offered to make their spectra available for the development of analysis software for CDS and SUMER.

8. Acknowledgements

Financial support for this project has been provided by the Science and Engineering Research Council.

SUMER SCIENCE DATABASE

J. ABOUDARHAM*
Institut d'Astrophysique Spatiale
Bât. 121
Université Paris-Sud
F-91405 Orsay cedex
France

Abstract. The complete description (ie Master Directory) of SUMER observations will be available using the relational database management system ORACLE, giving the opportunity to make complicated queries and cross references. Data will be stored using FITS format. At least the last 28 days of observation will be on-line while older data will be stored on off-line supports (such as CD-ROMs), accessible only on request. We present here an overview of the SUMER science database.

Key words: SOHO/SUMER – Database

1. Introduction

SUMER observations can be grouped in observing programs, corresponding to a main scientific interest. Each program is (generaly) made of several sequences (a set of quite similar observations), and each sequence is formed by series of images of the same kind (e.g. n line profiles along the instrument slit, then again after moving the slit a few steps, and so on). Hereafter, we'll call *elementary images* the smallest decomposition of an observing program (for instance, one line profile, one spectrum, ... corresponding to an instant image of the detector). The word *image* will be a set of elementary images of the same kind, coming from the same observing sequence. This term will be the basis of the conception of the SUMER science database. As a matter of fact, we can estimate from present observing proposals that the amount of elementary images each day for SUMER can reach 2000! It is not reallistic (and, in fact, not useful) to manage such a tremendous amount of data. The use of images (as defined before) reduces this volume to about 75 images per day. The goal of the database is to allow scientists to scan observing programs, sequences and images, to have the opportunity to make queries in order to look for the data they are interested in (for example : observations of a special target, or at a given wavelength, ...)

* On leave from Observatoire de Meudon

Space Science Reviews **70**, 115–117.

2. Filling of the database

There are two kinds of data which are important for the scientific use of observations : images themselves, of course, but also the complete description of them.

2.1 IMAGES

The results of observations will arrive through the telemetry and then be reformatted in images, after being calibrated, then stored in files (one file corresponding to one image). We shall keep at least 28 days of observation on-line while older data will be stored off-line, being available on request, after a manual load. There will also be data from coordinated ground-based observations. Data from SOHO inter-instrument observations will be also available from the involved instruments.

2.2 IMAGES DESCRIPTION

The complete description of images will also be found in the telemetry. It fully describes the observation, giving the target expected (which could be different from the target achieved), its exact position on the Sun, wavelength coverage, observing time, and several minor informations concerning the data. These housekeeping data will be used in two ways: to fill the tables of the database and to prepare FITS headers for each image.

2.3 PHYSICAL ORGANIZATION

In order to optimize queries velocity and to minimize costs, we do not plan to include image files in the database. As mentioned earlier, only recent data will be available on line and older ones will be loaded only on request. They will be presumably stored on long duration media such as CD ROMs. The database will be managed using ORACLE7 software. User interface will be implemented using SQL (Standard Query Language), or menus (with SQL*MENU), or 'Window type' environment (presumably ORACLE Data Browser, when available for DEC WINDOW or MOTIF). Routines will copy the files chosen in user's personal environment. The data processing will be possible using IDL, together with some standard facilities.

3. Users queries

Users will have the opportunity to make various kinds of queries, such as scanning a target, an observing time, a scientific program, coordinated observations, a wavelength, etc... Each image will be described in one row of the main table of the database. There will be about 70 fields (columns) in one row, each field corresponding to an elementary description of the image; only directly useful parts of them will be accessed unless otherwise specified. Some columns (such as wavelength, time, target) will be indexed in order to accelerate queries, but all these fields can be accessed for any kind of query. Then, a list of the images asked for will be displayed and, on user request, data are copied in user's working area, in FITS format. The queries could also be made from various levels : globally in observing programs, or in observing sequences, as well as in images.

4. Database 'queries'

We do not want the SUMER database to be a passive archive. So we encourage informations from users after they processed data. This might concern several items such as the observational target actually achieved if different from the one expected, or complementary targets discovered in images, description of the physical data obtained from images, reference of papers published using those data, etc... Therefore, we plan to develop facilities concerning this 'feedback'.

5. Optional possibilities of the database

A few additional possibilities may be included in the database conception if time and manpower can be found for it. First of all: to have debased images (i.e. to lower spatial or spectral resolution of original images), for the most important ones, in order to be able to have them on-line, so it will be possible for the user to plot quickly some of the images he or she selected and then make a more precise choice.
Another interesting point (depending on the future evolution of software) concerns the possibility to send directly ORACLE requests from IDL , to plot selected data (if they are on-line) without having to copy it in user's working area.

6. Conclusion

A detailed Master Directory will be developed, using ORACLE software, containing all useful informations concerning images. This will be the basis of users' queries. Various ways to make queries will be possible: Window-type interface, menus, or elementary SQL. Image files will be copied in user's working area and then processed with IDL; then new informations concerning data will be inserted in the database.

TRACE – THE TRANSITION REGION AND CORONAL EXPLORER

K. STRONG, M. BRUNER, T. TARBELL, A. TITLE, and C. J. WOLFSON
Lockheed Solar & Astrophysics Laboratory, Palo Alto, CA 94304, USA

Abstract. TRACE is a single-instrument solar mission that will be put into a Sun-synchronous polar orbit and will obtain continuous solar observations for about 8 months per year. It will collect images of solar plasmas at temperatures from 10^4 to 10^7 K, with 1-arcsec spatial resolution and excellent temporal resolution and continuity. With such data, we expect to gain a new understanding of many solar and stellar problems ranging from coronal heating to impulsive magnetohydrodynamic phenomena.

Key words: Sun – Instrumentation - Coronal Heating - Coronal Structure

1. Introduction

The solar magnetic field and plasma are roughly in pressure equilibrium in the photosphere whereas magnetic fields dominate the structure and distribution of the coronal plasma. The transition between these two regimes is extremely difficult to model. The data that will be obtained from TRACE will be used to quantitatively explore the connections between fine-scale magnetic fields and associated emission throughout the solar atmosphere.

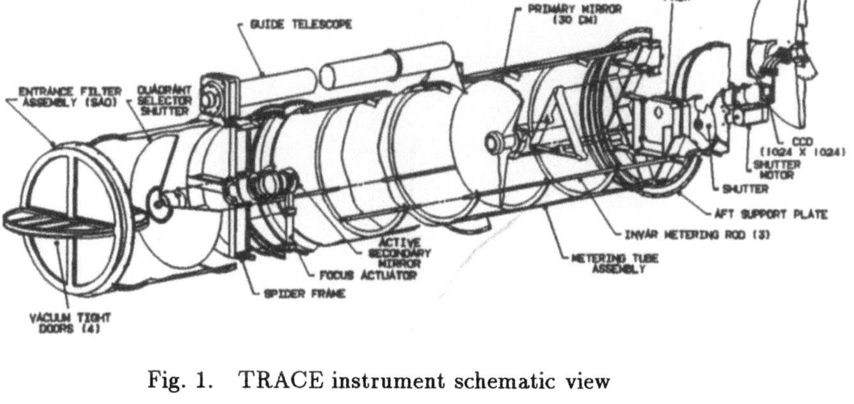

Fig. 1. TRACE instrument schematic view

Space Science Reviews **70**, 119–122.

2. The TRACE Instrument

TRACE will take images of the solar plasma from the photosphere to the corona. The 30-cm-aperture TRACE telescope has a Cassegrain design (see Fig. 1) with the four quadrants of the primary and secondary mirrors coated differently so that they can produce narrow-band XUV and UV images. The segmented coatings on the mirrors form identically sized and perfectly coaligned images, since they are on a common substrate. Filter wheels isolate the desired wavelengths and a shutter sets the exposure times. The 1024 × 1024 CCD camera allows TRACE to take images with a high cadence over an area of 8.5 × 8.5 arcmin. A separate aspect telescope produces an error signal that is fed into an articulated secondary mirror, stabalizing the pointing to $\leq$ 0.1 arcsec. A data-handling computer enables flexible use of the CCD array including adaptive target selection, data compression, and faster operation (0.2-s cadence) for a more limited field of view. Table I lists the main characteristics of the TRACE telescope.

TABLE I

TRACE Characteristics

Primary Mirror Diameter	30 cm
Effective Focal Length	8.66 m
Pixel Size	21 × 21 μm
	0.5 × 0.5 arcsec
CCD Size	1024 × 1024 pixels
CCD Read-out Noise	20 e^-
Field of View	8.5 × 8.5 arcmin
Image Stabilization	± 0.1 arcsec
Telescope Dimensions	37 × 37 × 187 cm
Instrument Mass	44 kg
On-board Memory	1 Gbit
Data Rates	3.6 Gbits/day
RAL	2.0 Gbits/day
Wallops Flight Facility	1.6 Gbits/day

TRACE is to be launched in 1997-1998 on a Pegasus rocket into a Sun-synchronous orbit This will allow continuous observing for about 8 months of the baseline 1-year mission. TRACE will produce complementary data with SOHO, enhanced by the coordinated planning of the TRACE daily observations with those of the SOHO instruments. The TRACE quick-lookdata will be processed at the SEOF and made available to the SOHO teams.

3. TRACE Science and Observational Objectives

The main observational objective of TRACE is to obtain 1-arcsec-resolution, multithermal images of the solar atmosphere with a cadence and continuity of coverage commensurate with the variability and lifetimes of primary solar targets, such as X-ray bright points, active regions, filaments, and flares. To observe all levels of the solar atmosphere, we must observe the Sun simultaneously at many different wavelengths. TRACE does this by using a combination of UV and XUV multilayers and filters. Table II shows the wavelengths that will be covered by the various TRACE passbands and the ions to which they correspond. However, just observing these wavelengths is insufficient for interpreting the data; the images taken at different times and in different wavelength (temperature) bands must be coaligned to within the size of a pixel for unambiguous interpretation of the relation between the magnetic structures. This is achieved by a combination of the secondary coating design and the articulated secondary mirror, as discussed above.

TABLE II

TRACE Image Wavelengths

Wavelength (Å)	Ion	Region	Log (T_{max})	Log(T) Range
2500	UV cont.	Photosphere	3.7	3.6 – 3.8
1700	UV cont.	T_{min}/Chromosphere	3.8	3.6 – 4.0
1570	C I, Fe II	T_{min}/Chromosphere	3.8	3.6 – 4.0
1550	C IV	Transition	5.1	4.8 – 5.4
1216	H Lyα	Chromosphere	4.2	4.0 – 4.5
284	Fe XV	Corona	6.4	6.1 – 6.6
195	Fe XII	Corona	6.2	5.7 – 6.4
171	Fe IX	Corona	6.0	5.2 – 6.3

The primary scientific objective of TRACE is to explore the connection between fine-scale magnetic fields from the photosphere up into the corona. It will do this with spatial resolution of about 700 km on the Sun (i.e., 0.5-arcsec pixels). Hence we can resolve the fine structure of the solar magnetic field as it emerges through the photosphere and changes its character as it expands into the corona. Such interactions have not previously been observed consistently owing to:

- spacecraft orbit limitations, which, at best, limit the observation of the evolution of a region to 60-min periods every orbit.
- on-board data storage limitations, which have meant interrupted temporal coverage during even the optimum observing times.

– low instrumental spatial resolution, which blurs the geometry of loops.
– the lack of wide temperature coverage, which has made it impossible to follow the field lines throughout the solar atmosphere.

TRACE will change all this by providing multiwavelength images at a cadence as short as every 0.2 s for a partial frame image and with exceptional spatial resolution continuously over a period of at about 8 months.

Coordinated observations with SOHO, other spacecraft, and ground-based observatories will increase the scientific value of the TRACE data. For example, MDI on SOHO and TRACE will be able to see the fine-scale changes in the magnetic and velocity fields in the lower parts of the solar atmosphere. There is evidence that the small-scale and large-scale fields evolve in antiphase with one another. Whereas active regions emerge in increasing numbers and strength at solar maximum, the small-scale bipolar fields (e.g., X-ray bright points) seem to be more numerous at solar minimum. TRACE's high spatial resolution can look at the details of the small-scale field in both active regions and emerging bipoles with unprecedented detail. Hence, for the first time, we should be able to build a comprehensive model of how the Sun evolves from a quiet to an active configuration on all spatial scales from 1 arcsec to several solar radii. We should be able to see the detailed interactions of the emerging flux and the existing large-scale field. Some tantalizing examples of this exist in the Yohkoh data, but observational and instrumental limitations of the Soft X-Ray Telescope make it impossible to determine the nature of these interactions.

4. Conclusions

TRACE will produce images of all levels of the solar atmosphere with perfect coalignment, providing insight on their heating and dynamics. By flying at the same time as SOHO, TRACE can gather coordinated data that are far more valuable than any data taken alone. SOHO and TRACE provide the unique opportunity to obtain an extended set of simultaneous digital measurements of all the temperature regimes of the solar atmosphere with high spatial and spectral resolution.

Acknowledgements

A. Title is the principal investigator on TRACE, T. Tarbell, M. Bruner, and K. Strong are the deputy PIs for operations, instrument design, and science, respectively. The authors are supported by NASA contract NAS5–38099 and the Lockheed Independent Research Programme.

WORKING GROUP 1: SMALL SCALE FEATURES

O. KJELDSETH-MOE
Institute of Theoretical Astrophysics, University of Oslo, Norway

Abstract. A review is given of the activities of the working group on small scale features at the 2nd SOHO Workshop on Elba, 27 September – 1 October 1993. The small scale and filamentary structure of the solar transition region, and possibly also the corona, was pointed out. The reported observations furthermore demonstrated that the upper solar atmosphere is strongly dynamical, containing rapidly flowing gas and with features changing with time. Theoretical concepts and simulations of conditions in the transition region and corona were presented and discussed. Finally some ideas on future observations and modelling were put forward.

1. Introduction

The 2nd SOHO Workshop at Elba clearly demonstrated the importance of small scale features for the understanding of the solar atmosphere and the solar wind. The significance of small scale features was emphasized, not only in the session devoted to them, but papers proposing and discussing such features were presented in the other sessions as well.

The group had the advantage of some very good reviews and contributed talks and a number of interesting posters touching upon several relevant subjects. Taken together the oral presentations and posters demonstrated how wide and far ranging the subject of small scale features actually is.

In this paper I will sum up an overall impression of the relevant results from the presentations and conclude with some remarks on how to deal with observing small scale phenomena using the SOHO coronal instruments, as well as mention some future efforts at theory and modelling.

The article is not intended as a review on the subject of small scale features. Thus, only a few references are given. Neither are the various papers in these proceedings referred to explicitly. This is not to depreciate the contributions of the workers in the field, which are many and important.

2. Geometrical Extent of the Small Scale Features

The invited and contributed papers and posters dealt with a range of small scales. This raises the question of what is the significant scale for the important features and physical processes in the upper solar atmosphere.

The features that were considered may perhaps be divided into "macro" features and "filamentary" structures. Macro features would comprise such observed structures as:

– Spicules, which are seen at the solar limb.

Space Science Reviews **70**, 123–130.

- Mottles, which may be the same physical feature on the disc, observed in the chromospheric canopy structure.
- Macro spicules, seen in Hα, He^+ 30.4 nm, and other lines.
- Spicular features seen at the limb in C^{+3} 155 nm, i.e. in the transition region around T = 10^5K.
- Explosive events and jets, which also are well observed in the same transition region lines.
- Bright points, which apparently are associated with small magnetic loops.

The macro structures are characterized by a typical angular size of $\phi \approx 1''$–$10''$, i.e. they are clearly to be considered as small scale structures in the usual sense.

The filamentary structures are, however, much smaller. By considering volume emission measures, the observed apparent extent of the emitting features, and the electron (or gas) densities derived from density sensitive line ratios, Dere *et al.* (1987) derive filling factors of the transition region in the range 0.01 to 0.0001. This correspond to filamentary structures with diameters d $\approx$ 10–30 km.

Obviously such structures cannot be directly observed. They must be inferred from comparing different measured parameters, as is described above. However, if they exist they may also explain why the transition region emission extends to altitudes of several thousand kilometers above the solar limb.

Another characteristic property of the transition region and corona which may be explained by filamentation, is the co–existence of structures at different temperatures within a small geometrical region. This appears to be the general case, whether one observes in quiet regions, coronal holes or active regions.

The macro structures listed above may themselves consist of smaller filamentary structures. Thus, we may not be referring to different structures, but to characteristic sizes in the transition region and corona.

While the filamentary character of the transition region appears to be reasonably well established, the situation is not clear in the corona. Recent X–ray images taken with instruments equipped with multi–layer coated mirrors, show evidence of coronal features of sub–arcsecond size, although filamentary structures cannot of course be distinguished. Thus, the picture of a corona filled with loops of diameters a few thousand kilometers, which emerged from the Skylab observations, cannot be maintained. This is definitely a subject for further investigations with the SOHO instruments.

3. Time Variations, Flows and Dynamics

Another result of the posters and presentations in the working group sessions was the realization that the upper solar atmosphere is in a dynamical state and is characterized by temporal changes.

Thus, observed line intensities clearly change with time. Generally the emission is associated with interaction of photospheric magnetic fields at network boundaries. Not all interaction will necessarily produce detectable emission. The emission varies, however, on a well defined pattern independent of a large scale field.

Structures and time variations of spicule flows in the chromosphere were described. In the future it would be of interest to pursue the connection between these flows and the fine structure dynamics at higher altitude, using simultaneous ground based and SOHO observations.

Attempts have been made to explain the observed average net redshift of transition region lines in terms of dynamics of small scale features. This redshift correspond to an apparent average downflow with velocity v $\approx$ 5 km s^{-1}. The downflow must be apparent, or the corona will be emptied of gas in a short time.

Cheng (1992) showed that net downward velocities is obtained from upward running waves used to simulate spicules. This may not, however, give a net redshift, since his calculation neglects the connection between compression, velocity and emitted radiation in the waves (i.e. Hansteen and Wikstøl, 1994). A mechanism by Hansteen (1993) has redshifts caused by downward running waves generated by "nano–flares" in the coronal part of loops. This results in a net redshift in the 155 nm C^{+3} transition regions lines, but not in the high values actually observed, for any reasonable energy flux in the waves.

Another characteristic property of the fine structure flows in the transition region is the presence of multiple velocities. This is the appearance of several distinct velocities observed simultaneously within a small area of $1'' \times 1''$, i.e. the resolution of the best UV instruments.

The multiple flows are seen particularly in active regions, but there is also a sufficient amount of such flows in quiet areas to characterize them as a general feature of the transition region. The multiple flows would tie in with a filamentary transition region, where each velocity would occur in a separate flow channel or filament strand.

A special case of particularly strong flows is occurring above or near sunspots. Here one often (but not always) observe downflows with velocities v $\approx$ 100–200 km s^{-1}. These are multiple flows which may contain several downflow velocities as well as less prominent upflows and gas nearly at rest. They last for hours, although their detailed structure may change on shorter time scales

Finally one has the rapid flows in transient events. Typical flow velocities of explosive events in the transition region amount to 100 km s^{-1}, extending over a $2'' \times 2''$ region. A characteristic property of these events are their short duration, $\tau \approx$ 1 minute, and their rapid onset and decay, i.e. a few seconds at most. Explosive events have caused considerable interest. They have been tied to primary heating and acceleration mechanisms for the solar atmosphere, and alternatively to secondary effects of the primary mechanisms, as mentioned below.

An attempt has been made to extend observations of explosive events to the solar corona, using data from the slitless spectrometer, SO82A, on Skylab. The result suggest that they may occur simultaneously at a range of temperatures in the transition region up to 1 MK. Furthermore, the corona may radiate strongly in the same regions. A separate class or temporal stage of these events occur at coronal temperatures, but is not prominent below 1 MK. The data material is, however, scarce, and firm conclusions cannot be drawn.

A task for the SOHO instruments would be to extend the observations of all the above features and properties to cover simultaneously, or near simultaneously, a range of temperatures, as well as look at variations with time.

4. Theoretical Ideas and Simulations

Theoretical ideas of a broad conceptual kind have been proposed. These would include having the corona heated (and gas accelerated) by "nano–flares", caused by the release of magnetic energy of a twisted and twisting magnetic field in the corona. The release mechanism could be reconnection. It has been hypothesized that the "nano–flares" might give rise to explosive events.

The numerical modelling of an idea falling in this general class was presented. Here, the magnetic field was pulled around by footpoint motions, resulting in the creation of small regions with high current densities. With very small spatial scales one might heat the corona by Joule dissipation of currents of short duration ($\tau \leq 100$ s), using classical resistivity. The small scales might be created through a cascade mechanism. Alternately non–classical resistivity might apply. This would require drift velocities close to the velocity of sound and would still imply high electrical currents.

Another interesting result of the simulation was that flux tubes may not be connected over large distances on the Sun. In different regions the same magnetic field line belongs to different tubes.

It would be of interest to pursue theories of this kind, particularly to develop their dynamical consequences in order to check their observability.

An alternate conceptual idea was the "chromospheric furnace", proposed by Axford in another session of this workshop. These would generate waves which could dissipate further out in the corona. Dissipation of waves in the corona at several solar radii is needed to account for the wave signature and velocity of the high speed solar wind. The furnaces may of course also generate heat for warming the corona and transition region, and may cause the explosive events as a secondary effect.

A third idea was presented, pertaining to the heating of the transition region and generation of explosive events by proton beams from the corona. The existence of such beams is inferred from observations of radio bursts. The energy source might be reconnection in the corona. It was pointed out that MeV protons could heat the top of the transition region. The question would then be what the properties of explosive events could tell us about the primary mechanism, which could play an important role in the energy budget of the corona.

Other papers dealt with different specific mechanisms, such as:

- A spicule driving mechanism consisting of a train of slow MHD shocks in a tube, initiated by compressing the tube by increased external pressure.
- Numerical simulation of instability of a jet in a MHD shear layer providing transition to turbulence with a sufficiently short decay time to account for observed surges.
- Simulation of the coronal response to foot point motions that may produce small reconnection events as well as catastrophic loss of equilibrium, depending on the helicity injected into the field.

In all the cases it would be interesting to see the theoretical simulations pursued to derive observable consequences, which might distinguish between different theoretical mechanisms.

5. The Means: Instruments and Analysis Tools

The study of small scale features in the solar atmosphere will require data from all the coronal instruments. Furthermore, some of the particle instruments, notably CELIAS, are very useful. However, the two spectrometers, CDS and SUMER, will be the most important instruments. Their combined spectral coverage, 15.5 nm to 160 nm, allows study of the structure and dynamics of the entire temperature region from 10^4K to 3 10^6K.

To optimize use of the spectrometers, essential support is needed from EIT, SOI and from ground based observatories. A main role of SOI and ground based instruments will be to provide magnetograms and photospheric and chromospheric velocities, as well as participate in detailed, collaborative studies of fine structure phenomena and processes.

Above we have stressed the significance of feature sizes of $1''$, and less than that in the case of filamentation. It is then sobering that the angular resolution of the two spectrometers are approximately $4''$ and $2''$, respectively for CDS and SUMER. Thus, many conclusions about the small scale structures will have to be inferred from indirect evidence, as is the case today.

The spectral resolution of CDS and SUMER correspond to a capability for determining velocities of 20 km s^{-1} and 2 km s^{-1}, respectively. Thus, SUMER is the instrument for detailed dynamical studies of small scale phenomena. CDS will, however, be capable of registering rapid multiple flows and explosive events. Its spectral resolution exceeds that of previous stigmatic instruments covering the same wavelength region, i.e. 15.5–78.7 nm.

The great advantage of the SOHO spectrometers over UV instruments on many earlier missions is the possibility of observing long time series, which may be virtually uninterrupted for their designed length of operations. From the evidence of YOHKOH, it seems safe to predict that this added temporal dimension to the observations will bring about a new understanding of the processes going on in the solar atmosphere.

One may also add that the time resolution, which may amount to a few seconds for strong emission lines, will be adequate for studying many rapidly varying phenomena. This combination of a reasonable angular resolution with good spectral and time resolutions is of essential value in the study of small scale phenomena.

Using the data for diagnostic purposes also requires good analysis software. There is presently an on–going effort to equip the ground based software for the spectrometers with good analysis tools, including improved atomic data, codes for emission measure analysis and density and velocity determination, more elaborate analysis programs not assuming ionization equilibrium, and perhaps some MHD simulation codes.

6. Discussion and Some Future Directions

It seemed obvious to the working group that small scale structures in the solar atmosphere are of fundamental to our understanding of the transition region and corona and the processes that generate and transport energy and momentum in the upper solar atmosphere. However, when it came to the actual physical mechanisms for energy generation, there was no general agreement. Particularly, the importance, and indeed the presence of the mechanism of reconnection was strongly debated. This only serves to emphasize the need for high resolution observations of improved quality, particularly with regard to simultaneous coverage in temperature and extending over longer time intervals.

In preparation for SOHO, three of the most important themes to be addressed in the context of small scale features are:

- The connection to the transition region and corona from features and processes observed in the photosphere and chromosphere,
- The filling factor in the upper solar atmosphere, and
- Energy dissipation mechanisms, of which there are several proposed, and maybe many different ones actually occurring.

Since both the mass and the energy for the corona and the solar wind comes from below, it is important to look for connections between physical structures and processes going on in the lower and upper parts of the solar atmosphere.

One question addressed both at this meeting and often elsewhere in this connection, is the role of spicules in the supply of mass and momentum to the corona. SOHO gives the opportunity of addressing such studies in great detail. This requires that extended collaborations be set up between space and ground based observations. The ground work for such collaborative studies are being prepared and should be encouraged, since practical procedures must be worked out well in advance.

A particularly important parameter is the magnetic field. It probably governs the structure of the transition region and corona, serves as an energy source and acts as a conduit for the transport of mass, momentum and energy into the corona and the solar wind. There is no dedicated instrument on board SOHO for measuring magnetic fields. Measurements from SOI will be available, but probably not to the extent required for the study of the detailed and time variable conditions in the solar transition region. Thus, it is important to co–observe with ground based magnetographs in order to map the magnetic field and its variation with time in the areas where the SOHO instruments study processes in the chromosphere, transition region and corona.

The emission filling factor appears to be very small around $T \approx 10^5$ K. One should attempt a theoretical study of the effect of a filamentary transition region on the transport and dissipation of energy, mass and momentum. This, however, may require better knowledge about the geometry of the filling factor than is presently available. Thus, observational studies of emission and density in the transition region should be designed to cover a larger range in temperature and to be extended into the corona. The question of whether the corona also has a filamentary structure, needs very much to be addressed through observations. (The point may be made that, like every other region in the solar atmosphere, the corona has become increasingly complex with improved angular resolution of the observing instruments.)

In these studies, theoretical as well as observational, one should continue to keep in mind the dynamic character of the upper solar atmosphere. In order to map the processes going on in the solar gas, observations of non–thermal and multiple velocities are just as important as extending our knowledge about the geometrical structure.

Of theoretical studies one would like to see numerical simulations of processes that might lead to filamentation and give rise to multiple velocities, that are either resolved, or unresolved as a non–thermal line broadening of the transition region lines. Such velocities might perhaps result from interaction of filament strands moving around inside a wider magnetic tube

(Dahlburg 1994, private communication).

With regard to the present ideas for energy dissipation, both the "nano-flares" and the "furnace" mechanisms may be questionable. Part of the problem may be that the theories are not mature enough to make sufficiently precise predictions to distinguish between different ideas. However, one should certainly continue to collect partial evidence on these and other possible mechanisms from a number of sources.

Among many ideas to pursue one might mention the following:

- Cascade or avalanche processes of flaring structures may be important in connection with the energy dissipation and should be studied.
- The validity of extrapolating flare properties to "nano-flares".
- Whether one needs to consider several energy generating mechanisms, considering the existence of many different types of structures and temperature regimes.
- How to ascertain the presence of energetic particle beams, and to model their effect.
- The presence of various types of waves should be assessed through observations, and theoretical models should account for their propagation and observability in the context of small scale and filamentary structures.
- Explosive events, their dynamics, time variations and densities, should be studied at a number of temperatures in order determine the generating mechanism and assess whether they are primary or secondary (i.e. derived) processes with respect to the supply of mass, energy and momentum to the corona. Models should address the fast rise and decay of explosive events, as well as the fact that they may appear simultaneously over a range of temperatures.

The list is obviously far from complete. Instead it represents some of the ideas that were presented and discussed by a limited group in the short period of the workshop. It is to be hoped that a large segment of the solar physicists in the world will be drawn into thinking about observational programs, as well as pursuing theoretical investigations, in connection with the future SOHO activities. A multitude of good ideas will be required to best use the SOHO instruments in the study of a part of the solar atmosphere which may appear less well understood today than 20–25 years ago.

References

Cheng, Q.Q.: 1992, *Astronomy and Astrophysics* **262**, 581

Dere, K.P., Bartoe, J–D.F., Brueckner, G.E., Cook, J.W. and Socker, D.G.: 1987, *Solar Physics* **114**, 223

Hansteen, V.: 1993, *Astrophysical Journal* **402**, 741

Hansteen, V. and Wikstøl Ø.: 1994, *Astronomy and Astrophysics* , accepted

II Loops and Prominences

OBSERVATIONS OF LOOPS AND PROMINENCES

KEITH T. STRONG AND THE YOHKOH TEAM
Lockheed Solar & Astrophysics Laboratory, Palo Alto, CA 94304, USA

Abstract. We review recent observations by the Yohkoh-SXT in collaboration with other spacecraft and ground-based observatories of coronal loops and prominences. These new results point to problems that SoHO will be able to address. With a unique combination of rapid-cadence digital imaging ($\geq$32 s full-disk and $\geq$2 s partial-frame images), high spatial resolution ($\geq$2.5 arcsec pixels), high sensitivity (EM $\leq 10^{42}$ cm^{-3}), a low-scatter mirror, and large dynamic range, SXT can observe a vast range of targets on the Sun. Over the first 21 months of Yohkoh operations, SXT has taken over one million images of the corona and so is building up an invaluable long-term database on the large-scale corona and loop geometry. The most striking thing about the SXT images is the range of loop sizes and shapes. The active regions are a bright tangle of magnetic field lines, surrounded by a network of large-scale quiet-Sun loops stretching over distances in excess of 10^5 km. The cross-section of most loops seems to be constant. Loops displaying significant increase in the ratio of the footpoint to loop-top diameter (Γ) are the exception, not the rule, implying the presence of widespread currents in the corona.

All magnetic structures show changes. Time scales range from seconds to months. The question of how these structures are formed, become filled with hot plasma, and are maintained is still open. While we see the propagation of brightenings along the length of active-region loops and in X-ray jets with velocities of several hundred km/s, much higher velocities are seen in the quiet Sun. In XBP flares, for example, velocities of over 1000 km/s are common. Active-region loops seem to be in constant motion, moving slowly outward, carrying plasma with them. During flares, loops often produce localized brightenings at the base and later at the apex of the loop. Quiescent filaments and prominences have been observed regularly. Their coronal manifestation seems to be an extended arcade of loops overlying the filament. Reliable alignment of the ground-based data with the X-ray images make it possible to make a detailed intercomparison of the hot and cold plasma structures over extended periods. Hence we are able to follow the long-term evolution of these structures and see how they become destabilized and erupt.

Key words: Sun – Corona - Magnetic Structure

1. Introduction

The recent flood of results from Yohkoh on the nature of coronal structures is changing our view of the solar atmosphere. The picture that is emerging is one of an extremely dynamic and complex corona. The high-cadence imaging available from the Soft X-ray Telescope (SXT) on Yohkoh makes it possible for the first time to observe the detailed evolution of coronal structures. These data are leading to new questions about how the corona is heated and supplied with mass, how active regions emerge and evolve, and how flares are produced. However, perhaps the most surprising aspect of these data is how they are causing us to reevaluate the basic building block of coronal structure, the magnetic loop. Coronal loops are not the stable, semicircular, potential structures that are often used in models. Rather, they seem to be asymmetric and dynamic structures with their own com-

Space Science Reviews **70**, 133–142.

plex internal structure. This poses a problem that cannot be solved with existing data, namely that the processes that produce a system of loops are occurring on rapid timescales and over a broad range of temperatures. So to understand this complex structure, we will need to adopt a multiwavelength approach. SoHO, with supporting observations, offers the best opportunity to probe the magnetic structure of the solar atmosphere at different heights simultaneously with high-resolution imaging and a wide range of plasma diagnostics.

In this paper, we illustrate the level of complexity and the range of geometries of coronal loops. Then we will show some of the advantages of the multiwavelength approach that SoHO will be able to take, especially in conjunction with other space-based missions such as Yohkoh and with the network of ground-based instrumentation. To do this we will discuss joint observations between Yohkoh and several recent rocket-borne payloads that observed filaments.

2. Coronal Currents

When we look at a full-disk image from SXT, the striking impression is that there is a wide variety of loops. In Figure 1 we can see some post-flare loops on the northwest limb (upper right). Several bright and complex active regions show a tortured weave of loops. The quiet-Sun loops tend to be larger, fainter, and low-lying. X-ray bright points (XBPs) are tiny, short-lived bipolar regions. They seem to be smaller versions of active regions, but when examined closely, they are found to fluctuate and flare (Strong et al. 1992). One of the first things noticed about the SXT images were that the loops seemed to be uniform in cross-section. This discovery was quantified by Klimchuk et al. (1992), who showed that the upper limit on the ratio of the footpoint to loop-top diameter (Γ) was about 1.1. They attributed this effect to currents flowing along the loops, producing a toroidal field which maintains a near-constant cross-section. This result is in direct contrast to the submerged dipole model of coronal loops which came out of the Skylab era.

This interesting result implies the presence of coronal currents, whose dissipation has long been proposed as a source of coronal heating. This conclusion is supported by Lang et al. (1993), who used an entirely independent method. They compared SXT images with VLA and RATAN-500 radio maps and were able to calculate the coronal magnetic field strength and compare it with the extrapolated potential field determined from a magnetogram. The measured coronal fields were significantly stronger than predicted by the model, indicating that the divergence of the field was less than expected.

These results imply that coronal currents are a significant factor in determining the geometry of the loops; however, that is not proof that they have a

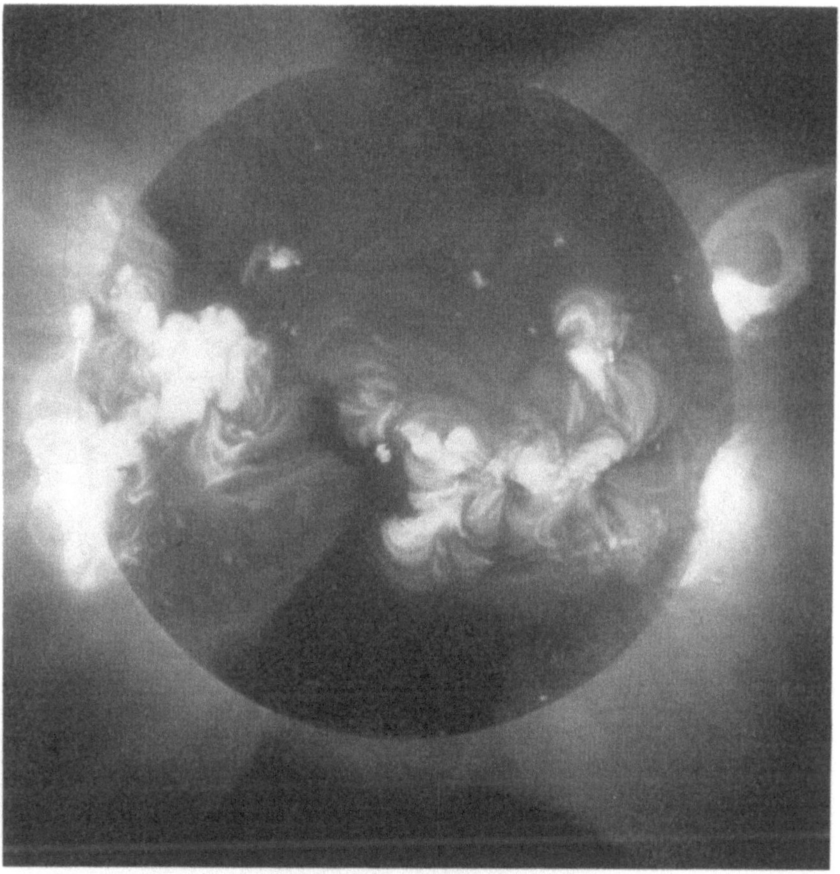

Fig. 1. SXT image of the corona (log scaling of a 2.556-s exposure taken in the Al 1400E filter).

significant role in heating the corona. If that were the case, we would expect that the brightest and hottest coronal loops would be well correlated with the strongest currents in the corona.

Metcalf et al. (1993) have found a way of checking this hypothesis by using SXT data in conjunction with Mees vector magnetograms. From these data they can derive the location and strength of the coronal currents and find no significant correlation with bright, soft X-ray loops. At first glance this seems to eliminate the possibility of currents heating the corona. It does eliminate coronal heating by large-scale, organized current systems. However, there still remains the possibility that a chaotic, randomly directed system of filamentary currents exists in the corona and heats it.

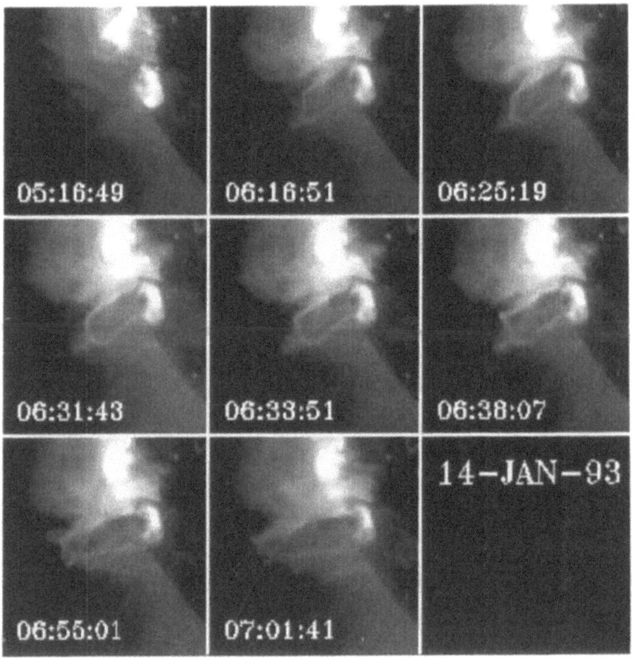

Fig. 2. The development of a possible kink instability in a loop on the southeast limb. Initially this loop has an odd angular shape, but it slowly develops over the next 20 min to a more smooth, curved shape. Suddenly a kink clearly forms in the northeast corner of the loop, after which it rapidly, almost explosively, expands and dissipates. Note the fine twisted structure of the field revealed as the loop expands.

3. Loop Geometry

There are other indications that currents exist in the corona and play a significant role in determining loop geometry. Evidence for currents can be found in the existence of sheared and twisted loops. While sheared loops abound in SXT images, remaining stable for hours to days, twisted loops are quite rare and seem to be unstable. Figure 2 shows one of the clearest examples of the development of a twisted loop.

One type of loop geometry is much more common in the Yohkoh images than we expected, and that is the cusped loop. We see many different sizes of cusped loop, ranging from a few arcseconds to over a solar radius. Several examples are shown in Figure 3. The cusped loops observed by Yohkoh associated with coronal mass ejections (CMEs) and flares are also discussed elsewhere in this volume (see Hiei 1994 and Cheng 1994, respectively). Both of these types of cusped loops seem to be associated with an arcade of loops

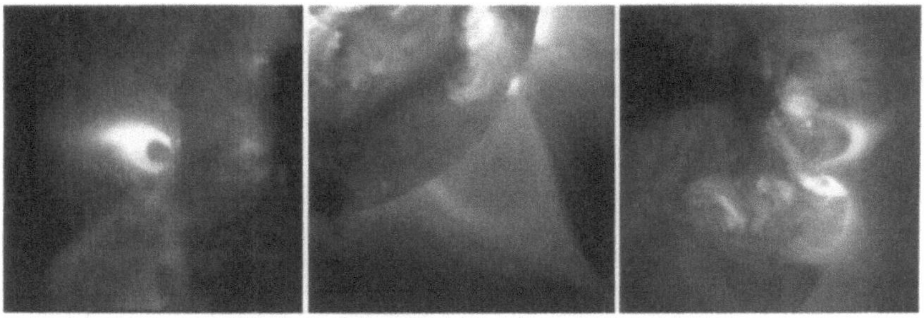

Fig. 3. Examples of cusped loops from flares (left), filament eruptions (centre), and active regions (right).

that can stretch over several arcmin.

The small, bright cusped loops particularly associated with long-duration flares are dynamic and last for a few hours. They grow at a rate of 5 to 10 km/s and generally have a bright apex (Figure 3a). The Yohkoh data on these seem to be consistent with the idea that these structures are formed by the successive reconnection of open field lines higher and higher in the corona. The larger-scale cusped loops are very reminiscent of helmet streamers and also expand upwards slowly (Figure 3b). They often seem to be associated with the aftermath of a filament eruption. They can last for several days and are often explained in similar terms to the smaller, flare-related structures. However, there are some important differences. They are very rare; we have observed only about 10 such events clearly in over two years. CMEs and filament eruptions are far more common than these large-scale cusped loops. Unlike their flare-related cousins, they are rarely bright at the apex of the helmet, implying that reconnection is not going on there, if it were, one would expect a somewhat elevated temperature due to local heating, which would produce a significant increase in the brightness at the apex. The explanation of these events by the reconnection of open fields was needed to explain the slow growth of the helmet. However, other loops have been seen to grow through the same region prior to the event. None of these had a cusp-like structure. Further, we observe expanding loops in active regions (Uchida et al. 1992) that are not necessarily explained in terms of reconnection as the driving mechanisim (although it might be caused by the expanding loops interacting with existing field lines). Although none of these objections to the current reconnection models for large scale X-ray helmet streamers are in and of themselves a fundamental problem, they should make us pause to reexamine our assumptions about the mechanism that forms these structures.

There is another type of cusped loop often seen in the quiet Sun or in active regions. These share the general geometry but seem to have a fundamentally different structure (Figure 3c). They seem to be thin structures, i.e., not associated with an arcade of loops. They do not seem to grow rapidly but can last from a few minutes to over a day. One of their most surprising aspects is that they can often be seen to be reconnected back to the Sun via a fine loop-like structure to another cusped loop. If this is the actual structure of these loops, then the filamentary loop joining the two cusps is probably a current sheet. This leads one to speculate how many of the other loop-like structures we see in the corona are in fact current sheets. The answer seems to be not many, but we should be careful when interpreting the properties of any given loop, because it might be a current sheet.

So far, Yohkoh is providing us with a complex picture of the solar corona. It is highly dynamic, changing on time scales from seconds to days. These changing loop structures have spatial scales from the practical limit of the instrument (5 arcsec) up to about a solar radius. The complexity of the geometry that is observed is often daunting.

4. The Multiwavelength Solution: SoHO

The real problem with interpreting the SXT data lies in that there is no information about the cooler ($\leq$2 MK) plasma and that the continuity of the data is broken by regular, long periods of spacecraft night. SoHO will remove these limitations. To illustrate this point, we shall describe collaborative observations of filaments and prominences in this section.

5. Prominences and Overlying Coronal Arches

In Figure 4 we show simultaneous observations by SXT, the Transition-Region Camera (TRC) on the SPDE rocket, and 10830 Å images (KPNO) of a filament channel. Note that in the C IV image, several clear, dark channels run across the image where the bright network structure is suppressed. Comparing this with the 10830 Å image, we can identify areas of dark filaments, but they exist only along portions of each channel. It is clear that the coronal loops are most sheared along (i.e., parallel to) the filament channel where the darkest and most prominent filaments can be seen. There is little evidence of any overlying magnetic field lines in Figure 4.

To observe these overlying loops more clearly, we have to look at prominences. Figure 5 shows a set of images from NIXT, BBSO, and Yohkoh, taken just after the SPARTAN flight on 13 April 1993. A prominence can be seen in the Hα image on the southeast limb. The same prominence is clearly visible in absorption in the Fe XII and Fe XV images from NIXT. Such absorption features are not usually seen in coronal images, but as NIXT operates in the XUV region of the spectrum, the light is more easily absorbed than at soft X-rays wavelengths. The SXT image shows a dark

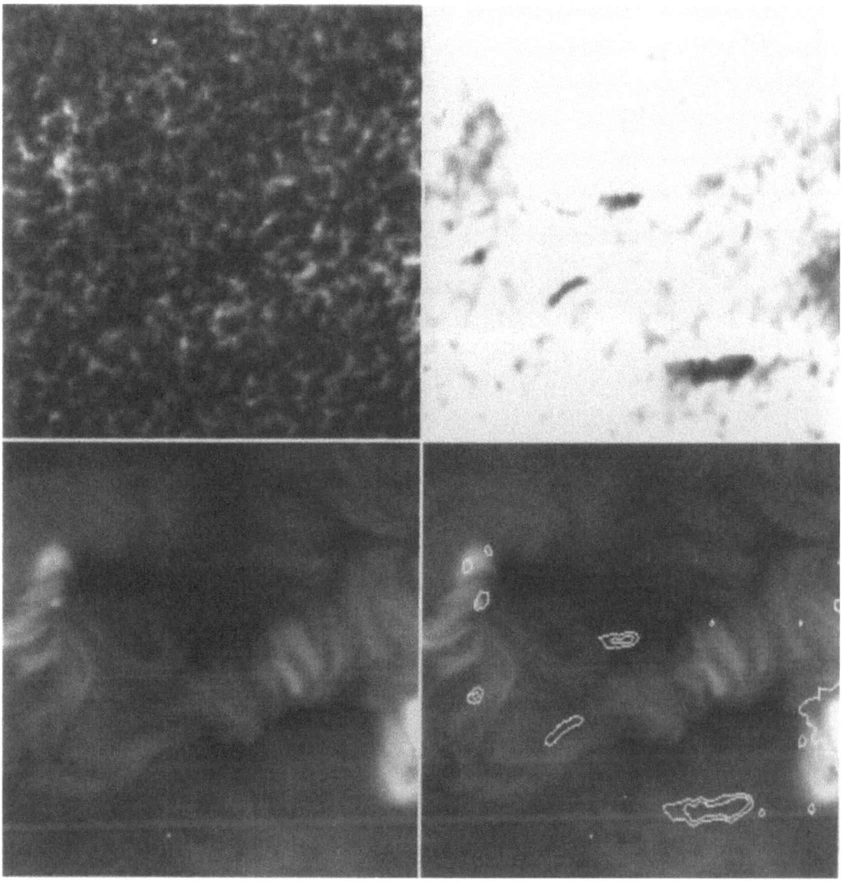

Fig. 4. Filament channels observed simultaneously at several different wavelengths: C-IV (1550 Å) from SPDE (upper left), 10830-Å from KPNO (upper right), and a soft X-ray image from SXT images of filament channels (lower left). In the lower right we overlay the darkest filament matterial contours on the soft X-ray image. Note how the dark filaments correspond to the most sheared coronal loops

cavity where the prominence is situated, but a faint arch of overlying magnetic field lines can be seen above the cavity. These two data sets seem very reminiscent of the model discussed by S. Antiochos at this conference, in which highly sheared field lines along the neutral line support the filament material and overlying (near potential) fields confine it.

These data illustrate the structure of a filament in the corona. What happens on a larger scale? We need to compare images from the lower corona with data from white-light coronagraphs. The only large-scale coronagraph observations that have been available during Yohkoh have been those from

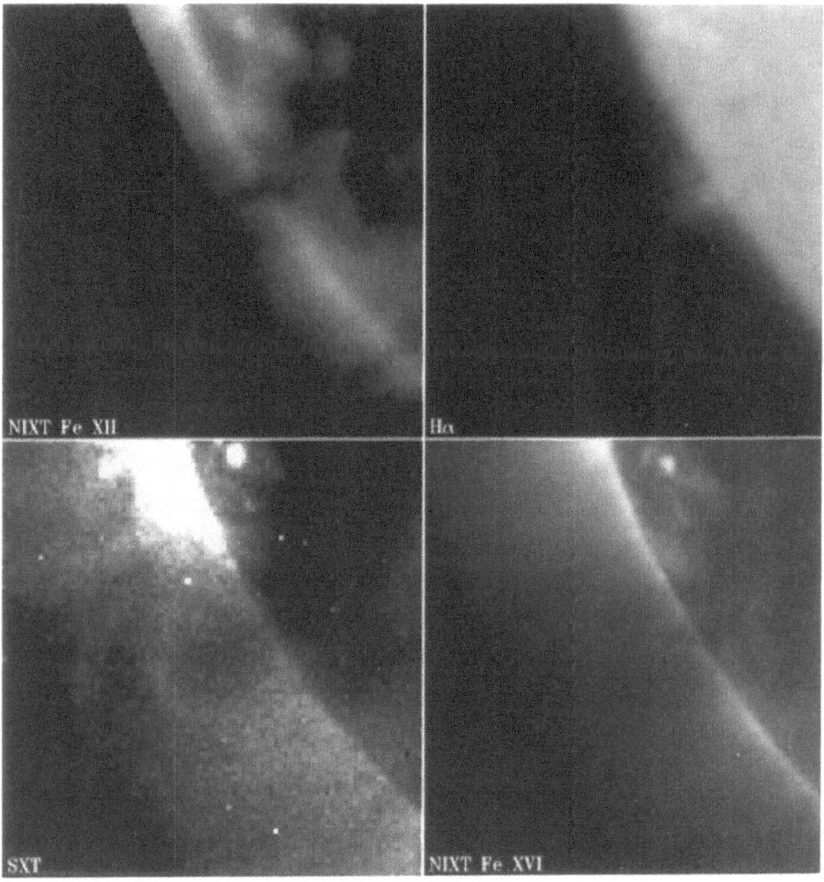

Fig. 5. Comparison of Hα (BBSO), XUV (Fe XII and Fe XV from NIXT, courtesy L. Golub), and SXT images of a prominence on the SE limb.

the recent SPARTAN mission (e.g., Kohl et al. 1994). Figure 6 shows a composite SPARTAN and SXT image from 12 April 1993.

Note that there are three prominent helmet streamers visible: two along the equator of the Sun and one in the southeast. Of the two equatorial helmet streamers, the one on the west limb is by far the brighter, probably because of a coronal transient and flare from the western active region below it. In fact, an arcade of postflare loops can be seen about 5 arcmin above the west limb. However, the brightest and most prominent helmet does not lie above the active regions but, rather, over the filament channel on the southeast, which is the same one as shown in Figure 5. So, as SoHO will be able to do with its unique combination of instruments, we have been able to

Fig. 6. A white light coronagraph image (SPARTAN, courtessy R. Fisher) of the streamer above the prominence shown in Figure 5, taken on 12 April 1993. Solar north is rotated about 45 degrees to the right from the virtical.

investigate the magnetic structure of a prominence from the chromosphere out to several solar radii.

6. Conclusions

All magnetic structures are seen by SXT to change on time scales that range from seconds to months. The solar corona as seen by Yohkoh has a complex and dynamic structure. We may have to wait for observations from SoHO for these to be understood better. SoHO will be able to look much further out into the corona than the above Yohkoh observations and have available UV and EUV spectral diagnostics. It has uninterrupted viewing of the Sun and so it can see the formation, development, and decay of most solar structures. Hence we look forward to SoHO to shed further light on the magnetic structure of the solar atmosphere.

The ideal situation would be to have Yohkoh still operating during SoHO, because its observations will complement those of SoHO, not only from the

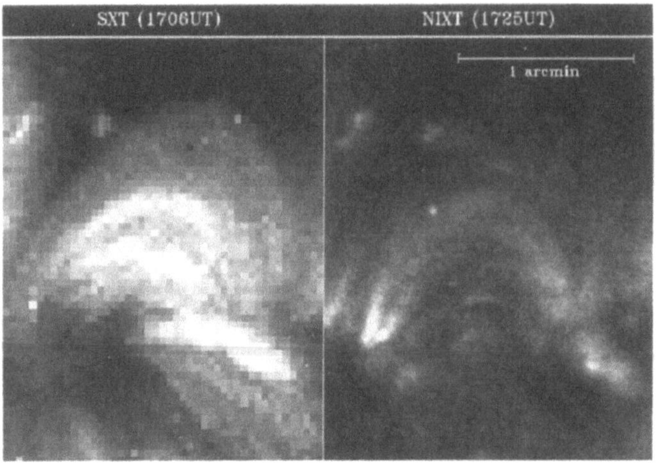

Fig. 7. Comparison of broad-band (left) and multilayer (right) images of quiescent loops. Comparison of observations of a quiescent active region seen through SXT and NIXT (courtesy L. Golub).

differences in the response of a broad-band X-ray telescope compared with multilayer optics (Figure 7) but also in the sense that the EIT full-disk images will be very limited in number owing to telemetry limitations of SoHO, whereas Yohkoh can provide rapid-cadance data on flares, active regions, and the quiet Sun. Hence SoHO and Yohkoh can provide a powerful set of imaging and spectroscopic tools that should be able to answer many of the questions that have perplexed us.

Acknowledgements

This work is supported by NASA contract NAS8 32334 with MSFC and the Lockheed Independant Research Programme. SXT was built at Lockheed Solar and Astrophysics Laboratory in collaboration with the National Astronomical Observatory of Japan, the University of Tokyo, and ISAS.

References

Cheng, C.-C.: 1994, these proceedings
Hiei, E.: 1994, these proceedings
Klimchuk, J. et al.: 1992, *PSAJ* **44**, L181
Kohl, J. et al.: 1994, these proceedings
Lang, K. et al.: 1993, *ApJ* **419**, 298
Metcalf, T. et al.: 1994, *ApJ*, in press
Strong, K.T. et at.: 1992, *PSAJ* **44**, L161
Uchida, Y. et al.: 1992, *PSAJ* **44**, L155

MASS FLOWS IN CORONAL LOOPS

SPIRO K. ANTIOCHOS
E.O. Hulburt Center for Space Research, Naval Research Lab, Code 7675, Washington, DC 20375, USA

Abstract. Although static loop models are often used to describe the structure of coronal loops, it is evident on both observational and theoretical grounds that mass motions play a crucial role in the physics of the corona and transition region. First we review the observations of emission-line broadening and wavelength shifts, which imply the presence of random motions and systematic downflows in coronal loops. Some discrepancies in the observations are discussed. It is argued that velocities due to gas pressure gradients are the most likely explanation for the observed flows. A number of models that have been proposed for these motions are reviewed. The implications of the various models on observations of the corona and transition region by SOHO are discussed.

Key words: SUN – CORONA – UV, X-rays

1. Observational Results

Since the observations from SKYLAB, it has become widely accepted that the bulk of the plasma in the solar corona and transition region is contained in loops. Each loop is presumed to correspond to a magnetic flux tube. The field both constrains the bulk flow of the material and impedes any diffusion of mass or heat out of the flux tube. Consquently, the corona is usually approximated as a collection of one-dimensional loops.

The lifetime of coronal loops as observed in X-ray or EUV is often long compared to the timescale for evolution of the plasma by radiative or conductive cooling; consequently, static models are frequently assumed for the loop plasma. These have proved to be quite successful at explaining several features of the observations, specifically, the intensities of emission lines formed in the corona and upper transition region. The line intensities are conveniently reported in terms of the differential emission measure. The standard hot loop models agree well with the observed form of the emission measure for temperatures above 10^5K, but fail badly for lower temperatures (e.g., Pallavicini *et al.* 1981). This discrepancy is generally explained by invoking a different structure for the lower transition region, such as cool loops. Note that this lower transition region structure may still be static, so that the observations of line intensities by themselves do not necessitate dynamic models.

Compelling evidence for dynamic effects in the corona and transition region is obtained from spectrally-resolved emission lines. Observations with high-spatial and temporal resolution often exhibit a large variety of small dynamic phenomena such as spicules and macrospicules, surges, jets, coronal bullets, etc.; however, observations with large spatial or temporal averaging

Space Science Reviews **70**, 143–148.

show only two types of persistent motions. From the position of the centroids of transition region lines with respect to the position of photospheric lines, a systematic redshift is deduced to be present throughout the solar surface (e.g., Doschek, Feldman and Bohlin 1976). The redshifts occur in all types of solar regions; they are even observed in the UV emission from other late-type stars (Ayres, Jensen and Engvold 1988). These line shifts imply velocities of approximately 10 km/sec at 10^5K. Another type of persistent motion that is observed in the data is a nonthermal broadening of the emission lines. The broadening is usually interpreted as due to unresolved motions that are approximately randomly oriented with respect to the line-of-sight.

Athay and Dere (1989) have performed a careful analysis of available spectral line data and have derived the following forms for the persistent motions as a function of temperature:

$$V_{red} \approx 10(T/10^5)^{1.5} \text{ km/s}, \quad \text{and} \quad V_{broad} \approx 20(T/10^5)^{0.5} \text{ km/s}, \tag{1}$$

where, V_{red} is the downward velocity deduced from the redshifts and V_{broad} is the random velocity deduced from the linewidths. Note that these results are valid only for $10^4 < T < 10^5$K. Above this temperature there are few available data with high spectral resolution, which has resulted in conflicting estimates of the magnitudes of the persistent motions in the upper transition region and corona. SOHO should clear up this confusion by providing us with accurate spectral data on lines formed above 10^5K.

Another area where there appears to be observational confusion is in the longitudinal dependence of the persistent motions. Since the nonthermal velocities are believed to be randomly oriented, they should not exhibit any center-to-limb variation and, indeed, this is observed to be the case. The redshifts, however, must be oriented downward toward the chromosphere; hence, they should exhibit a sinusoidal dependence with longitude, vanishing at the limb. Some observers do find from their data that the redshifts vanish at the limb as predicted (Hassler, Rottman, and Orrall 1991), but others find very little center-to-limb variation (Feldman, Cohen, and Doschek 1982; Brekke 1993). From a careful study of SMM observations, Klimchuk (1987) concluded that in quiet regions the redshifts do show a center-to-limb dependence, but that in active regions the redshifts are approximately constant out to the limb. It should be noted that this conclusion is at odds with the emission measure observations, because the differential emission measure appears to have the same shape in quiet and active regions, implying a similar structure for both regions. Again this is an important area of controversy that, hopefully, we will be able to resolve with the SOHO observations.

2. Theoretical Considerations

In order to determine the possible origins of the observed motions, we can examine the momentum equation for the loop plasma,

$$\frac{D\mathbf{V}}{Dt} = -\frac{\nabla P}{\rho} + (\nabla \times \mathbf{B}) \times \frac{\mathbf{B}}{\rho} + \mathbf{g}. \tag{2}$$

In equation (3), D/Dt indicates the convective derivitive, $\partial/\partial t + \mathbf{V} \cdot \nabla$, and all the variables have their usual meaning. There are three possible forces accelerating the plasma, pressure gradients, the Lorentz force, and gravity. Each is associated with a characteristic velocity determined by the magnitude of the force: pressure gradients generally produce motions of order the sound speed C_s, the magnetic field produces motions of order the Alfven speed V_A, and gravitational acceleration results in the free-fall velocity V_f. For typical solar coronal parameters, temperature $T \sim 10^6$K, magnetic field strength $B \sim 50$G, electron density $n \sim 10^{10}\text{cm}^{-3}$, and loop height $H \sim 10^4$km, the three characteristic velocities have magnitudes: $C_s \propto T^{1/2} \sim 150$ km/s, $V_A \propto B/n^{1/2} \sim 2,000$ km/s, and $V_f \propto H \sim 100$ km/s.

There are several important points in these results. One is that plasma motions at the Alfven speed are rarely seen in the solar atmosphere. Occasionally surges or coronal mass ejections may reach velocities of order the Alfven speed, but these events are infrequent. Even in highly dynamic events such as flares, the typical velocities deduced from lineshifts or broadenings are of order hundreds of km/s rather than thousands. This result is somewhat puizzling since many of the theoretical models for flare energy release and for coronal heating involve MHD instabilities such as kinking or reconnection, which are expected to produce motions at or near the Alfven speed. It can be argued that Alfvenic velocities are indeed common in the corona, but that the high speed plasma is cool and of low density. Once this plasma slows down, it compresses and heats up, thereby raising its emission measure and becoming visible as the subAlfvenic coronal plasma. Although this argument appears reasonable, it still seems unlikely that motions at the Alfven speed can play a major role in the coronal magnetic-energy release process, but are unobservable. This is a key issue that SOHO with its high spatial resolution and sensitivity must resolve. If we find that even at the highest resolution and sensitivity, Alfvenic motions are not observed, this would argue strongly that the magnetic energy release process is not due to an MHD instability but more likely to some enhanced difussion mechanism such as anomalous resistivity.

Another important point is that the persistent motions are probably not due to simple free-fall. The observed motions have a clear temperature dependence, whereas the free-fall velocity depends only on height. Hence, the motions are most likely due to pressure gradients. Note that the sound speed does have the right temperature dependence and the correct magnitude to account for the broadening velocities, at least in the lower transition region. It may be that the broadening can be explained as due to flows generated by small-scale impulsive heating events in loops. Impulsive energy release would be expected to give rise to expansions and contractions of the plasma at the local sound speed.

A different possible explanation for the broadening velocities is that they are due to waves propagating through the transition region and corona. For a constant flux of either Alfven ($\rho v^2 V_A = const.$) or acoustic waves ($\rho v^2 C_s = const.$) in a constant pressure atmosphere, one obtains the same dependence of wave amplitude on temperature, $v \propto T^{1/4}$. This dependence is somewhat weaker than what is observed, but is probably within experimental error. If either the bulk flow or the wave model is correct, then either the $T^{1/2}$ or $T^{1/4}$ dependence for the nonthermal velocities should be observed to extend to coronal temperatures, a prediction that SOHO can test.

3. Models for the Redshifts

Although the persistent broadenings have yet to be explained definitively, there do appear to be good candidate models. The redshifts, on the other hand, have proved to be more problematical, especially if they have no center to limb variation. One obvious possibility is that the redshifts are due to steady flows in coronal loops. A truly static loop is unphysical, in general, since it requires complete symmetry about the loop midpoint. Any difference in geometry or heating rate or boundary conditions between the two legs of a loop will result in flows. If the heating is time independent, these flows may be steady. For a constant mass flux we expect that, since the pressure is approximately constant in the transition region, $V \propto T$, which is in rough agreement with equation (1).

The magnitude of the steady flows can be estimated from the energy equation for the loop plasma. We expect that if the flows are to have a significant effect on redistributing the plasma energy, then the enthalpy flux 5/2 PV should be of order the input energy flux to the loop. The coronal heating flux is estimated to be $\sim 10^5$ergs/cm^2/s; implying a steady velocity of 10 km/s for a typical coronal plasma pressure of 10^{-1} ergs/cm^3. However, this is the magnitude of the velocity in the corona. If the transition region temperature is an order of magnitude smaller, then the velocities there are proportionally smaller and are too low to match the observed redshifts.

Another problem with the steady state model is that for a loop at disk center, one would expect to observe redshifts from only one leg of the loop, that with the downflowing material. The other leg, which contains upflows, would produce blueshifts. Hence, in order to agree with that data, the observations must contain spatially unresolved loops, and the emission from the downflowing plasma must be more intense than that from the upflowing. For models with small asymmetries or with flows due to differences in the chromospheric boundary conditions, the blueshifts are expected to dominate (Antiochos 1984), in direct contrast with observations.

One way to circumvent the problems with the steady state models for the redshifts is to invoke extreme asymmetries in the heating. A number of authors have considered such models (e.g. Mariska and Boris 1983; McClymont and Craig 1987). The difficulty with these models is that very stringent assumptions must be placed on the heating magnitude and location, and on the loop length in order to minimize the blueshifted emission and to obtain large flows at 10^5K. An interesting variation on the steady state model has been proposed by McClymont (1989). He argues for heating asymmetries in cool loops. The key point here is that for the same heating asymmetry, a cool loop will have substantially larger flows than a hot one because its pressure is much lower (Antiochos and Noci 1986). There are still problems, however, with the predicted presence of blueshifts, especially at lower temperatures. Note that since the cool loops do not exist for temperatures above $\sim 10^5$K, this model predicts that the shifts should disappear abruptly above this temperature.

Even if steady flows are not the explanation for the redshifts, it would still be very interesting to search for them with SOHO. The predicted lineshifts are small; but have a distinctive signature, red on one side and blue on the other, so that some procedure such as differencing the emission from the two legs of a loop may be able to detect steady flows.

A different possible model for the redshifts is time-dependent flows produced by impulsive heating. In this case the redshifts would be due to cooling flows that result from an abrupt decrease in the coronal heating rate. Both loop (Mariska 1987) and spicule geometries (Athay and Holzer 1982) for this model have been proposed. The situation is similar to the decay phase of flares, which typically do show significant redshifts during their decay phase. Cooling flows in loops have been investigated in detail by Mariska (1987). He found from his simulations that although redshifts of the magnitude observed could be produced, they tended to occur late in the cooling phase when the emission measure was low; hence the time-averaged emission did not exhibit redshifts of the required magnitude.

A related model has been proposed recently by Hansteen (1993) who argues that small impusive heatings in a loop will create sound waves due to the initial expansion of heated plasma. These will propagate down to the transition region creating a net redshift. Again the main difficulty is producing large enough redshifts; however, the model raises the interesting point that if coronal heating is highly impulsive, then a significant flux of acoustic waves may be produced in the corona. It may be possible to detect these waves with SOHO. In addition, this whole class of models would predict a correlation between changes in coronal emission with redshifts in the transition region. SOHO, with its broad complement of instruments will be able to search for such correlations.

A final point concerns the variation of redshifts with longitude. If the result that in active regions the redshifts are constant out to the solar limb is confirmed by SOHO, it would argue very strongly against any model in which the redshifted emission originates in an optically thin region. The only explanation that we can imagine for the observation of a net redshift at the limb is that there is preferential absorption of blueshifted emission there. We have proposed a model for this absorption in which the UV emission lines originate in high pressure coronal loops whose transition region occurs *below* the top of the surrounding chromosphere (Antiochos 1984). This model predicts that the redshifted emission should occur at the footpoints of bright X-ray loops, and that there should be obscuration of UV emission near the limb. It may be possible to detect this obscuration as a decrease in the expected limb brightening of active regions. These are predictions that SOHO should easily be able to confirm or deny.

In summary, there are many aspects of coronal and transition region structure and dynamics that are presently in a state of controversy. Hopefully, SOHO will resolve all this observational confusion and raise our knowledge of these regions to a new level, so that we can finally begin to understand the basic physics of the upper solar atmosphere.

Acknowledgements

This work has been supported in part by grants from NASA and ONR.

References

Antiochos, S.K.: 1984, *Astrophys. J.* **280**, 416
Antiochos, S.K. and Noci, G.: 1986, *Astrophys. J.* **301**, 440
Athay, R.G. and Holzer, T.E.: 1982, *Astrophys. J.* **255**, 743
Athay, R.G. and Dere, K.P.: 1989, *Astrophys. J.* **346**, 514
Ayres, T.R., Jensen, E., and Engvold, O.: 1988, *Astrophys. J. Suppl.* **66**, 51
Brekke, P.: 1993, *Astrophys. J.* **408**, 735
Doschek, G.A., Feldman, U., and Bohlin, J.D.: 1976, *Astrophys. J. Lett.* **205**, L177
Feldman, U., Cohen, L., and Doschek, G.A.: 1982, *Astrophys. J.* **255**, 325
Hassler, D.M, Rottman, G.J., and Orrall, G.Q.: 1991, *Astrophys. J.* **372**, 710
Hansteen, V.: 1993, *Astrophys. J.* **402**, 741
Klimchuk, J.A.: 1987, *Astrophys. J.* **323**, 368
Mariska, J.T. and Boris, J.P.: 1983, *Astrophys. J.* **267**, 409
Mariska, J.T.: 1987, *Astrophys. J.* **319**, 465
McClymont, A.N. and Craig, I.J.D.: 1987, *Astrophys. J.* **321**, 489
McClymont, A.N.: 1989, *Astrophys. J. Lett.* **347**, L47
Pallavicini, R., Peres, G., Serio, S., Vaiana, G.S., Golub, L., and Rosner, R.: 1981, *Astrophys. J.* **247**, 692

SOHO CONTRIBUTION TO THE UNDERSTANDING OF MASS SUPPLY AND FLOWS IN THE SOLAR CORONA

E. ANTONUCCI
Istituto di Fisica, University of Torino, Torino, Italy

Abstract. We expect a variety of dynamic phenomena in the 'quiescent' non-flaring corona. Plasma flows, such as siphon flows or convective flows of chromospheric material evaporating into the corona, are expected whenever a pressure differences is established either between the footpoints or between the coronal and chromospheric segments of a coronal loop. Such flows can induce phenomena of spatial and temporal brightness variability of the corona. In particular, evaporation induces a net mass input into the corona and consequently coronal density enhancements. Flows are also expected in the regions where energy is released during magnetic reconnection. From the observational point of view the dynamics of the solar atmosphere has been investigated in great detail mostly in the lower transition region with the HRTS, and during flares with the *Solar Maximum Mission* and *Yohkoh*. The high spectral, temporal and spatial resolution of the *SOHO* ultraviolet spectrometers should enable us in the near future to fill the gap providing a continuous coverage from the chromosphere to the corona, in the 10^4–10^6 K domain, and therefore to best study the dynamics throughout the solar atmosphere.

Key words: corona of the Sun – atmosphere of the Sun – hydrodynamics

1. Introduction

Systematic downflows at a few tens of km s^{-1} are observed at all locations of the solar atmosphere in the transition region lines. Although observed in localized areas, blue-shifts are absent on the large-scale. The picture of the dynamics of the solar atmosphere changes completely during flares, when considerable large-scale blue-shifts are observed in the hot flare lines formed in the soft X-ray region ≤ 3 Å. Upflows with velocities of a few 100 km s^{-1} were discovered at the onset of solar flares during the *Solar Maximum Mission*. These are high-velocity bulk motions at a few hundreds of km s^{-1} and they persist throughout the impulsive phase, as long as the intensity of the soft X-ray emission in the flare region is increasing. They have been identified with the evaporative convective flows of chromospheric plasma induced by the impulsive energy release which transiently heats up the chromosphere (Antonucci et al. 1982).

The picture of the dynamics in the non-flaring solar atmosphere remains still incomplete, since though emerging from a considerable sample of observations, these have been performed mostly in the ultraviolet and they cover predominantly the transition region. The normal-incidence slit spectrograph on *Skylab* measured red-shifts corresponding to downward velocities ≤ 15 km s^{-1}, in the temperature range from 7.1 10^4 K to 2.2 10^5 K in coronal holes and in the quiet sun (Doschek et al. 1976). The data collected up

Space Science Reviews **70**, 149–160.

to now show that the average large-scale downflow velocity decreases from the quiet sun values within 5 km s^{-1} and 12 km s^{-1} to about 2 km s^{-1} in coronal holes where a proportion of the area greater than in the quiet sun is affected by localized outflows (Dere et al. 1989).

In active regions, the High Resolution Telescope and Spectrograph (HRTS) measured velocities within 10–60 km s^{-1} in C IV at 10^5 K confined in areas 3"–5" wide (e.g. Brueckner 1981). Velocities up to 70 km s^{-1} in the Ne VII line emitted at 5 10^5 K were found by Mariska and Dowdy (1992), by re-examining *Skylab* data. Downflow velocities over sunspots can be much higher ($\sim$ 150 km s^{-1}) than in the surrounding active region (e.g. Brueckner 1981), but also low values of about 10 km s^{-1} are measured (e.g. OSO VIII observations in C IV, Bruner et al. 1976). Transition region downflows exhibit a characteristic differential velocity pattern as a function of temperature. The downward velocity reaches its maximum near 10^5 K and decreases monotonically below and above this temperature. Flows become negligible around 2.5 10^5 K.

According to the active region observations, redshifts either disappear or are somewhat reduced in coronal lines; for example, downflows of 13 km s^{-1} are observed in the Fe XII line at 1.7 10^6 K (Brueckner, 1981). The quiet sun observations obtained by Hassler et al. (1991) support the tendency for the directional flows to cancel approaching the corona: they find no Doppler shifts in the Ne VIII line formed at 8 10^5 K. Coronal lines however are also observed to be blue-shifted indicating outflow velocities within 10 - 20 km s^{-1} over coronal holes (e.g. Cushman and Rense 1976) and sunspots (Neupert et al. 1982). While the presence of pervasive downflows in the transition region is a well-established fact, coronal spectroscopic observations, with the exception of flare observations, are however still too scarce and contradictory to draw definite conclusions on the absence of flows in the non-flaring corona.

The permanent and ubiquitous presence of downflows in the transition region poses two fundamental questions. The first one is related to the origin of downflows. The second one concerns the mass input required to maintain the corona, owing to the fact that the observed continuous draining of matter toward the chromosphere is altering the mass balance in the solar atmosphere.

Concerning the origin of flows, one possible hypothesis is to interpret downflows as draining flows of plasma diverging from the top of magnetic arcades which consist of a mixture of cool and hot loops (Athay and Dere 1989). An alternative possible approach is to investigate whether downflows can be explained in terms of siphon flows which are expected in the solar atmosphere, since in principle they can be easily induced in a coronal loop whenever a pressure difference between the footpoints is established. The models developed to investigate the behavior of siphon flows however tend to predict, for spectral lines typical of the transition region regime, a net predominance of blue-shifts in contradiction with the observations (Spadaro et al. 1990a; Antiochos 1984; Peres et al. 1992). It has been suggested that siphon models might explain red-shifts in the limiting case of cool, asymmetrically heated loops (e.g. Mariska 1988, McClymont 1989).

The second problem arising from the presence of downflows concerns the identification of the source of the mass input required to balance the persistent draining of plasma through the transition region. The downward mass flux inferred in the transition region at 10^5 K is estimated to be sufficient to empty the corona on a time-scale of a few minutes. Brueckner (1981) has suggested that 'to maintain the mass balance it must be assumed that a large portion of the upward-moving gas is invisible in transition region lines.' Several phenomena have been considered as possible candidates for the required mass input, e.g. from spicules to the explosive events observed in the ultraviolet and characterized by velocities of about 200 km s^{-1}, without reaching firm conclusions.

Both the origin of downflows and the mass required to balance the permanent draining associated with downflows remain therefore open problems which do not find yet a definitive answer. This is in part due presumably to the fact that the phenomenology of flows in the solar atmosphere is derived from sporadic observations, obtained mostly during short space missions or rocket experiments. Furthermore, observations have not provided a sufficient coverage of the entire range of temperatures existing in the solar atmosphere, being mainly focused on transition region emission lines. In view of the *SOHO* mission, designed to overcome both kinds of drawbacks in the investigation of the large-scale flows in the solar atmosphere, in this paper we consider the signatures of siphon flows and convective flows present in the evaporation-condensation process, in connection with the origin of downflows and mass supply to the transition region and corona. We also consider the role of the outflows from a reconnecting current sheet in transforming cool slow flows in hot flows at the Alfvén velocity. The discussion is then focused, on the basis of one example of flows, on the capabilities of the *SOHO* experiments of detecting unambiguously the expected observational flow signatures in the solar atmosphere.

2. Flows in Coronal Loops

Macroscopic bulk motions such as siphon flows, evaporation-condensation flows and outflows from a reconnecting current sheet are quite different in nature and they contribute differently to the mass input into the solar corona. Any pressure imbalance between the footpoints of a coronal loop induces a flow along the loop which persits until the pressure difference is canceled. Therefore if the pressure difference is not a transient effect, it is possible to induce steady siphon flows. No mass supply is however involved in siphon flows, but only a circulation of mass within the loop.

Any change in the heating rate of a coronal loop or in the length of the loop has the effect of inducing a thermal imbalance. The plasma will respond by re-adjusting its physical conditions through the mechanism of evaporation of chromospheric plasma, or condensation of coronal plasma, depending on the sign of the change. The process of evaporation involves a net input of mass from the chromosphere into the coronal portion of the loop. On the contrary the net result of condensation is a draining of mass from the corona to the chromosphere.

Flows of plasma are expected out of a reconnecting current sheet whenever the fluid pressure between two oppositely directed magnetic fields is insufficient to keep the fields separated. The fluid pressure reaches its maximum where the magnetic field goes through zero. As a result plasma is expelled out of the reconnecting current sheet at the Alfvén velocity. In this process relatively cool and slow plasma flows are transformed into high velocity flows of relatively hot plasma. Therefore the increase in mass at high temperature occurring during magnetic reconnection is at the expenses of cooler plasma already present around the region of the reconnecting fields and does not imply an effective new mass input into the corona or the transition region.

The energy released in the corona at the expenses of the reconnecting magnetic fields is partially converted in form of thermal energy and kinetic energy of the plasma inside the reconnecting current sheet, which is heated and expelled at the Alfvén velocity. The thermal energy acquired by the plasma inside the current sheet is predominantly transferred to the chromosphere in form of downward heat flux. Providing the chromosphere can dissipate only a negligible amount of the energy deposited, the downward heat flux is almost completely transformed in an upward enthalpy flux. The resulting convective upflows supply a net mass and energy input into the corona. Because of the mass input, the coronal radiation increases considerably being proportional to n^2, where n is the electron density. When evaporation occurs in a coronal loop, the only effective energy dissipation process of the coronal loop system is in the end radiation in the ultraviolet and in soft X-rays.

2.1 Chromospheric evaporation

The variability observed in the coronal loops of active regions suggests significant changes in the heating rate, although the observed excursions of the average loop temperature remain within relatively narrow limits. According to the *Skylab* observations the values of the average coronal temperature in a coronal loop fall within a quite restricted range, from 2 to 3 10^6 K (e.g. Webb 1981). The *Yohkoh* observations have indicated a wider range of variations for the loop temperatures: a range from 2 to 6 10^6 K is obtained with the

Soft X-ray Telescope (SXT) measurements (Hara et al. 1992), and from 3.5 to 4.5 10^6 K, with the Bragg Crystal Spectrometer (Watanabe et al. 1992). These results are consistent with the average temperature of the whole corona, of 4.0 10^6 K, measured with the GOES satellite (Thomas et al. 1985). The temperature excursions induced by variations in the heating rate can remain within relatively narrow limits if the plasma is re-adjusting through processes which allow a mass exchange between the chromosphere and the corona; that is, through chromospheric evaporation or coronal condensation (e.g. Antiochos and Sturrock 1978; Antiochos and Krall 1979).

Krall and Antiochos (1980) have compared a static model of a coronal loop, which assumes that the chromosphere can dissipate the heat flux received from the corona in response to a variation in the heating rate, with an evaporative model which assumes that the chromosphere is able to dissipate only a negligible part of the heat. For the static model with the same increase in the heating rate, the final coronal temperature in the loop departs significantly from the initial one. For a typical loop, 5 10^9 cm long, an increase by a factor of 10 in the quasi-static heating rate, which is 2.5 10^{-3} erg cm^{-3} s^{-1}, induces a variation by a factor of 3.5 in the final temperature, reached when the conductive losses at the base become comparable to the new heating rate. For the evaporative model a similar increase in temperature is reached only transiently. In fact, if a mass exchange between chromosphere and corona is allowed, an upward enthalpy flux of the same order of the downward conductive flux balances the new heating rate. The coronal loop transiently heated cools radiatively due to the density enhancement in the loop, approaching the initial temperature. The final density increase caused by evaporation is by a factor of 5.5. Therefore the effect of evaporation is in the end that of controlling the excursion of the temperature variations within the loop induced by a transient energy release in the corona. The loop brightness depending on n^2 is sensitive to a larger extent to density variations.

Recently frequent transient brightenings have been discovered in active regions with the Soft X-ray Telescope (SXT) on the *Yohkoh* satellite by Shimizu et al. (1992). These phenomena are evidence for localized energy release in the corona with heating of the plasma up to $\approx$ 7 10^6 K, a value close to flare temperatures. The localized variations in the heating rate can induce localized heat conduction losses from the corona exceeding 10^7 erg cm^{-2} s^{-1}, that is, well above the average conduction losses estimated in the active regions from the *Skylab* results (e.g. Withbroe and Noyes 1977). These transient heat fluxes can trigger local episodes of chromospheric evaporation.

2.2 Outflows from a reconnecting current sheet

In coronal plasmas rapid reconnection at some fraction of the Alfvén velocity V_A should be a general occurrence. During steady reconnection net inflows of plasma with density n_o and velocity v_d into the reconnecting current sheet are balanced by flows, with density n_s, out of the current sheet: $n_o\, v_d\, b = n_s\, V_A\, (a + \xi\, b)$, where a and b are the thickness and width of the sheet, respectively, and $\xi = B_y/B_o$ is the ratio of the transverse component to the reconnecting component of the magnetic field (Somov 1992). The presence of a transverse field B_y increases the amount of plasma flowing out of the reconnection region. If pressure equilibrium is established across the current sheet, $B_o{}^2/(8\,\pi) = n_s\, k\, (T_e + T_i)_s$, where $T_{e,s}$ and $T_{i,s}$ are the electron and ion temperature inside the current sheet, the velocity of expulsion V_A is a function of the temperature inside the current sheet: $V_A \propto f(T_e/T_i)_s \sqrt{T_{e,s}}$ (Antonucci and Somov 1992). The fraction $(T_e/T_i)_s$ depends on the state of the plasma turbulence within the current sheet.

A reconnecting current sheet acts as a source of high-temperature plasma in the corona in two ways: directly, by converting cooler in hotter plasma in the reconnection process, and indirectly, owing to the fact that the energy released can heat the chromosphere and induce chromospheric evaporation. The amount of heated plasma ejected by the current sheet in a time τ is a simple function of the geometrical parameters of the current sheet and the Alfvén velocity ($N = 4\, n_s\, V_A\, (a + \xi\, b)\, l\, \tau$, where N is the total number of particles ejected and l is the length of the current sheet, e.g. Antonucci and Somov 1992). The expelled plasma will undergo a cooling process by radiation and thermal conduction, accompanied by a deceleration process. Since the plasma is expelled from a current sheet mainly in opposite directions, the spectral lines emitted from the outflows are expected to be symmetrically broadened according to the projection along the line of sight of the distribution of the plasma velocity. The plasma flowing out of a current sheet in quasi-steady conditions, is therefore expected to emit spectral lines which are typically non-thermally broadened and the observed non-thermal velocity is a signature of the thermal and turbulent conditions within the reconnection region (Antonucci and Somov 1992, Antonucci et al. 1994).

No definitive evidence for outflows from reconnection regions in the non-flaring corona has been obtained yet. Evidence for outflows from reconnecting current sheets, although not conclusive, has been instead found for flares. At the onset of flares large non-thermal broadenings are observed in the soft X-ray spectral lines at wavelengths ≤ 3 Å, with non-thermal velocities of the order of 100 km s^{-1}. These observations are consistent with the outflows expected from a reconnecting current sheet with physical properties typical of the hot phase of flares (Antonucci et al. 1993). Magnetic reconnection has also been suggested to be at the origin of the explosive events observed in the transition region (e.g. Dere 1992).

2.3 Siphon flows

Siphon flows are driven by a pressure difference between the two footpoints of a coronal loop. The first models of siphon flows have been developed for steady siphon flows, which can be established for a constant pressure difference within hot coronal loops. Noci (1981) has studied the case of polytropic flows with polytropic index $\alpha = 1.1$, assuming a symmetric loop with uniform cross-section. Flows set up in loops with non-uniform cross-section have been first investigated by Cargill and Priest (1980) and Priest (1981). Of special physical interest are the continuous solutions corresponding to subsonic flows obtained for moderate pressure differences, and the discontinuous solutions obtained for a relatively higher pressure difference, corresponding to supersonic flows with the formation of a stationary shock (e.g. Noci 1981). In the latter case flows, subsonic in the ascending branch of the loop, become supersonic through the critical point, which for a symmetric model is at the apex of the loop, and again subsonic in the descending branch of the loop through the stationary shock which decelerates the plasma.

3. Observable Characteristics of Flows

As an example, the signatures of a specific kind of flows, namely siphon flows, will be discussed in detail in view of deriving the observational requirements for unambiguously detecting their presence in the solar atmosphere with *SOHO*.

3.1 Asymmetries and discontinuities in the velocity, density and temperature profiles in a siphon flow

The density and velocity variations expected in siphon flows appear to be large enough to give an observable effect. Hot loops are the best candidates for observing siphon flows, since they are expected to be the site of higher velocities. Subsonic flows with velocities of the order of 150 km s^{-1} can be established, for instance, in uniformly heated loops with coronal temperatures of 1.6 10^6 K (Noci et al. 1989). Even higher velocities are expected for supersonic flows in hot loops (Noci 1981). Lower velocities, about 32 km s^{-1}, are predicted for non-uniformly heated cold loops, at a temperature of 2.5 10^5 K (Mariska 1988). Robb and Cally (1992) have shown that the velocity of steady siphon flows is an increasing function both of the pressure difference between the footpoints and of the coronal temperature of the loop, by modeling hot symmetric loops with temperatures within 1.8 10^6 K and 2.5 10^6 K, uniformly heated.

Steady siphon flows are characterized by distinct signatures which should be easily detectable. The velocity and density profiles along subsonic flows set up in a loop are considerably asymmetric. The velocity is increasing along the loop and is higher in the downflowing leg. The density which is inversely proportional to the velocity, v, because of the law of mass conservation ($n \propto 1/v$), on the contrary, is higher in the upflowing leg. The degree of asymmetry expected for the electron temperature is instead generally small and it might not be detectable (e.g. Noci et al. 1989). Asymmetries in the profiles of the flow parameters are more significant for supersonic flows. The presence of a shock introduces in the profiles a characteristic discontinuity which should be easily observed even for the electron temperature (e.g. Noci 1981).

The UV brightness for steady subsonic flows is not expected to depart significantly from the static case, providing the emission in the UV lines is computed under the assumption of ionization equilibrium. Owing to the formation of the shock significant asymmetries and discontinuities in the emission are however expected for supersonic flows. In this case, the downflowing leg is significantly less bright than the other one, while the footpoints emit almost at the same level (Noci 1981).

3.2 Signatures of departure from ionization equilibrium in a siphon flow

The large temperature and velocity gradients occurring in the legs of a coronal loop, site of a siphon flow, induce large departures from ionization equilibrium. The degree of ionization of an ion can differ considerably from equilibrium where, in presence of steep gradients, the time scale of the ionization-recombination process is much longer than the time of the ion passage through a temperature scale height.

The expected signatures for the departure from ionization equilibrium obtained with the siphon flow models are the following. A very large overabundance of the higher ionization stages of ions is expected in the downflow region. The consequence is an enhancement of the coronal lines in the descending branch of the loop. On the other hand, the models predict a large overabundance of the lower ionization stages of ions in the upflowing region, with a consequent enhancement of the transition region lines in the ascending branch of the loop. These effects are discussed in detail by Noci et al. (1989), who have analysed for steady subsonic flows the variations of ion abundances along the loop for the complete set of ionization states of carbon, forming in the interval from 2.2 10^4 K to 10^6 K. The analysis has been extended to the ionization states of oxigen by Spadaro et al. (1990b).

Because of the departure from ionization equilibrium, the emissivity of ions such as C III is enhanced in the upflows and drastically reduced in the downflows. On the contrary, the emissivity of O VI is reduced in the upflows and considerably enhanced in the downflows (Peres et al. 1992). In non-equilibrium ionization conditions, the asymmetry of the UV brightness along the loop is significantly enhanced, and large brightness differences are expected for the two legs. The effects due to non-equilibrium ionization are even larger for supersonic flows (Borrini and Noci 1982).

The non-ionization equilibrium conditions for siphon flows in cool loops, $\leq 6\ 10^5$ K, have been studied by Mariska and Boris (1983). Spadaro et al. (1991) have shown that even in cool loops, where the temperature gradients are less steep and the non-equilibrium effects less marked, the enhancement of the transition region lines in the upflow region leads to predict predominantly blue-shifts, showing that according to the present models siphon flows are not consistent with the observations of the transition region red-shifts.

3.3 Signatures for non-steady siphon flows

A steady flow can be established in a time of the order of several l/c_s, where l is the loop length and c_s is the sound speed. Robb and Cally (1992) have modeled the development of a steady siphon flow, showing that this condition is reached in a few 10^3 s. For a relatively high pressure difference between the footpoints, $> 6\%$, and moderate heating, $\sim 10^{-4}$ W m^{-3}, the authors reach a non-steady nearly periodic solution. This non-steady flow, which has been named as 'surge flow', shows a quasi-periodicity of about 10^3 s.

4. Soho Contribution to the Study of Plasma Flows Through the Solar Atmosphere

In absence of unambiguous experimental evidence for siphon flows, the predictions of the models developed up to now become essential to design observing programs apt to reveal their existence. The observational requirements derived for siphon flows, which can be easily generalized to include flows caused by evaporation-condensation and reconnection processes, are fully met by the *SOHO* coronal instruments.

4.1 Requirements for observing siphon flows

Considering that the identification of siphon flows should be easier in hot coronal loops, where the expected velocities are higher, the range of observations, up to now obtained preferentially in the transition region, has to be extended to include coronal lines. An almost continuous coverage of the temperature from 10^4 K up to 2–2.5 10^6 K should therefore permit the identification of the sites of upflows at higher velocity. The same temperature coverage should also enable us to trace flows continuously from the chromosphere up to coronal heights.

Spectral lines should be measured at a spectral resolution allowing to detect flow velocities at least within 1 km s^{-1} and 300 km s^{-1}. The highest spectral resolution is required for the cooler lines in order to investigate the slower flows which can develop in cool transition region loops. The largest asymmetries and departures from ionization equilibrium are expected in the legs of the coronal loops, where the highest spatial resolution, of approximately 1", is required; while, in the coronal portion of the loop a spatial resolution of 2"–3" should be sufficient to trace the smoother variations in the physical parameters of the flow. High space resolution is also needed to explore the region of the shock and to detect the discontinuity of the flow parameters through the shock. A temporal resolution of about 1 minute should be sufficient to trace the dynamic evolution of siphon flows toward steady conditions, which is taking place on a time-scale of the order of 10^3 s, and to study the stationary phase. A field of view FOV $\leq$ (2.5' x 2.5') is sufficient to cover the full range of coronal loops, from the compact to the extended ones up to a maximum length of $\sim 10^{10}$ cm. Smaller FOV's might be chosen to achieve a better temporal resolution over a specific segment of the loop.

The detection of a complete set of density-sensitive lines, on a relatively wide range, e.g. from 10^8 to 10^{11} cm^{-3}, can provide direct information on the density along the entire flow. At the same time a complete set of temperature sensitive lines, covering the interval 10^4–2.5 10^6 K, gives direct information on the different temperature regimes in the flow. Combining these data the profiles of density, temperature and pressure along the loop can then be derived with great accuracy. An important aspect for the observation of flows concerns the possibility of detecting spectral lines emitted from the same element in different ionization conditions, preferably forming a complete set of ionization stages, to reveal possible departures from ionization equilibrium in the different parts of the loop.

Exactly the same selection of spectral windows, spectral and spatial resolution and field of view desirable for a complete investigation of siphon flows is required to search for other kinds of flows in the solar atmosphere, such as evaporation - condensation flows or outflows from reconnecting current sheets. Only the requirement on the temporal resolution should be more stringent ($\sim$ 10 s) when searching for the onset of evaporation - condensation flows and reconnection in a current sheet. Evaporation is intrinsically a transient process through which the plasma within a coronal loop is readjusting to new conditions in the heating rate or changes in the geometry of the loop and this can occur on a short time-scale. On the other hand, outflows due to magnetic reconnection can in principle reach quasi-steady conditions and may pose less stringent requirements for temporal resolution.

4.2 CAPABILITIES OF THE SOHO INSTRUMENTATION TO DETECT PLASMA FLOWS

The *SOHO* coronal instruments are designed to meet the observational requirements discussed above, thus allowing the development of observing programs fully suitable for a complete investigation of flows in the solar atmosphere. This goal can be achieved by combining the capabilities of three of the *SOHO* coronal instruments: the Coronal Diagnostic Spectrometer (CDS), SUMER, and the Extreme-Ultraviolet Imaging Telescope (EIT). The CDS spectral coverage, within 155 Å and 787 Å, is fully sufficient to include the temperature and density regimes expected in the plasma flows. This instrument is best suited to derive the profiles of temperature within $\sim 10^4$–$2\ 10^6$ K, density within 10^7–10^{11} cm^{-3}, and pressure along the plasma flows in hot coronal loops. The simultaneous coverage of large wavelength bands does not only provide the capability to measure these parameters along the flow, but also to test the abundance of elements and different ionization states for the same element. The maximum CDS FOV of 4' x 4', with high spatial resolution of $\sim$ 3", can easily include an entire coronal loop, but the choice of a large FOV is at the expenses of the temporal resolution, which can be of 1 s at best. The CDS can measure flow velocities ≥ 10 km s^{-1}.

SUMER is best suited for investigating the rapidly changing physical conditions expected in the legs of coronal loops sites of plasma flows, owing to its high spectral resolution, which allows measurements of flows velocities ≥ 1 km s^{-1}, and spatial resolution of 1". The instrument is optimized to measure the lines emitted in the transition region and coronal lines up to $1.8\ 10^6$ K. As the CDS, SUMER allows an ample choice of lines sensitive to density and temperature, and lines emitted from different ions of the same element to study departures from ionization equilibrium.

The evolution of the region of the solar atmosphere where the flows are observed is monitored by the EIT, which is imaging the corona up to $3\ 10^6$ K in narrow bands centered on three iron lines, Fe IX (171 Å), Fe XII (195 Å) and Fe XV (284 Å), with a spatial resolution of 2.6" and a temporal resolution of 10 s, if a field of view of 4' x 4' is chosen.

The data provided by the joint observations of the *SOHO* coronal telescopes should therefore enable us to establish in detail the geometry of the flow region, the velocity, temperature, density and pressure fields throughout the transition region/corona, the ionization conditions in the flows, and the spatial and temporal evolution of all the relevant parameters. An accurate knowledge of the flow phenomenology then allows the identification of the distinctive signatures of the observed flows. Hence, the capability of measuring for the first time all the essential physical properties of flows and their evolution with the required resolution and coverage should in the end provide the key to understand the origin of flows as well as to define their role in supplying mass to the corona.

Acknowledgements

The author wishes to thank the support by the Agenzia Spaziale Italiana.

References

Antiochos, S.K.: 1984 *The Astrophys. J.* **280**, 416.

Antiochos, S.K., and Sturrock, P.A.: 1978 *The Astrophys. J.* **220**, 1137.

Antiochos, S.K., and Krall, K.R.: 1979 *The Astrophys. J.* **229**, 788.

Antonucci, E. et al.: 1982, *Solar Phys.* **78**, 107.

Antonucci, E. and Somov, B.V.: 1992, *Coronal Streamers, Coronal Loops and Coronal and Solar Wind Composition, ESA*, **SP 348**, 293.

Antonucci, E., Dodero, M.A. and Somov, B.V.: 1993, *Physics of Solar and Stellar Coronae*, eds. J.F. Linsky and S. Serio Kluwer Academic Publishers, 179.

Antonucci, E., Dodero, M.A. and Somov, B.V.: 1994, *X-ray Solar Physics from Yohkoh*, eds. Y. Uchida, H.S. Hudson, K. Shibata and T. Watanabe Universal Academy Press, Tokyo, in press.

Athay, R.G., and Dere, K.P.: 1991 *The Astrophys. J.* **381**, 323.

Borrini, G. and Noci, G.: 1982 *Solar Phys.* **77**, 153.

Brueckner, G.E.: 1981 *Solar Active Regions. A Monograph from Skylab Solar Workshop III*, (ed. F.Q. Orrall) Colorado Associated University Press 113.

Bruner, E.C. Jr. et al.: 1976 *The Astrophys. J.* **210**, L97.

Cargill, P.J. and Priest, E.R.: 1980 *Solar Phys.* **65**, 251.

Cushman, G.W. and Rense, W.A.: 1976 *The Astrophys. J.* **207**, L61.

Dere, K.P.: 1992 *AIP Conference Proceedings* **267**, 63.

Dere, K.P., Bartoe, J.-D., Brueckner, G.E. and Recely, F.: 1989 *The Astrophys. J.* **345**, L95.

Doschek, G.A., Feldman, U. and Bohlin, J.D.: 1976 *The Astrophys. J.* **205**, L177.

Hara, H., Tsuneta, S., Lemen, J.R., Acton, L.W. and McTiernan, J.M.: 1992 *PASJ* **44**, L135.

Hassler, D.M., Rottman, G.J. and Orrall, F.Q.: 1991 *The Astrophys. J.* **372**, 710.

Krall, K.R. and Antiochos, S.K.: 1980 *The Astrophys. J.* **242**, 374.

Mariska, J.T.: 1988 *The Astrophys. J.* **334**, 489.

Mariska, J.T. and Boris, J.P.: 1983 *The Astrophys. J.* **267**, 409.

Mariska, J.T. and Dowdy Jr., J.F.: 1992 *The Astrophys. J.* **401**, 754.

McClymont, A.N.: 1989 *The Astrophys. J.* **347**, L47.

Neupert, W.M., Brosius, J.W., Thomas, R.J. and Thompson, W.T.: 1982 *The Astrophys. J.* **392** , L95.

Noci, G.: 1981 *Solar Phys.* **69**, 63.

Noci, G., Spadaro, D., Zappalà, R.A. and Antiochos, S.K.: 1989 *The Astrophys. J.* **338**, 1131.

Peres, G., Spadaro, D., and Noci, G.: 1992 *The Astrophys. J.* **389**, 777.

Priest, E.R.: 1981 *Solar Active Regions. A Monograph from Skylab Solar Workshop III*, (ed. F.Q. Orrall) Colorado Associated University Press 213.

Robb, T.D. and Cally, P.S.: 1992 *The Astrophys. J.* **397**, 329.

Shimizu, T., Tsuneta, S., Acton, L.W., Lemen, J.R., and Uchida, Y.: 1992 *PASJ* **44**, L147.

Somov, B.V.: 1992, *Physical Processes in Solar Flares*, Kluwer Academic Publ., Dordrecht, London.

Spadaro, D., Noci, G., Zappalà, R.A. and Antiochos, S.K.: 1990 a *The Astrophys. J.* **355**, 342.

Spadaro, D., Zappalà, R.A., Antiochos, S.K., Lanzafame, G. and Noci, G.: 1990 b *The Astrophys. J.* **362**, 370.

Spadaro, D., Antiochos, S.K. and Mariska, J.T.: 1991 *The Astrophys. J.* **382**, 338.

Thomas, R.J., Starr, R. and Crannel, C.J.: 1985 *Solar Phys.* **95**, 323.

Watanabe, T. et al.: 1992 *PASJ* **44**, L141.

Webb, D.F.: 1981 *Solar Active Regions. A Monograph from Skylab Solar Workshop III*, (ed. F.Q. Orrall) Colorado Associated University Press, 165.

Withbroe, G.L. and Noyes, R.W.: 1977 *Ann. Rev. Astr. Astrophys.* **15**, 363.

FEATURES OF MASS SUPPLY AND FLOWS RELATED WITH RECONNECTION IN THE SOLAR CORONA

B.V. SOMOV
Astronomical Institute, Moscow State University, Moscow, Russia

Abstract. Symmetrical broadening in the emission spectral lines is the ultimate observational effect of magnetic reconnection in the solar corona. Reconnection can create plasmas of very different temperatures and, hence, very different electrical conductivities in the corona. The electrodynamical effect of such a mass supply is considered. Electromagnetic expulsion force, different from Parker's well-known magnetic buoyancy force, can effectively balance gravity in prominences and generate fast vortex flows in the vicinity of fine threads inside prominences. The possibility of observing this effect from SOHO is discussed.

Key words: Magnetic Fields – Prominences – MHD flows

1. Introduction

It was believed for a long time (e.g., Syrovatskii and Somov, 1980) that reconnection (i.e., interaction of magnetic fluxes having antiparallel components) plays a key role in the global dynamics of coronal plasma. Now, this concept seems to be confirmed by Yohkoh observations. The theory of magnetic reconnection in high-temperature turbulent current sheets (HTTCSs) explains the total amount of accumulated energy in solar flares, the power of energy release and some other parameters of solar flares (Somov, 1992).

One of the distinctive features of magnetic reconnection in the solar corona is the presence of plasma outflows whose velocity depends on the electron and ion temperatures inside HTTCSs. The outflowing plasma is characterized by an anti-symmetric velocity distribution along the line-of-sight. Therefore the ultimate observational effect of reconnection is a suprathermal symmetrical broadening in the emission spectral lines. If reconnection occurs in a pre-flare active region one can expect a broadening in the XUV lines formed at a few millions degrees, which would be observable with SOHO (Antonucci and Somov, 1992; Antonucci, 1993).

Another feature of reconnection is related with the condensation mode of thermal instability. During the nonlinear stage of this mode, cold dense filaments can be formed, surrounded by high-temperature low-density plasma inside HTTCSs. The reconnection process then becomes essentially three-dimensional. Coronal plasma flows into the HTTCS with a drift velocity and is heated to high temperatures by magnetic field dissipation under conditions of anomalous conductivity. The high-temperature plasma is a source of X-ray emission and of intense heat flows oriented alond the field lines out of the current sheet.

Space Science Reviews **70**, 161–166.

Moving in both perpendicular and parallel directions to the HTTCS (see fig. 1.2.7 in Somov, 1992), the high-temperature plasma penetrates inside the filaments, cools by radiation and compresses. The cold dense plasma flows out of the current sheet along the filaments as in classical two-dimensional models of reconnection. Therefore, reconnection in the corona creates plasmas of very different temperatures and, hence, very different electrical conductivity. In the present paper, the electrodynamical effect of such a mass supply is considered. Nonuniform distributions of conductivity result in excitation of fast vortex flows which can be observed by SOHO.

2. Archimedian force in MHD

Let us consider the equation of magnetostatic equilibrium

$$-\nabla p + \frac{1}{c}\,\mathbf{j} \times \mathbf{B} + \rho\,\mathbf{g} = 0$$

or

$$\nabla p = \rho\,\mathbf{g} + \mathbf{f}\,, \tag{1}$$

where

$$\mathbf{f} = \frac{1}{c}\,\mathbf{j} \times \mathbf{B}$$

is the density of the volume Lorentz force.

Consider an incompressible conducting fluid situated in uniform magnetic field $\mathbf{B}_0$ and electric field $\mathbf{E}_0$. Provided the current flowing in the fluid is uniform, the created force is uniform as well:

$$\mathbf{f}_0 = \frac{1}{c}\,\mathbf{j} \times \mathbf{B}_0\,. \tag{2}$$

By virtue of (1), the force makes the fluid heavier or lighter. In both cases the volume force is potential and, much like the gravity force, will be balanced by an additional pressure gradient appearing in the fluid. This allows us to create a regulated expulsion force analogous to the Archimedian force in hydrodynamics.

A body plunged into the fluid is acted upon by the force

$$\mathbf{F} = \int\limits_{V} (\,\rho_1\,\mathbf{g} + \mathbf{f}_1\,)\,dV + \oint\limits_{S} p_0\,\mathbf{n}\,dS\,. \tag{3}$$

Here ρ_1 is the density of the body, which is generally not equal to that of the fluid ρ_0;

$$\mathbf{f}_1 = \frac{1}{c}\,\mathbf{j}_1 \times \mathbf{B}_1$$

is the volume Lorentz force, $\mathbf{j}_1$ is the current inside the body, $\mathbf{n}$ is the inward normal to the surface, p_0 is the pressure on the body from the fluid, resulted from (1):

$$\nabla p_0 = \rho_0 \, \mathbf{g} + \mathbf{f}_0 \, . \tag{4}$$

Once the current $\mathbf{j}_0$ is uniform, the right-hand side of (4) is a uniform force, and (3) may be rewritten as

$$\mathbf{F} = \int_V (\rho_1 - \rho_0) \, \mathbf{g} \, dV + \frac{1}{c} \int_V (\mathbf{j}_1 - \mathbf{j}_0) \times \mathbf{B}_0 \, dV \, . \tag{5}$$

The first term in (5) corresponds to the usual Archimedian force in hydrodynamics. It equals zero once $\rho_1 = \rho_0$. The second term describes the magnetic expulsion force. It vanishes once $\mathbf{j}_1 = \mathbf{j}_0$, i.e. $\sigma_1 = \sigma_0$.

Equation (5) shows that the *magnetic* expulsion force appears provided $\sigma_1 \neq \sigma_0$. This fact has been used to construct MHD-devices for the separation of mechanical mixtures.

Note, however, that Equation (5) is of purely illustrative value, since the electric field and current density are not uniform in the presence of a body with a conductivity σ_1 which is not equal to that of the fluid σ_0. In this case, the Lorentz force is generally not potential. Hence it cannot be balanced by potential forces. This is the reason why vortex flows of the fluid must be generated by the Lorentz force.

3. Expulsion force and vortex flows

Consider the stationary problem for an incompressible fluid having viscosity ν and magnetic diffusivity ν_{m}. Let both the usual and magnetic Reynolds numbers be small:

$$\mathrm{Re} = \frac{vL}{\nu} \ll 1 \, , \tag{6}$$

$$\mathrm{Re}_{\mathrm{m}} = \frac{vL}{\nu_{\mathrm{m}}} \ll 1 \, . \tag{7}$$

The freezing-in condition can be rewritten in the form

$$\Delta \mathbf{B} + \mathrm{Re}_{\mathrm{m}} \, \mathrm{curl} \, (\mathbf{v} \times \mathbf{B}) = 0 \, , \tag{8}$$

where, in view of (7), Re_{m} is a small parameter. In a zero approximation in this parameter, the magnetic field is potential; moreover, it will be assumed to be uniform, in accordance with the formulation of the problem discussed in Section 2.

Assuming the stationary flows occurring in the fluid to be slow, the inertial force (proportional to v^2) will be ignored in the equation of motion as compared to the other forces: pressure gradient, Lorentz force, viscous force. The term describing the gravity force will be dropped, since its effect has already been studied in Section 2. Finally, on multiplying the equation

$$0 = -\frac{\nabla p}{\rho} - \frac{\mathbf{B} \times \operatorname{curl} \mathbf{B}}{4\pi\rho} + \nu\,\Delta\,\mathbf{v}$$

by the fluid density $\rho = \rho_0$, it is rewritten in the form

$$\eta\,\Delta\,\mathbf{v} = \nabla p - \mathbf{f}\,. \tag{9}$$

Here $\eta = \rho_0\,\nu$ is the *dynamic* viscosity coefficient, and

$$\mathbf{f} = \frac{1}{c}\,\mathbf{j} \times \mathbf{B}_0 \tag{10}$$

is the Lorentz force.

The solution of the problem for a ball of the radius a can be written in spherical coordinates

$$\mathbf{v} = \{\, v_r,\, v_\theta,\, v_\varphi \,\}$$

(cf. Syrovatskii and Chesalin, 1963):

$$\begin{aligned} v_r &= 0\,, \\ v_\theta &= \frac{\beta j_0 a^3}{4c\eta\, r}\left(1 - \frac{a^2}{r^2}\right)\left(-B_\perp \cos\theta\, \sin\varphi + B_\parallel \sin\theta\, \sin 2\varphi\right), \\ v_\varphi &= \frac{\beta j_0 a^3}{4c\eta\, r}\left(1 - \frac{a^2}{r^2}\right)\left(B_\perp \cos 2\theta\, \cos\varphi + B_\parallel \sin 2\theta\, \cos^2\varphi\right). \end{aligned}$$

Here the constant

$$\beta = \frac{\sigma_0 - \sigma_1}{2\sigma_0 + \sigma_1}\,. \tag{11}$$

Thus, in the case of a ball, the flow lines of a conducting fluid are situated at $r =$ const surfaces. The force acting on the ball is

$$\mathbf{F} = -V\,\mathbf{f}_0 \cdot \frac{3}{2}\,\beta\,. \tag{12}$$

Here $V = 4\pi a^3/3$ is the volume of the ball, $\mathbf{f}_0$ is the Lorentz force (2) in the conducting fluid with uniform magnetic $\mathbf{B}_0$ and electric $\mathbf{E}_0$.

4. Expulsion force for large magnetic Reynolds numbers

In the previous section, we have considered the solution to the MHD problem concerning the expulsion force in the limit of small (usual and magnetic)

Reynolds numbers (6) and (7). As a rule, the opposite limiting case is applicable for astrophysical use. In this case, the problem on the expulsion force is difficult and hardly can be solved completely, especially given

$$\mathrm{Re} \ll 1\,, \qquad \mathrm{Re_m} \gg 1\,.$$

The situation of this kind takes place in prominences (see Section 5). Litvinenko and Somov (1993) have found the following expression for the electromagnetic expulsion force

$$\mathbf{f} = -\frac{c}{4\pi v_0\, a}\,\mathbf{E}_0 \times \mathbf{B}_0\,. \tag{13}$$

Here

$$v_0 \approx \left(\frac{cE_0 B_0 a}{4\pi\eta}\right)^{1/2} \tag{14}$$

is the characteristic velocity of vortex flows.

The approximate expression (13) valid in the limit of large magnetic Reynolds numbers. The physical sense of Equation (13) is obvious. Comparison of (13) with (2) for the electromagnetic force, which holds then a uniform current flows in the medium, shows that for $\mathrm{Re_m} \to \infty$ $(\sigma \to \infty)$ the plasma in the vicinity of the body possesses, as it were, an effective conductivity

$$\sigma_{\mathrm{ef}} \approx \frac{c^2}{v_0 a}\,.$$

This finite conductivity is a result of the energy losses to the generation of macroscopic vortex flows. It goes without saying that this mechanism of electric conductivity of plasma is different from the usual microscopic one, in which energy losses result from Coulomb collisions of current-carrying electrons with thermal electrons of the plasma.

5. Observable characteristics of prominences

The solar corona is a natural 'astrophysical laboratory' where Equation (13) can be tested. Recall several of its characteristics: density $\rho_0 \approx 10^{-16}\,\mathrm{g\,cm^{-3}}$, temperature $T_0 \approx 10^6\,\mathrm{K}$, dynamic viscosity $\eta \approx 1\,\mathrm{g\,cm^{-1}}\,s^{-1}$, magnetic field $B_0 \approx 10 - 100\,\mathrm{G}$, electric field $E_0 \approx 10^{-5}$ CGSE units. According to observational data (e.g. Demoulin *et al.*, 1987), prominences consist of numerous fine threads—cold dense formations having the transverse scale $a \approx 10^7\,\mathrm{cm}$ and temperature $T_1 \approx 10^4\,\mathrm{K}$. Hence the ratio $\sigma_1/\sigma \approx 10^{-3} \ll 1$, as applied to prominences in the solar corona. In the vicinity of the threads, as well as near a prominence as a whole, rather fast plasma flows are really observed.

According to the model under discussion, these flows are generated by the vortex component of the electromagnetic force. For Re $\ll 1$, their velocity, given by (14), is $v_0 \approx 10 - 30\,\mathrm{km/s}$, that, generally speaking, corresponds to the characteristic values of observed velocities. However the spatial resolution of modern optical, UV and X-ray observations is not sufficient for the model to be confirmed or refuzed. Consider another possibility.

Anti-symmetric distribution of velocities on line-of-sight projection (i.e., in the direction towards the observer) is a distinguishing feature of the model since it predicts the presence of a large number of vortex flows of plasma inside the prominence. Such a distribution will be observed by SOHO as a symmetric broadening of spectral lines (Antonucci, 1993).

The gravity force acting on prominences is supposed to be balanced by the electromagnetic expulsion. The equilibrium condition makes it possible to evaluate the characteristic value of the plasma density inside the fine threads forming the prominence

$$(\rho_1 - \rho_0)\, g = f\,. \tag{15}$$

Here the specific gravity on the Sun $g \approx 3 \cdot 10^4\,\mathrm{cm\ s^{-2}}$. Equations (13) and (15) result in

$$\rho_1 \approx \left(\frac{cE_0 B_0\, \eta}{4\pi\, g^2 a^3} \right)^{1/2} \approx 3 \cdot 10^{-13}\,\mathrm{g\,cm^{-3}}\,,$$

in accordance with observational data.

Even faster motions with characteristic velocities $10^2 - 10^3\,\mathrm{km\,s^{-1}}$ in eruptive prominences are probably a consequence of the fact that the coronal fields $\mathbf{E}_0$ and $\mathbf{B}_0$ can change (in magnitude or direction) in the course of evolution. As this takes place, the equilibrium described by Equation (15) can be violated.

Observations with high spectral resolution in UV and soft X-ray ranges are necessary to study the effect of electromagnetic expulsion stimulated by the presence of plasma regions with considerably different conductivity in the solar atmosphere.

References

Antonucci, E.:1994, these proceedings

Antonucci, E., Somov, B.V.: 1992, in *Coronal Streamers, Coronal Loops, and Coronal and Solar Wind Composition*, Proc. 1st SOHO Workshop, ESA SP-348, p. 339.

Demoulin, P., Raadu, M.A., Malherbe, J.M., Schieder, B.: 1987, *A&A* **183**, 142

Litvinenko, Yu.E., Somov, B.V.: 1993, *Astron. Letters* **19**, 27

Somov, B.V.: 1992, *Physical Processes in Solar Flares*, Kluwer, Dodrecht, 280

Syrovatskii, S.I., Chesalin, L.S.:1963, *Questions of Magnetohydrodynamics*, Zinatne: Riga, **17** (in Russian)

Syrovatskii, S.I., Somov, B.V.: 1980, 'Solar and Interplanetary Dynamics' in M. Dryer and E. Tandberg-Hanssen, ed(s)., *Reidel*, 425,

CORONAL HEATING DUE TO THE EMERGENCE OF MAGNETIC FLUX

S. T. WU and M. T. SONG*
Center for Space Plasma and Aeronomic Research
and Department of Mechanical Engineering
The University of Alabama in Huntsville
Huntsville, AL 35899 USA

C. C. CHENG
Naval Research Laboratory, Washington, DC 20375 USA

and

M. DRYER
Space Environment Laboratory/ERL
National Oceanic and Atmospheric Administration, Boulder, CO 80303 USA

Abstract. A self-consistent time-dependent, two-dimensional MHD model with a realistic energy equation is developed to understand the origin of bright coronal emission accompanying the occurrence of a new bipolar magnetic region. The motivation for this study is the interpretation of anticipated observations to be made by the SOHO mission.

Key words: Coronal Heating

1. Introduction

It has been shown observationally that the appearance of the emergence of a new bipolar magnetic region (BMR) is always accompanied by bright coronal emissions (Sheeley, 1976; Meyer *et al.* 1979; Chou and Fisher, 1989). Recently, Shibata *et al.* (1989) used a nonlinear, time-dependent, two-dimensional MHD, two-temperature simulation model to represent the transition region (i.e., photosphere, chromosphere and corona) to study the dynamical responses of this part of the solar atmosphere due to emergence of magnetic flux. The purpose of this simulation study is attempt to explain the bright coronal emission associated with the appearance of a BMR. Because the earlier model did not model the realistic features of transition, in this study we use a newly developed numerical technique (Song, Wu, and Dryer, 1993) which enables us to construct a transition region that includes the Harvard-Smithsonian standard atmosphere in the MHD model. We use this model to simulate the atmospheric responses due to the emergence of the magnetic flux across the transition region up to the edge of the corona with the objective to exhibit observed features. The fundamentals of the model are given in Section II. Numerical results are presented in Section III; finally, concluding remarks are given in Section IV.

* Purple Mt. Obser., PRC

Space Science Reviews **70**, 167–170.

2. Fundamentals of the Model

The fundamentals which describe this physical system contain two parts:

2.1 Mathematical Description

The mathematical model used for this study can be described by the first principle of the MHD theory, but the realistic energy equation was used as follows:

$$\frac{\partial p}{\partial t} + (\mathbf{u}\cdot\nabla)p + \gamma p(\nabla\cdot\mathbf{u}) + (\gamma-1)[\nabla\cdot\mathbf{Q} - (\frac{J^2}{\sigma}) + G_{\mathrm{mech}} + L_{\mathrm{rad}}] = 0 (1)$$

where p is the plasma pressure, $\mathbf{u}$ is the velocity vector, γ is specific heat ratio, $\mathbf{Q}$ is the thermal flux, $\mathbf{J}$ is the current, σ is electric conductivity, G_{mech} is the wave heating term and, finally, L_{rad} is the radiation loss term. The terms of $\mathbf{Q}$, G_{mech} and L_{rad} are determined by specific models.

2.2 Boundary and Perturbed Conditions

The boundary conditions set for this calculation are divided into two parts; (i) computational boundary conditions and (ii) physical boundary conditions. It is realized that the left, right and top boundary conditions are computational boundary conditions which are set as non-reflecting boundary conditions at all times. The bottom boundary condition is the physical boundary condition. At the initial time (i.e. $t = 0$), the quiet photospheric conditions (i.e. $\rho_0 = 3.5 \times 10^{-7}\mathrm{gm/cm^3}$, $T_0 = 6.5 \times 10^4\,{}^\circ$ K and $B_0 = 500$ gauss) are given. The perturbed boundary conditions are prescribed at the bottom boundary when $t > 0$, which are:

a. $[p + \frac{B^2}{8\pi}]|_{\mathrm{outside}} = 1.5[p + \frac{B^2}{8\pi}]|_{\mathrm{inside}}$
b. In order to make sure that $\nabla\cdot\mathbf{B} = 0$ is satisfied, we set $[B_{\mathrm{n}}]_{\mathrm{outside}} = [B_{\mathrm{n}}]_{\mathrm{inside}}$.

It should be noticed that this is a Lagrangian calculation. Hence, the boundary conditions and perturbed conditions are functions of time.

3. Numerical Results

In this study, three cases of perturbed conditions are employed. These three cases are:

Case I: The strong magnetic flux emergence strength is seven times the background of the field strength. In this case, the magnetic field outside the computational domain is pushed into the region by the pressure difference between the outside and inside of the computation domain.

Case II: The strength of the emergence of the magnetic field is the same as the background field strength but with a vertical velocity of 0.2 km s^{-1}

at the lower boundary. This velocity is required to carry the field into the computational domain.

Case III: The strong magnetic field emergence with isothermal atmosphere case is the same as Case I, but the transition region structure is ignored.

We briefly summarize the results in the following: The velocity vectors and the fractional density, temperature, and pressure contours at various times for Case I are shown in Figures 1 and 2 respectively. Figures 2(b) and 2(c), respectively, show temperature depletions and enhancements relative to the initial temperature profile from the chromosphere, through the transition region and into the corona. The schematic description of the physical results is shown in Figure 3. From these results we notice that the induced plasma flow oscillates at the Brunt-Vaisala frequency with a period of $\sim$ 200 s (see Fig. 1). In general, we observed from density and temperature contours (see Fig. 2) that the cool region surrounded by two hotter regions appeared in the chromosphere and is identified as the arch filament system surrounded by bright coronal emissions (see Fig. 3). Upward movement of the transition region (T.R.) is indicated by a vertical arrow in Figure 3.

Case II, not only induced the "Brunt Vaisala" oscillation, but also induces horizontal plasma flow which may be identified as the source of a Morton wave.

4. Concluding Remarks

We have presented a two-dimensional, time-dependent MHD model with realistic energy equation which not only can be used for interpretation of UVCS/SOHO data analysis, but also could be utilized for planning SOHO observations.

Acknowledgements

The work by STW is supported by NASA grant (NAGW-9) and NOAA (RANR000104). The work by MD is supported by U.S. Naval Research Laboratory Work Order N0017392WR20046.

References

Chou, D. Y. and Fisher, G. H.: 1989, *Ap. J.* **341**, 533.
Meyer, F., *et al.*: 1979, *Astron. and Astrophys.* **76**, 53.
Sheeley, N. R.: 1976, *Solar Phys.* **47**, 173.
Shibata, K., *et al.*: 1989, *Ap. J.* **345**, 584.
Song, M. T., Wu, S. T. , and Dryer, M.: 1993, *AIAA Paper No. 93-3177*, American Institute of Aeronautics and Astronautics: Washington, DC

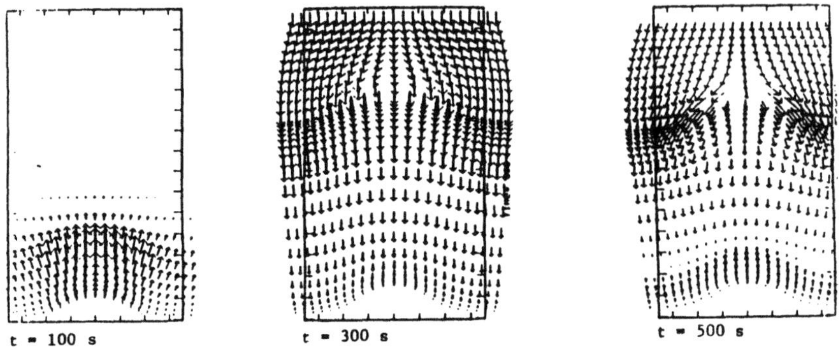

Figure 1. Velocity vectors at various times after appearance of the emerging magnetic flux of Case I.

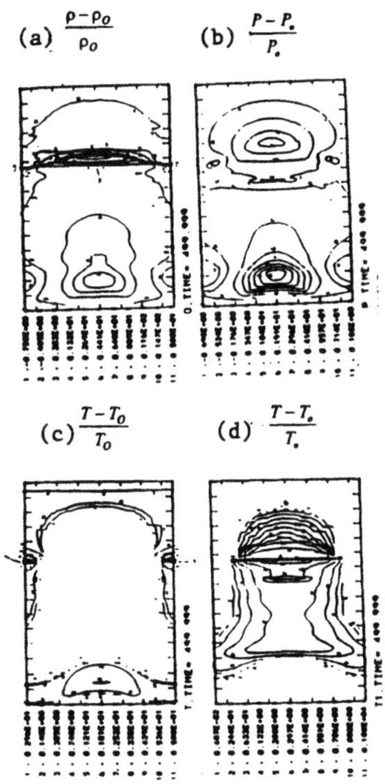

Figure 2. Plasma parameters of Case I at 500 s. Note that (c) represent temperature increases relative to the ambient values and (d), decreases.

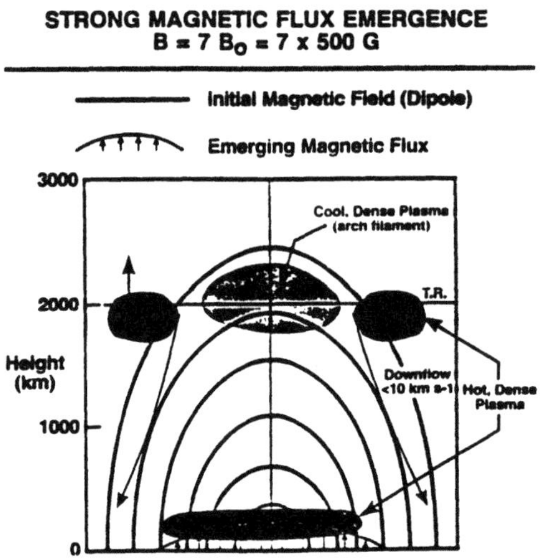

Figure 3. Schematic representation of the physical behavior of the simulation of Case I. Upward movement of the transition region (T.R.) is indicated by the short vertical arrow. Note the regions of cool dense plasma (arch filament) and hot dense plasma as indicated by Figures 2(a) - 2(c).

MODELING OF DYNAMIC EVOLUTION OF CORONAL LOOPS

T.V. STEPANOVA
Department of General Physics and Astronomy
Russian Academy of Sciences, Moscow, Russia

and

A.G. KOSOVICHEV
Crimean Astrophysical Observatory, Nauchny, Ukraine
Institute of Astronomy, University of Cambridge, U.K.

Abstract. Parameters of expanding magnetic loops and arches and of mass flows generated by them in the corona have been computed in a 1D two-fluid approximation. Two possible trigger mechanisms of the coronal transients have been considered: (i) sudden increase of the background magnetic field strength, and (ii) heating and compression plasma inside these magnetic structures. We discuss the formation of shock waves and their dependence on dynamics and geometry of the magnetic structures.

Key words: Sun: corona – Sun: activity

1. Models

The model is based on loss of magnetohydrostatic equilibrium of magnetic loops and arches as a result of an increase in magnetic field strength of the active region or as a result of flare-induced heating of plasma inside them. The expanding magnetic structures act as a piston on coronal plasma and produce flows and shock waves in the heliosphere. The flows result in back reaction on the structures. In other words, an attempt is made to construct a self-consistent model for the dynamics of magnetic loops and arches and for flows generated by them. Mathematically, the model is approximated by two coupled groups of one-dimensional gas-dynamic equations: the first of them, the 'internal' group, describes dynamics inside the magnetic structures, and their shapes and motions; and the second, 'external', describes flows induced outside them. The systems are coupled by boundary conditions on the magnetic surfaces. We suggest the use of a 'thin-tube' approximation for the internal group and, for the external group, the two-fluid Navier-Stokes equations for fully ionized hydrogen gas, taking account of the modified coefficients for electron and ion heat conduction, ion viscosity and energy exchange between ions and electrons (Wolff et al., 1971; Korzhov et al., 1984).

Following the model of coronal transients developed by Pneuman (1980) we consider magnetic flux tubes of two geometrical forms: loops and arches, and assume that their shapes change self-similarly in the course of the expansion. It is further assumed that the magnetic field is frozen into the

Space Science Reviews **70**, 171–174.

plasma and that all parameters inside the tube vary smoothly and monotonically. The basic equations of the model have been formulated by Kosovichev & Stepanova (1991).

2. Results

We have investigated the dynamics of the magnetic structures and the coronal flows in the case of initial disturbances of two types: (i) with an increase of the background magnetic field strength beneath the structures, which corresponds to the model of coronal transients, and (ii) with an increase of temperature and density inside the structures, believed to occur due to solar flares. Figures 1a and 1b show the results for the loop and arch models

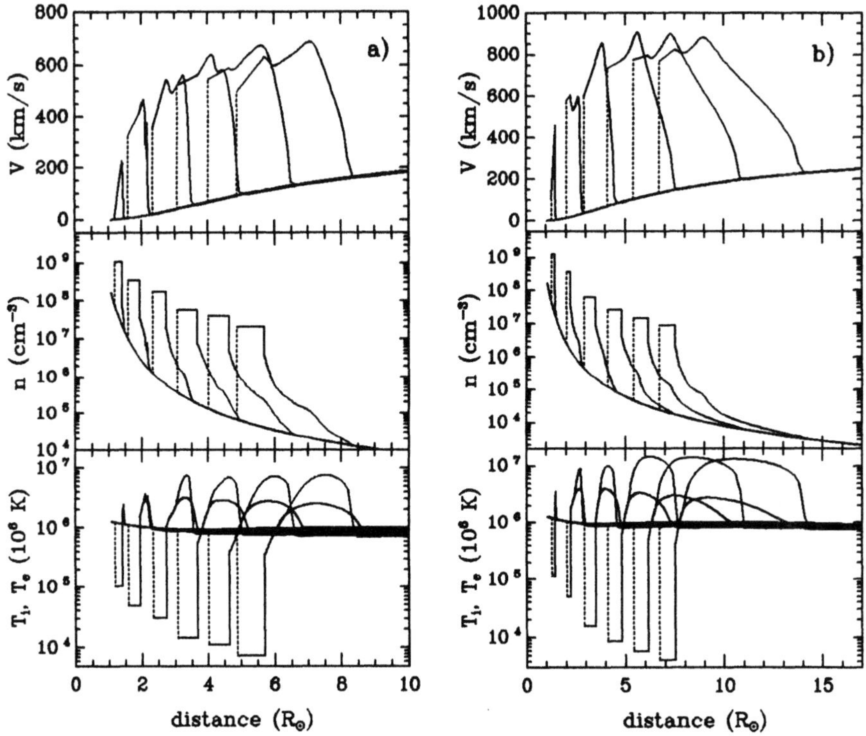

Fig. 1. The dependence on the distance from the solar centre of the velocity V, the density n and the ion temperature T_i (thin curves) and the electron temperature T_e (thick curves) inside the magnetic structures [a) for a loop and b) for an arch] and above them at successive moments of time: 10, 30, 50, 70 and 90 min after a sudden increase of the magnetic field strength beneath the structures from 7 to 10 G. The dashed lines show the lower boundaries of the magnetic structures. Their upper boundaries correspond to sharp decreases of n and sharp increases of both T_i and T_e. These parameters are assumed to be uniform inside the structures, whereas the velocity there is approximated by a linear function. The initial radial distance of the structures from the solar centre is 1.2 $R_\odot$, their width is 0.24 $R_\odot$; the initial internal density is $n = 10^9$ cm^{-3}, temperature $T_i = T_e = 10^5$ K, magnetic field 5 G.

respectively when the magnetic field beneath the structures increased from from its initial equilibrium value of 7 G to 10 G.

In front of the rising loops and arches there form shock waves that propagate in the corona with velocities $\simeq 10^3$ km/s. The electron temperature is several times lower than the ion temperature in a wide relaxation zone behind the shock front, of an extent of several solar radii. Near the upper boundary of the structures, a spherical expansion results in an increase of density and a decrease of temperature, so 'heaping-up' of the material occurs. The arches propagate faster than the loops under the same conditions, because the density inside the arches decreases more rapidly. Thus, they produce stronger shocks with wider relaxation zones.

If the background magnetic field increases to 15 G, then the processes in the corona are substantially more intensified. Thus, the transient velocities reach $\simeq 1200$ km/s; the plasma flow velocities behind the shock front are $\simeq 1500$ km/s, and the ion temperature is $\simeq 4 \times 10^7$ K. The structure of the shock front in this case is determined by the large-scale turbulent viscosity. An important role here is played by viscous heating behind the shock front, which does not result, unlike the preceding case, in the formation of a dense

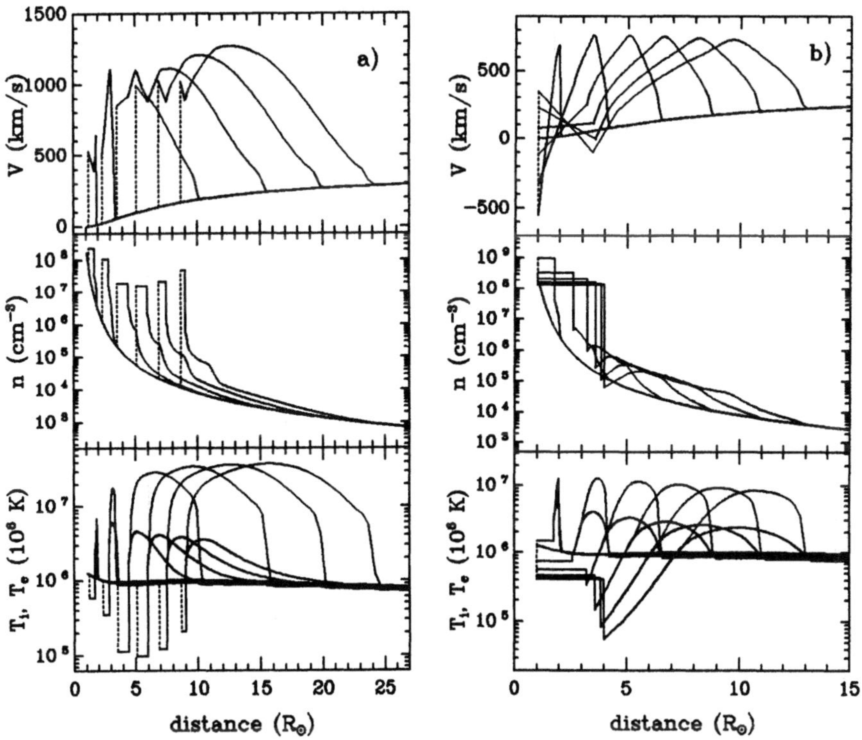

Fig. 2. Same as in Fig. 1 but for a simultaneous increase of the temperature from 10^5 to 10^7 K and the density from 10^9 to 1.6×10^{10} cm^{-3} inside (a) the loop and (b) the arch.

and cold layer near the upper magnetic boundary.

Another case of disturbance of the magnetic structures, of 'flare' type, (Fig. 2) simulates a simultaneous increase of the internal temperature from 10^5 to 10^7 K and the density from 10^9 to 1.6×10^{10} cm^{-3}, the background magnetic field being kept unchanged. In this case a substantial role in the resulting balance of forces in the structures is played by the gravity force. As a result, their upward motions have a pulsating character. The loop behaves similarly to the previous case of magnetic disturbance, whereas the arch almost does not rise, but increases its width significantly. The loop pulsations result in a series of relatively weak shock waves in the solar corona (Kosovichev and Stepanova, 1991).

3. Conclusions

Loss of equilibrium of magnetic loops and arches as a result of enhancement of the background magnetic field or due to plasma heating inside the structures leads to their rapid acceleration (in about 1 hour) to supersonic velocities. The arches are characterized by higher velocities and by more intensified coronal mass flows they produced, as compared with the loop. Shock waves in the flows can be divided into two types, depending on their intensity. In the case of relatively weak shock waves moving with velocities of $\lesssim 10^3$ km/s, the plasma behind the shock front is heated up to $\lesssim 5 \times 10^6$ K, and the electron temperature is lower than the ion temperature in the relaxation zone behind the front. A typical feature in this case is a dense and cold layer formed near the upper boundary of the magnetic structures. In the case of strong shock waves, having velocities of $\gtrsim 10^3$ km/s, the ion temperature can increase to $\simeq 5 \times 10^7$ K and can exceed the electron temperature by more than an order of magnitude. No appreciable piling-up of plasma occurs in this case. Further development of the model will take into account plasma dynamics along the coronal magnetic structures.

Studies of coronal shocks and flows produced by expanding magnetic structures are important for understanding the physics of coronal transients and coronal mass ejections. These processes can be studied by using the UV spectroscopic technique developed by Spadaro et al. (1994) for UVCS data.

References

Korzhov N.P., Mishin V.V. and Tomozov V.M., 1984, *Plan.Sp.Sci.*, **32**, 1169-1178
Kosovichev A.G., Stepanova T.V., 1991, *Soviet Astronomy*, **35**, 646-652
Pneuman G.W., 1980, *Solar Phys.*, **65**, 369-385
Spadaro D., Ventura R., Martin R., 1994, these proceedings
Wolff C.W., Brandt J.C., and Southwick R.C., 1971, *Astrophys.J.*, **165**, 181-194

HIERARCHY OF SPATIAL SCALES IN UV PROMINENCES

B. SCHMIEDER[1]
[1] *Observatoire de Paris, Section de Meudon, DASOP, F-92195 Meudon Principal Cedex, France*

J. E. WIIK[2]
[2] *ESA, Space Science Department - Sc, ESTEC Postbus 299, NL-2200 AG Noordwijk, The Netherlands*

and

K. P. DERE[3]
[3] *E. O. Hulburt Center for Space Research, US Naval Research Laboratory, Washington DC 20375-5352, U.S.A.*

Abstract. Ultraviolet spectra of a quiescent prominence observed with the *High Resolution Telescope and Spectrograph* (HRTS) are analyzed. Different techniques lead to greatly different spatial scales for the prominence structures. The UV spectra show strong variations in intensity and Doppler shift on scales larger than 1700 km. Spectroscopic diagnostics employing line intensity ratios indicate the existence of scales between 400 m to some hundred kilometers. We attempt to interpret various aspects of the prominence intensities and velocities with a multiple thread model.

Key words: Sun – Prominences – UV radiation

1. Introduction

Solar prominences are known to contain plasmas a hundred times cooler and a hundred times denser than the surrounding coronal plasma. Prominence plasmas with temperatures between 2×10^4 and 10^6 K are called the prominence transition region (PTR). The ongoing debate on the structure of the PTR is focused on a few central questions like: Does the PTR surround each cool thread as observed in Hα, or is it concentrated in separate iso-thermal threads? The answers to these questions help determine the formation mechanisms of prominences with respect to the energy balance.

During its first rocket flight the HRTS - *High Resolution Telescope and Spectrograph* (Bartoe and Brueckner 1975) - recorded ultraviolet spectra in the 1175 – 1715 Å range of a prominence. These spectra allow us to compute various physical parameters in the PTR such as the electron density, the emission measure and the thermal and non-thermal velocities on small spatial scales ($\sim 1''$).

2. UV Fine Structures in Prominences

More than ten resonance and intercombination lines from six elements were detected in the prominence. The temperatures of formation of these lines

Space Science Reviews **70**, 175–180.

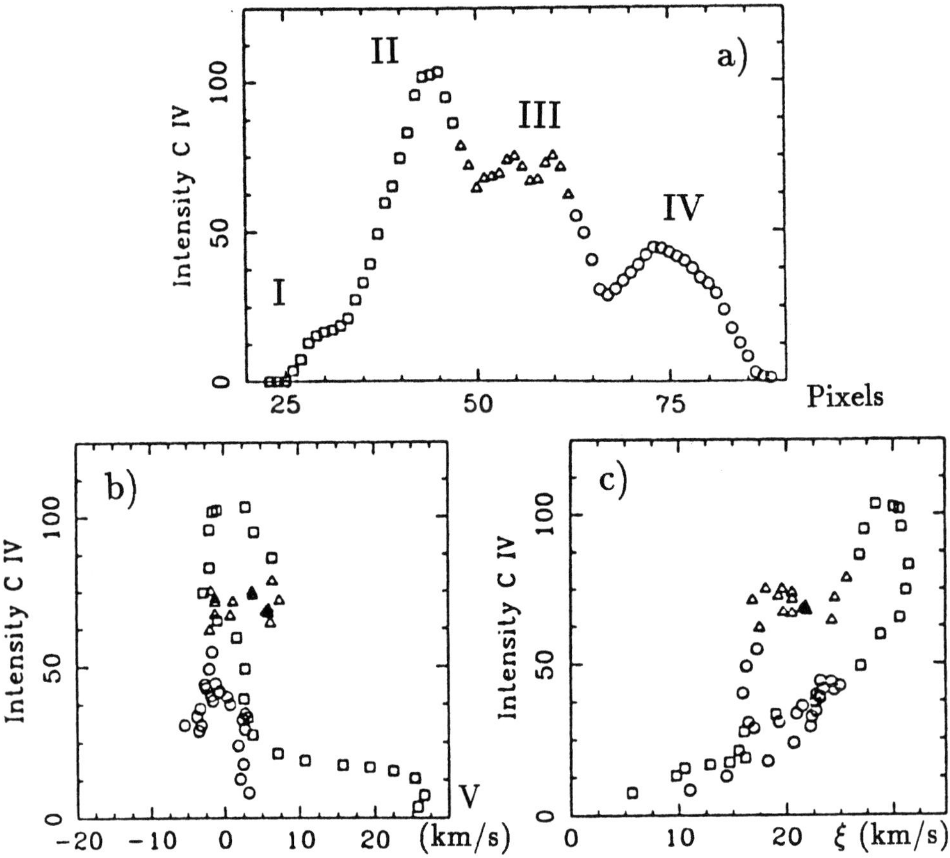

Fig. 1. (a) The C IV intensity along the slit. The symbols differentiate three of the structures (squares - structures I+II, triangles - structure III, and circles - structure IV). Each pixel corresponds to 0.5″. (b) Intensity versus Doppler velocity and (c) intensity versus non-thermal velocity.

span the interval 16 000 to 10^6 K. Many transition region lines are quite strong while usually strong chromospheric lines such as C I and Fe II are faint. A detailed analysis of these spectra has been presented previously by Wiik *et al.* (1993).

To analyse the structures of the prominence, we use the parameters derived from the three moments of the profiles (Dere *et al.* 1984): the integrated intensity (intensity), the net line shift (Doppler velocity), and the line width from which the non-thermal velocity has been derived. These parameters are presented, for a selected line (C IV), in Fig. 1. The intensity structures seen in Fig. 1a, labelled I-IV, appear to be discrete, separate structures from their appearance in the spectra. Their sizes are typically 2.5 – 5″ and are significantly larger than the spatial resolution of the HRTS (0.8″).

A good correlation is obtained between velocities from lines of different temperatures. Coherent motions over spatial scales of $\sim 5''$ along the slit are observed. The velocities are not distributed randomly in each structure suggesting real bulk flow. Due to the sign reversal of the velocities at the points of maximum intensity in the structures, shear or twisted motions are suspected. The high altitude component, structure I, shows velocities as high as $28\,\mathrm{km\,s^{-1}}$ while the observed non-thermal velocities in this structure are relatively low (Fig. 1b, c).

3. Physical Parameters of the PTR

From a number of pressure sensitive line ratios an average pressure of $N_e T_e \sim 10^{16}\,\mathrm{cm^{-3}\,K}$ has been derived (Wiik *et al.* 1993). The corresponding mean electron density is $N_e \sim 10^{11}\,\mathrm{cm^{-3}}$ at $T_e = 10^5\,\mathrm{K}$. Such high value is typical for active prominences (Poland and Tandberg-Hanssen 1983, Widing *et al.* 1986; Dere *et al.* 1982). Complementary information on the prominence plasma is contained in the emission measure $EM(T) = N_e(T)N_H(T)\Delta l$ which is derived from a spectral line intensity through: $I = \epsilon(T)EM(T)$, where the path length Δl is understood to the total path length along the line-of-sight where the temperature is within roughly a factor of 2 of the peak of the emissivity ϵ. Incorporating the derived densities, path lengths $\Delta l = 400\,\mathrm{m}$ are found for C IV, for example.

The intensity ratio of doublets from ions such as C IV and Si IV is very sensitive to the optical depth. In most pixels along the slit, the ratio of the C IV intensities is equal to 2.0 corresponding to optically thin conditions. Only in the central part of structure II, we measure a ratio equal to 1.8 corresponding to an optical depth at line center of 0.6 (Dere and Mason 1993). For a density of $10^{11}\,\mathrm{cm^{-3}}$, a path length Δl of 40 km is derived. This constitutes an upper limit on the path length based on the considerations of optical depth because most of the profiles appear to be optically thin. The path length is considerable larger than the path length derived from the emission measure analysis. The discrepancy between the path lengths may indicate non-uniform electron densities along the line-of-sight such that $< N_e >^2 \neq < N_e^2 >$ where the brackets $<>$ indicate an average along the line-of-sight.

4. Multiple Thread Model

Multiple thread models have been considered in the past to interpret optical spectra of prominences (Zirker and Koutchmy 1990, 1991; Mein and Mein 1991; Mein *et al.* 1994). The indication of subresolution spatial scales in the UV data suggests the applicability of such a model to the present observations. The appearance of a highly blueshifted structure with low non-thermal

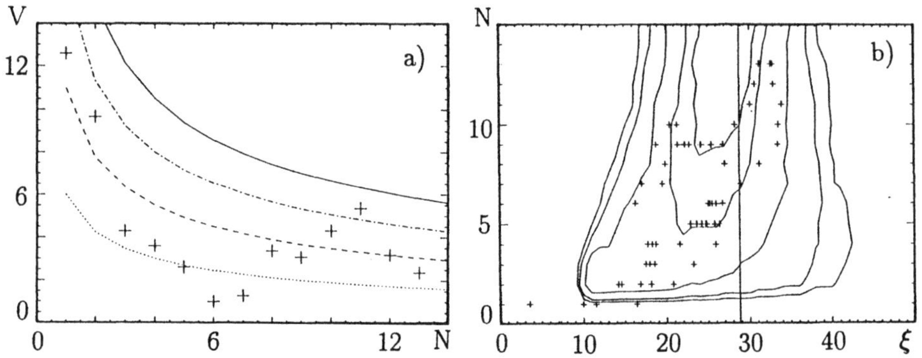

Fig. 2. a) The number of threads (C IV) in the line-of-sight versus the distribution of line shift (crosses). The curves correspond to a thread velocity distribution width of $6\,\mathrm{km\,s^{-1}}$ (lower curve), $9\,\mathrm{km\,s^{-1}}$, $12\,\mathrm{km\,s^{-1}}$, $15\,\mathrm{km\,s^{-1}}$, and $18\,\mathrm{km\,s^{-1}}$ (upper curve). b) The statistical distribution of non-thermal velocities versus the number of threads in the line-of-sight. The best fit to the C IV observations (crosses) was obtained with a thread velocity distribution width of $26\,\mathrm{km\,s^{-1}}$. The vertical line represents the sum of the thread velocity distribution width and the intrinsic non-thermal velocity ($10\,\mathrm{km\,s^{-1}}$)

velocity and low intensity (structure I in Fig. 1) lends further support to this idea. We now develop quantitative predictions of the correlation expected between line intensity, velocity and non-thermal velocity based on a multiple thread model and compare these with the observations.

We assume the prominence consists of a set of identical threads, each giving rise to the same intensity. Each thread has an intrinsic non-thermal velocity (the same for all threads) and the velocity of the threads is characterized by a random gaussian distribution. Each line profile is then integrated through a discrete number N of these threads. The observed line width is given by the combined width of the thermal $\Delta\lambda_{th}$, the instrumental $\Delta\lambda_{inst}$, the intrinsic non-thermal $\Delta\lambda_{ntv}$, and the random thread velocity distribution $\Delta\lambda_{thread}$. Weaker profiles should show the largest variations in bulk velocity and the most intense should show little net line shift and only small variations in line width about some average value. To derive an initial prediction, we assume that structure I in Fig. 1 consists of a single thread with an intrinsic non-thermal velocity for the thread of $10\,\mathrm{km\,s^{-1}}$ as observed. The C IV λ1548 intensity for a single thread is $10\,\mathrm{erg\,cm^{-2}\,s^{-1}\,sr^{-1}}$. If all the threads have the same intensity the number of threads would be simply the intensity of each pixel divided by the intensity of a single thread.

4.1 Gaussian distribution of threads

For a Gaussian distribution of threads, the rms distribution of velocities obtained from profiles that are the result of N threads in the line-of-sight is equal to $\Delta\lambda_{thread}/\sqrt{N}$ where we expect $\Delta\lambda_{thread}$ to be equal to the average non-thermal velocity. For C IV this is $18\,\mathrm{km\,s^{-1}}$. The rms distribution calculated from the above formula can be compared to the observed distribution of C IV bulk velocities. In Fig. 2a we have plotted the rms distribution of velocities for all pixels that correspond to a given number of threads. The solid line corresponds to $\Delta\lambda_{thread} = 18\,\mathrm{km\,s^{-1}}$ but a better fit to the data is found for values between 6 to $15\,\mathrm{km\,s^{-1}}$. In regions with many threads along the line-of-sight ($N > 10$) the velocity dispersion of the contributing pixels is greater than in regions of intermediate number of threads along the line-of-sight. This may indicate a variable $\Delta\lambda_{thread}$ or that the intensity of single threads is not constant along the line-of-sight. The large velocity gradient across the high intensity large scale structure (structure II in Fig. 1) could also lead to large distribution of line shifts for pixels with many threads.

4.2 Random distribution of threads

It would be also possible to predict certain characteristics of the non-thermal velocities expected from a multiple thread model. By randomly generating a set of N velocity values corresponding to the N threads producing the profile, the distribution of non-thermal velocities expected for a given $\Delta\lambda_{thread}$ can be calculated. We have performed such calculations for N threads where N goes from 1 to 15 (Fig. 2b).

In order to make the model agree with the observed data (the observed non-thermal velocities) it was necessary to assume that $\Delta\lambda_{thread} = 26\,\mathrm{km\,s^{-1}}$ (instead of $18\,\mathrm{km\,s^{-1}}$). It was again assumed that the intrinsic non-thermal velocity was $10\,\mathrm{km\,s^{-1}}$. This discrepancy may be due to the fact that the fine structures are not randomly distributed as far as bulk velocities are concerned in the structures.

4.3 Number and dimension of structures in the line-of-sight

An estimation of the number of structures in the line-of-sight could be obtained by assuming that the non-thermal velocities are due to unresolved moving structures in the line-of-sight with a Gaussian distribution of velocity. The relation between the line shift V and the non-thermal line broadening ξ is then, $< V^2 >^{1/2} \sim \xi/\sqrt{N}$. Using this relation we find for profiles in the intermediate to bright regions, an average number of threads equal to 30 in the hot line C IV. Likewise, we find a mean value of 15 structures in the line-of-sight for the brighter parts of the prominence emitting in cool lines such as C II.

We interpret these results as follows: the line-of-sight in the central part of the structures crosses a number of micro-structures, apparently twice more of hot plasma than of cool plasma. If we assume that the structures have a circular form with diameter equal to $D = 3500\,\mathrm{km}$, and that the fine structure is distributed homogeneously, the upper limit of the size of these micro-structures d may be simply derived from the ratio between the volume contributing to one pixel ($\pi(0.8''/2)^2 D$) and the volume occupied by the subresolution structures ($N\pi(d/2)^2 l$), $d = 0.8''\sqrt{D/Nl}$, where l is the effective length of the subresolution structures. In the case of 30 hot lines in the line-of-sight we get $d \sim 100$ km for threads parallel to the line-of-sight ($l = D$) and $d \sim 250$ km for threads normal to the line-of-sight ($l = 0.8''$).

5. Conclusion

Using the high quality spectra of a prominence observed with HRTS a number of physical parameters for the PTR plasma has been derived (Wiik *et al.* 1993). We have summarized in this paper the main results concerning the definition of scales of prominence fine structures, showing the uncertainty of the physical parameters involved in the filamentation of prominences. The spatial scales of the prominence structures are larger than 2.5″ (FWHM) which is greater than the spatial resolution of the instrument. A comparison of the densities derived from pressure sensitive line ratios with the emission measure leads to spatial scales of 400 m. A similar comparison with the optical depth in C IV indicate spatial scales 100 times greater. These small scale lengths seems a general characteristics of the transition region (Dere *et al.* 1987, Dere and Mason 1993). The analysis of the behaviour of the UV line profiles in the prominences suggest the presence of multiple subresolution threads along the line-of-sight. We have proposed a method based on a multiple thread model to interpreted these profiles. Considering the simplicity of the model, a number of aspects of the distributions of intensity, velocity and non-thermal velocity were reproduced, although not always self-consistently. The number of threads contributing to each pixel could be about 15 at temperatures near lines formed at 2×10^4 K and about 30 in transition region lines.

References

Bartoe J. D. F., Brueckner G. E., 1975, *J. Opt. Soc. Am.* **65**, 13
Dere K. P. and Mason H. E., 1993, *Solar Phys.* **144**, 217
Dere K. P., Bartoe J. D. F., Brueckner G. E., 1982, *Astrophys.J.* **259**, 366
Dere K. P., Bartoe J. D. F., Brueckner G. E., 1984, *Astrophys.J.* **281**, 870
Dere K. P., Bartoe J. D. F., Brueckner G. E., Cook J. W., Socker D. G., 1987, *Solar Phys.* **114**, 223
Mein P., Mein N., 1991, *Solar Phys.* **136**, 317
Mein P., Mein N., Wiik J. E., 1994, *Solar Physics*, in press
Poland A. I., Tandberg-Hanssen E., 1983, *Solar Phys.* **84**, 63
Widing K. G., Feldman U., Bhatia A. K., 1986, *Astrophys.J.* **308**, 982
Wiik J. E., Dere K. P., Schmieder B., 1993, *Astron. Astrophys.* **273**, 267
Zirker J. B., Koutchmy S., 1990, *Solar Phys.* **127**, 109
Zirker J. B., Koutchmy S., 1991, *Solar Phys.* **131**, 107

COORDINATED OPTICAL AND YOHKOH OBSERVATIONS OF 26 JUNE 1992 FLARE LOOPS

P. HEINZEL and P. KOTRČ
Astronomical Institute, 25165 Ondřejov, Czech Republic

B. SCHMIEDER
Observatoire de Meudon, DASOP, 92195 Meudon-Cedex, France

E. HIEI
Meisei University, Tokyo, Japan

and

B. ANWAR
National Institute of Aeronautics and Space, Bandung, Indonesia

Abstract. Optical spectra of large flare loops were detected by the Ondřejov Multichannel Flare Spectrograph (MFS) during coordinated observations with MSDP at Pic du Midi (Hα) and the soft X-ray telescope (SXT) on Yohkoh. The CCD video images taken by the MFS slit-jaw camera document the time-development of the flare loops as seen through the Hα filter. Preliminary analysis of the MSDP images shows the intensity structure of the cool flare loops and their velocity fields. From the spectra we can clearly see the intensity variations along the cool loops. SXT images show the structure of hot X-ray loops similar to that of cool loops. Special attention is devoted to the bright tops, simultaneously observed in X-rays, Hα and other optical lines. Based on a preliminary analysis of the optical spectra, we speculate about possible mechanisms leading to an observed bright emission at the tops of cool loops. We suggest that direct soft X-ray irradiation of cool loops at their tops could be, at least partly, responsible for such a strong brightening.

Key words: Flare loops – Heating mechanisms

1. Coordinated observations

A huge system of W-limb (post) flare loops was observed simultaneously by the Ondřejov Multichannel Flare Spectrograph (MFS), Multichannel Subtractive Double Pass Spectrograph (MSDP) at Pic du Midi and the soft X-ray telescope (SXT) on board Yohkoh on 26 June 1992. These coordinated observations span the time period from about 07:00 UT till 09:40 UT with overlaps between MFS, MSDP and SXT.

MSDP observations

Hα images of cool loops have been digitized on the MAMA microdensitometer at the Observatory of Paris and processed using the technique described by Mein (1977). Fig. 1a shows the Hα intensity structure at about 07:08 UT, while Fig. 1b displays the velocity field derived from MSDP data. Highly structured cool loops with rather bright tops clearly exhibit the downward motions along both legs (see light pattern - the blue shift and dark one - the

Space Science Reviews **70**, 181–184.

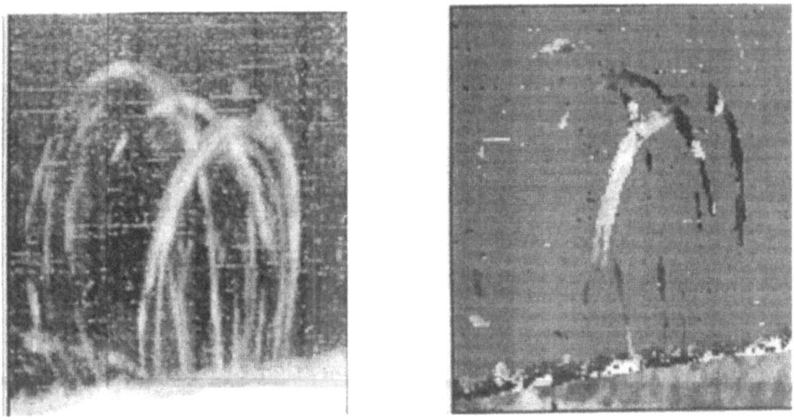

Fig. 1. Flare loops observed with the MSDP. a) left - intensity in the Hα line center, and b) right - velocity in Hα $\pm$ 0.3 Å(see text).

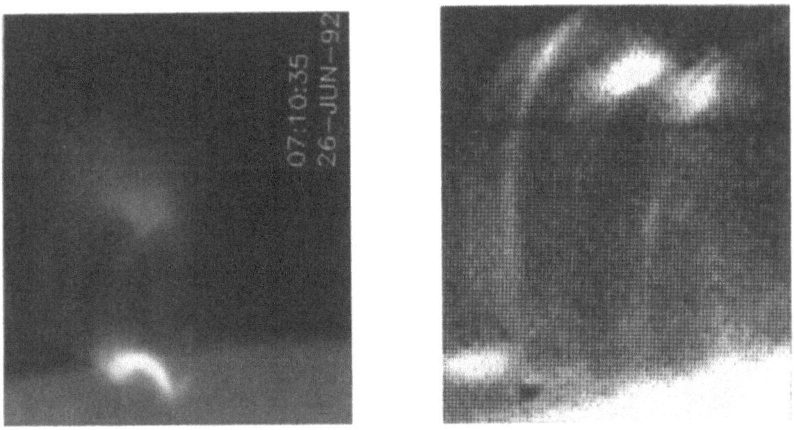

Fig. 2. Flare loops observed: a) in X-rays from SXT/Yohkoh (left) and b) by the MFS Hα slit-jaw camera (right). Note different scales in both pictures.

red shift in Fig. 1b). The Hα line profiles reconstructed from MSDP pixels show a complex behaviour which is to be compared with MFS spectra.

SXT images

The system of hot loops was observed by Yohkoh SXT almost continuously from 01:49:03 till 10:35:29 UT. Almost all of SXT images (e.g. Fig. 2a) exhibit bright tops of the loops. The temperatures of the hot loops were derived from the SXT images, each being observed with a different filter. They are of the order of 6×10^6 K (Anwar et al., 1993).

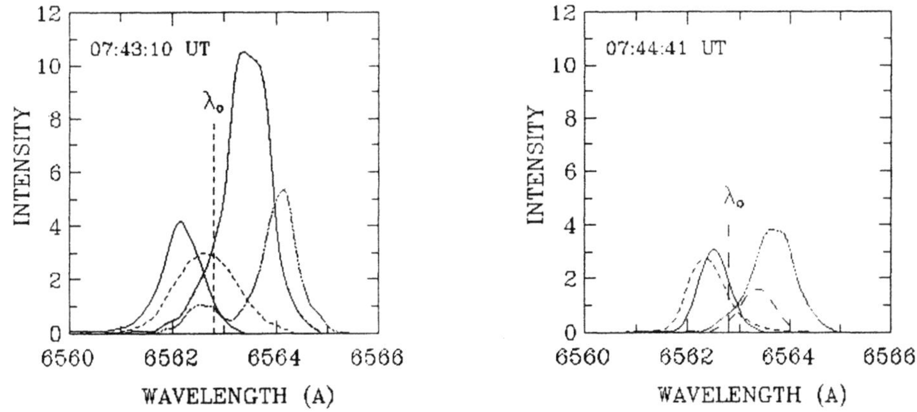

Fig. 3. Hα profiles of various knots observed at tops (left) and at middle heights (right) of the loops. Intensities are in % of the disk-center continuum.

MFS data

In Fig. 2b we show the Hα CCD slit-jaw picture with the indicated position of the spectrograph slit. Around 07:40 UT we obtained several sets of MFS spectra of the loop system at various heights above the limb, including the bright tops. The spectra have been taken in the hydrogen, helium and CaII lines with exposure times 5 or 9 sec and the dispersion 1 Å/mm. They were digitized by the Ondřejov Universal Microphotometer (Kotrč and Zicha, 1993). The intensities are generally higher at the tops as compared to loop legs (Fig. 3), which is consistent with Hα slit-jaw pictures (Fig. 2b). An important feature is the velocity pattern. In the legs we detect substantial Doppler shifts to the blue and to the red, depending on which leg actually intersects the spectrograph slit. At the top of the loops velocities of the same order are also present, but some plasma knots move rather slowly.

2. Heating of the loops

It is usually assumed that the enhanced brightness at the tops of both hot (X-ray) and cool (Hα) loops is due to the heating processes taking place at the reconnection site (Forbes and Malherbe, 1986, Démoulin - private communication). Here we propose an idea that the tops of cool loops are, at least partly, heated by a strong soft X-ray irradiation from close hot loops (such heating mechanism in the case of the chromosphere was studied by Hénoux and Nakagawa, 1977). Moreover, this X-ray irradiation will modify the hydrogen ionization thus leading to a change of the Hα brightness.

Using the non-LTE technique described in Heinzel et al. (1992), we have performed some test computations in order to see whether Hα can be enhanced at the loop tops due to strong soft X-ray irradiation, consistent with SXT images.
Two models have been considered:
T=10000 K, P=0.5 dyn cm^{-2}, D=1000 km
T=10000 K, P=5.0 dyn cm^{-2}, D=1000 km,
where T is the temperature, P is the gas pressure and D represents the thickness of the cool loop.
Lower-pressure model leads to lower Hα emission when X-ray irradiation is taken into account. However, the higher pressure one gives almost a factor of two brighter Hα than the same model without incident soft X-ray radiation.

3. Conclusions

Our simultaneous optical and SXT observations provide us with a unique opportunity to discriminate between various heating mechanisms in cool flare loops. A possibility of soft X-ray heating at the tops of cool loops was suggested, in analogy to studies of the chromospheric heating (Hénoux and Nakagawa, 1977, Hawley and Fisher, 1993). To check the proposed idea, we plan to proceed further in the following directions:
- the reduction of MFS spectra and MSDP images
- specification of soft X-ray fluxes from hot loops, using SXT data
- detailed non-LTE modeling of cool loops, irradiated by soft X-rays
- evaluation of synthetic hydrogen, helium and CaII line profiles.

One important output from our modeling is the predicted range of UV-line emissivities which can be compared to SOHO-SUMER observations.

Acknowledgements

MSDP observations have been done by J. E. Wiik and T. Roudier. We also thank E. Priest for valuable comments.

References

Anwar B., Hiei E., Hudson H. S., Acton L. W., Lemen J. and Metcalf T.: 1993, *Proc. Kofu Symp. "New Look at the Sun", in press.*
Forbes T.G. and Malherbe J.M.: 1986, *Ap. J.* **302**, L67.
Hawley S.L. and Fisher G.H.: 1993, *SSL-preprint, Berkeley; Ap. J., submitted.*
Heinzel P., Schmieder B. and Mein P.: 1992, *Solar Phys.* **139**, 81.
Hénoux J. C. and Nakagawa Y.: 1977, *Astron. Astrophys.* **57**, 105
Kotrč P. and Zicha J.: 1993, *Proc. IAU Symp. No 161 "Wide-field Imaging", in press.*
Mein P.: 1977, *Solar Phys.* **54**, 55.

CORONAL ELECTRON DENSITY MEASUREMENTS USING THE NEAR-IR [FE XIII] EMISSION LINES

M.J. PENN, J.R. KUHN
*National Solar Observatory/Sacramento Peak, National Optical Astronomy Observatories, * P.O. Box 62, Sunspot, NM 88349 USA*

J. ARNAUD
Observatoire Midi-Pyrenees, 14 Avenue Edouard Belin, 31400 Toulouse FRANCE

and

D.L. MICKEY, B.J. LABONTE
University of Hawaii, 2680 Woodlawn Drive, Honolulu, HI 96822 USA

Abstract. Observations made during the 1991 total solar eclipse and recent observations from NSO/Sac Peak are discussed. The ground-based density measurements will be complimentary to SOHO observations, particularly SOHO electron density measurements.

Key words: Sun - corona

1. Discussion

Eclipse data from from the University of Hawaii 61-cm South telescope of the Mauna Kea Observatories were obtained by CCD imaging of an active region corona using four near infra-red wavelength channels isolated with narrow-band interference filters (see Penn et al. 1993). Two long exposure images in each channel were obtained including the continuum (λ = 10690 Å), two [Fe-XIII] emission lines, (λ = 10747 and 10798 Å), and the He I line (λ = 10830 Å). The images are calibrated to Sun center intensity. The field-of-view (FOV) is about 180 × 270 arcseconds centered 140 arcseconds above the south-east limb; each image has 0.5 arcsecond pixels with ≈1.5 arcsecond seeing. Within the FOV are the edge of a coronal streamer, a section of a prominence, and coronal loops associated with active regions behind the limb (see Figure 1).

The ratio of emission of the two [Fe-XIII] lines provides a direct measure of the coronal electron density (see Flowers & Pineau des Forêts, 1973). From a very simple analysis, the electron density from the [Fe-XIII] emission is found to be $2.1 \times 10^9 \geq N_e \geq 1.6 \times 10^7$ cm^{-3} in the FOV. An enhanced continuum loop corresponds with a region depleted of [Fe-XIII] emission; this bright continuum loop must be significantly hotter or colder than the surrounding [Fe-XIII] loops. Analysis of the radial dependence of the density (Figure 2) suggests the continuum loop has a large density scale height and

* Operated by the Association of Universities for Research in Astronomy, Inc. (AURA) under cooperative agreement with the National Science Foundation

Space Science Reviews **70**, 185–188.

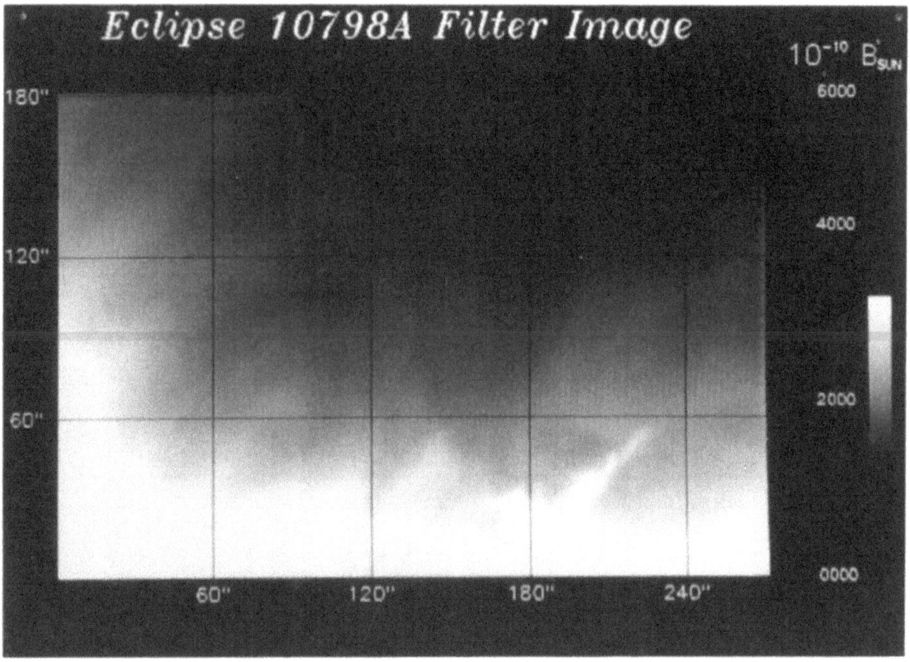

Fig. 1. Calibrated [Fe-XIII] infrared emission line (λ = 10798 Å) image from the 1991 eclipse. The field-of-view is about 180 × 270 arcseconds with 0.5 arcsecond pixels; the seeing was estimated to be about 1.5 arcseconds in the predawn sky. Coronal loops from the active region NOAA 6728 on the farside of the limb are the dominant structures in the field-of-view; a prominence is also visible in the lower right corner of this image.

is hotter than the [Fe-XIII] loops, although the assumption of hydrostatic equilibrium gives anomolously high temperatures for both continuum and [Fe-XIII] loops.

Preliminary observations of these two [Fe-XIII] emission lines have been made at the Evans Solar Facility at NSO/Sac Peak. The 40 cm aperture coronagraph and littrow spectrograph are used with an infra-red sensitive array. Wide-slit spectra of both [Fe-XIII] emission lines were obtained simultaneously at one coronal height, and then the telescope was pointed to a greater height for the next exposure. The mean electron density as a function of height (from 1.03 to 1.13 $R_\odot$) was measured in the corona over a weak active region. Low structures were seen with a mean density of $N_e = 1.7 \times 10^8$ cm^{-3} at a height of 1.06 $R_\odot$ and below, followed by a smooth drop in the mean electron density to $N_e = 6.3 \times 10^7$ cm^{-3} at a height of 1.13 $R_\odot$ (see Figure 3).

More [Fe-XIII] electron density measurements will be made from Sac Peak. Calibrated [Fe-XIII] (1213 Å) SOHO data (perhaps from the UVCS

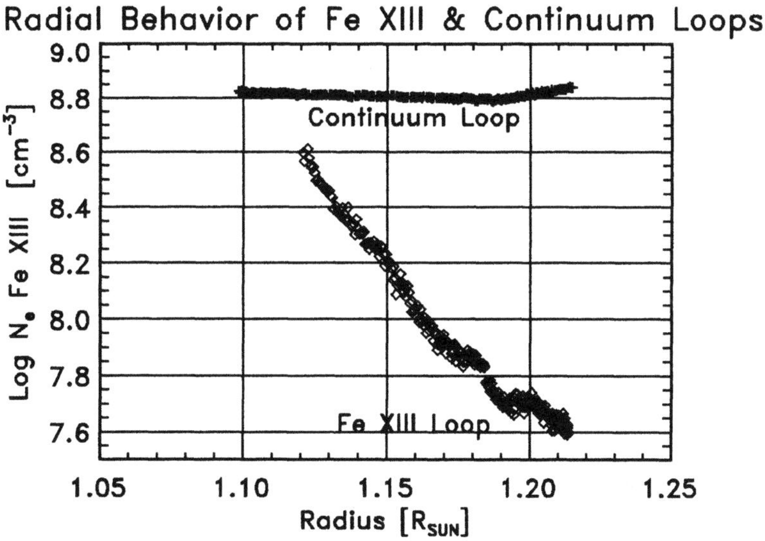

Fig. 2. Log of the electron density versus radius in a typical [Fe-XIII] loop in the 1991 eclipse FOV. The log of the electron density in the continuum loop, which has an arbitrary vertical offset, is also plotted. The continuum loop has a larger density scale height than the [Fe-XIII] loop; this suggests that the temperature of the continuum loop is greater than that of the [Fe-XIII] loops.

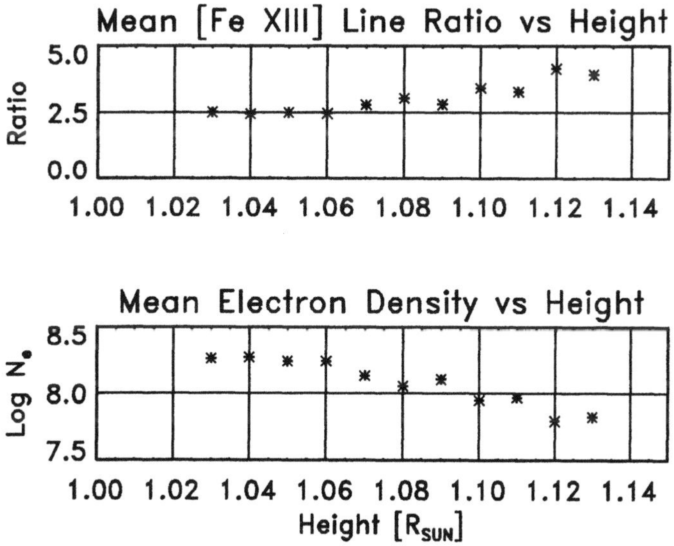

Fig. 3. The mean [Fe-XIII] line ratio at different heights above the limb for the preliminary NSO/Sac Peak observations is plotted in the upper figure; corresponding electron densities are plotted in the lower figure. Nearly constant densities occur in low structures at heights less than 1.06 $R_\odot$; above this height the electron density decreases in a way suggestive of the background corona.

experiment) could also be used in conjunction with ground-based near-IR [Fe-XIII] observations for coronal electron density measurements; the ratio of UV [Fe-XIII] emission to the 10747Å emission would provide a density diagnostic in the outer corona. Density measurements from S-X emission line ratios (from the SUMER experiment) can be compared to the ground-based Fe-XIII observations; density measurements from these two ions (at two temperatures) would be relevant to the study of gas pressure balance in the corona.

References

Flower, D.R., Pineau des Forêts, G.: 1973, *A&A* **24**, 181
Penn, M.J., Arnaud, J., Mickey, D.L., LaBonte, B.J.: 1993, *preprint*

STRUCTURE AND DEVELOPMENT OF QUIET LOOPS IN THE SOLAR CORONA

EIJIRO HIEI
National Astronomical Observatory and Meisei University Mitaka, Tokyo 181, Japan

Abstract. X-ray emission from solar coronal loops changes on two different timescales: a) flare loops and transient active region brightenings show a rapid variability, b) quiet region loops are quasi-steady and change only slowly with time. This different time behavior has been analyzed on the basis of Yohkoh SXT observations and we report here on the results from our analysis, mainly focussing on quiet loop variability.

Key words: Coronal Loop - Solar Corona - X-ray Image

1. Introduction

Soft X-ray images of the solar corona, taken with the SXT (Soft X-ray) Telescope onboard the Yohkoh spacecraft, show that the solar corona has a much higher and ubiquitous time variability than appeared from previous space experiments. It is never static and quiet, and transient phenomena occur on a wide variety of scale sizes; i) transient brightenings of small size (less than several tens of arcsec.) occur in X-ray bright points in active regions and coronal holes, ii) flare loops, jets and bright loops associated with prominence disappearances are typical examples of dramatic time variability in medium size (0.1 to 0.5 R_{sun}) features and iii) large scale restructuring of the solar corona on a size larger than 0.5 R_{sun} occurs, for instance, in CME events.

SXT images of the solar corona show several kinds of loops of different sizes projected onto a homogeneous background corona. The homogeneity of the background corona might in fact result from many fine low-contrast threads with a geometrical scale much smaller than the spatial resolution of the SXT instrument. However, in the following the background corona will not be discussed any further.

2. Coronal Loops Seen in SXT

Coronal X-ray loops show different characteristics. SXT images — appearing now also on Solar Geophysical Data — give fair examples of different loops, although on sizes larger than allowed by the instrument resolution ($\approx$ 2.5 arcsec FWHM at 8 Å), and lead to the following loop classification:
i) flare loops
ii) loops in active regions (AR)
iii) loops connecting two AR

Space Science Reviews **70**, 189–192.

iv) loops connecting between AR and their periphery
v) loops in a quiet region.

The brightest and most prominent loops are flaring loops (for further details see: Initial Results from Yohkoh, *Publ. Astr. Soc. Japan* 44, 1992). Active Regions usually contain many bright loops, some of which show transient brightenings (Shimizu, 1992), while some are steady (Tsuneta,1993). Larger scale loops may extend from one AR to its peripheral areas or even connect different ARs. In a quiet region there are many loops, hereafter called quiet loops, which show time changes both in intensity and shape, but on a much longer time scale than active regions loops.

The time scale of a loop brightness/shape change can be defined as the time over which its identity, as seen in SXT images, is lost. The following Table (Table I) shows the time scales of loop variability for loops of different classes, as estimated from an analysis of the temporal behavior of loops seen in SXT images.

TABLE I
Time scale of loops

classification	time scale	τ (Alfvén time-scale)	reference
i) flare loop	< 1 min.	30 x	Tsuneta (1993)
ii) transient brightening	1 min.	30 x	Shimizu (1993)
ii) steady loops in AR	1 h	1000 x	Tsuneta (1993)
iii) AR-AR loop	5 h	1000 x	
iv) AR-periphery loop	5 h	1000 x	
v) quiet loop	1 d	1000 x	

The Alfvén time-scale which appears in Table I has been evaluated assuming typical values for the magnetic field strength, the plasma density, and the characteristic scale length of loops of different classes. Table 2 gives the figures we adopted for this estimate.

It is apparent from Table I that the characteristic time scales given in column two define two classes: one representing a quite rapid change occurring on a time scale on the order of a few tens the Alfvén time-scale, the other representing slow changes over a time scale as long as a thousand times the Alfvén time-scale. Hence, there are two types of time changes in loops. It is worth noticing that, according to laboratory experiments, magnetic reconnection may occur over a time scale on the order of several tens of the Alfvén time-scale. The results of Table 1 suggest that the time change of flaring loops and/or transient loop brightenings may be due to magnetic

TABLE II
Physical parameters

	B (Gauss)	log n	L (km)
flare loop	500	11	10,000
transient brightening	100	10	5,000
steady loop in AR	100	10	5,000
AR-AR loop	100	10	50,000
AR-periphery loop	100	10	50,000
quiet loop	10	9	100,000

reconnection. On the contrary, time variability in loops of the other classes has probably to be ascribed to a different mechanism, such as, for instance, a change in the magnetic field strength/direction at the loop footpoints, and/or a change in the loop energy input.

3. Structure of Quiet Loops

Isolated quiet loops are not often observed: usually several loops, nearly parallel to each other, bridge over a well-defined neutral line of the magnetic field, forming an arcade structure. Generally, but not always, a prominence lies over the neutral line. Therefore, there is not a one-to one correspondence between prominences and quiet loops. Eclipse images have shown that quiet loops may also appear *above* prominences, as exemplified by coronal streamers.

The existence of small loops cannot be ruled out; however, SXT images show that, typically, the quiet loop length ranges in between 10000 and a few times 100000 km. A quiet loop keeps its shape for several hours, although gradually fading. In an arcade, there is a chain process where one of the loops slowly changes its intensity and eventually disappears, and then a nearby loop goes through the same process. The time scale for changes in the quiet loop intensity is of the order of 10 h - 1 day.

Quiet loops sometimes change their structure drastically. A good example is seen in an event of 12 November 1991 (Tsuneta, 1992), where the process of formation of closed loops propagates westward with a speed of 20 - 40 km/s, eventually leading to the formation of an enormous arcade which extends over a length of $\approx 1\ R_{sun}$ near the north polar region of the sun.

Quiet loops above a prominence brighten up at the time of a sudden disappearance of the prominence (disparition brusque (DB) (McAllister et.al, 1992, Hiei et. al, 1993). An interesting example of this behavior is seen in the event of January 12 - 16, 1993, whose SXT images have been published also in Solar Geophysical Data. A loop system, seen above a prominence on January 12, 1993, increases in width over successive days, and, following the disappearance of the prominence between 12:20 UT on 16 January and 05:00 UT on 17 January, grows in intensity as well. Hundhausen (1993) reported that a coronal streamer observed with the K-coronameter at Mauna Loa has expanded before the Coronal Mass Ejection. Because the change of the width of loops seen on January 12 - 16, 1993, was observed on disk images, and loops, not being at the solar limb, cannot be observed with the K-coronameter, it is not easy to associate the two phenomena, but we surmise that the growing loop width and the "swollen event" are part of the same scenario.

4. Concluding Remarks

We have shown that quiet loops, bridging over a neutral line, change in intensity over a time-scale of about 1000 Alfvén time-scale. The loop shape is mantained by the magnetic field supporting the loop and/or a constant energy input to the loop. The time scale for loop changes are therefore dictated by changes in these mechanisms.

Quiet loops may change drastically with/without a prominence disappearance. Also the large-scale magnetic configuration of the solar corona may change with time, affecting the lower lying loop structures as well. Coronal Mass Ejections are an obvious example of a phenomenon which modifies the large scale magnetic configuration of the corona. In order to understand fully the mechanisms which lead to loop time variability, it is important to compare the behavior of low and high loops, as Yohkoh SXT images have provided some evidence in favor of a similar behavior for a large range of loop sizes.

Acknowledgements

The author is extremely thankful to Dr. G. Poletto for her comments on an earlier version of this paper and for encouraging me to attend the meeting.

References

Hiei, E., Hundhausen, A.J., Sime, D.: 1993, *Journ. Geophy. Res. Letters*, Vol. 20, No. 24, p. 2785

Hundhausen, A.J.: 1993, Coronal Mass Ejections, in *The Many Faces of the Sun*, eds. K. Strong, J. Saba, and B. Haisch

McAllister, A. et. al: 1992, *Publ.Astr.Soc.Japan* **44**, L205

Shimizu,T. et. al: 1992, *Publ.Astr.Soc.Japan* **44**, L147

Tsuneta,S. et. al: 1992, *Publ.Astr.Soc.Japan* **44**, L211

Tsuneta, S.: 1993, private communication

CORONAL LOOPS AND PROMINENCES AS OBSERVED WITH RATAN-600

A.N.KORZHAVIN and V.M.BOGOD
Special Astrophysical Observatory RAS
Pulkovo, St.Petersburg 196140, Russia

and

V.N.BOROVIK, G.B.GELFREIKH and V.I.MAKAROV
Pulkovo Astronomical Observatory RAS
Pulkovo, St.Petersburg 196140, Russia

Abstract. This paper presents a short summary of observations of coronal structures at microwaves using an instrument with high spatial resolution and good wavelength coverage. The comparison of the RATAN-600 data with optical observations of coronal loops in the green line and with the Very Large Array maps at 21 cm has shown that the loops represent only a small part of coronal matter, although their role may be of great importance in the physics of the solar corona. Prominence (filament) associated sources, especially "peculiar" ones, are also reviewed.

Key words: Sun – Corona – Loops – Prominences - Radio Observations

1. Introduction

The important role of the radio observations in studying coronal structures, both of low and high temperatures, is widely accepted nowadays. One, however, should keep in mind that effective diagnostics of the parameters of the plasma and physical processes in the plasma structures essentially depend on the parameters of the radio telescopes used for the observations. We need both high spatial resolution and good spectral coverage and resolution, and reliable polarization measurements as well. These requirements are contradictory and are hardly met in a single antenna system.

The radio telescope RATAN-600 of the Russian Academy of Science is a reflector type telescope with an aperture of up to 600 meters in diameter and as such is used for detailed analysis of the spectra of different structures of the solar plasma (Pariysky et al, 1967). The main parameters of the instrument are given in Table I. Some ways of possible future development of the instrument for solar studies are also shown (Bogod, 1992, Gelfreikh, 1990, 1992). The main limitations (at present) are its moderate spatial resolution (about 15 arcsec at λ=1.7 cm in WE–direction), especially at decimeter wavelengths, and the transitional nature of the observations. The full coverage of the wavelength range 1.7-32 cm has been proven to be of great value, in particular when measuring coronal magnetic fields.

Space Science Reviews **70**, 193–198.

TABLE I
Solar observations at RATAN-600

Wavelengths range	0.8-32 cm, more then 30 frequency channels
Polarization measurements	Circular (I and V or R and L, $\delta p < 0.5\%$)
Spatial resolution	One-dimensional, 17 arcsec at $\lambda = 2.0 cm$
Time resolution	1 scan per 15 minutes, from 7:30 to 10:30 UT
Modernizations	Panoramic spectrograph with 80 frequency channels
(planned)	and 400 channels with frequency step 1.5 MHz
	Radioheliograph with 10 arcsec resolution at λ=2.0 cm

2. Active Region Observations

A typical well-developed active region (November 1980, SD548/1980) was studied at a number of radio wavelengths obtained with the RATAN-600. During disk passage we could see bright cyclotron sources above the main sunspots (strongly polarized) and an interspot component – rather weak at $\lambda < 4$ cm. When the region was on the limb (November 19) photographic observations made in the green coronal line at the Mountain Solar Kislovodsk Station have shown the loops reaching up to a height of about 100,000 km (see Fig.1). The radio sources observed with RATAN-600 have practically the same height above the photosphere. Their spectrum of fluxes is flat implying thermal bremsstrahlung as the most probable mechanism for the generation of the radio emission. The height of the cyclotron sources above the sunspots does not exceed a few thousands km and they are not seen on the radio scans of November 19, though clearly registered on the previous day. These may most probably be considered as the base of the above loops, seen in the corona in the green line.

We have used for this particuliar active region three methods of measuring magnetic fields developed in our group. They are based on the analysis of spectra and polarization of bremsstrahlung in magnetic fields (Bogod and Gelfreikh, 1980), the analysis of the spectra of the cyclotron emission (Akhmedov et al., 1982) and the interpretation of the observed inversion of the sign of circular polarization in a quasi–transversal magnetic field (Gelfreikh et al., 1987). Some parameters of this AR derived from these observations are summarized in Table II.

Fig. 2 illustrates another case (July 1982, SD228+SD229/1982). In the decimeter wavelength range the interspot component became a halo covering the whole AR (Akhmedov et al., 1987). We should note that the observations with higher spatial resolution of the region made with the VLA show the arc-type components but they are responsible only for a small fraction of the total flux at 20 cm. The physical nature of this decimeter halo is rather

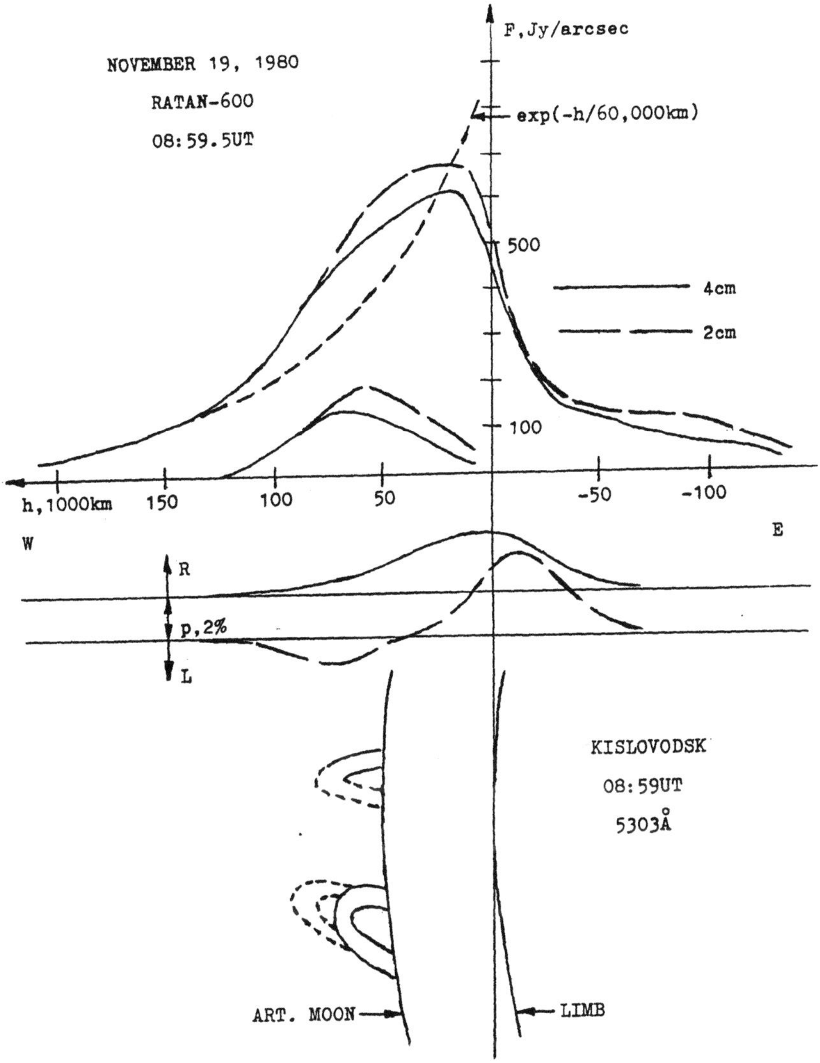

Fig. 1. Radio scans of an active region off the W-limb at the November 19, 1980, in total intensity (F) and circular polarization degree (p) at 2.0 and 4.0 cm wavelengths. The observations were made using the RATAN-600 at five short cm wavelengths (2.0, 2.3, 2.7, 3.2, 4.0 cm). Below is a sketch of optically observed coronal loops in the green coronal line made from coronograph spectroheliograms of Kislovodsk Solar Mountain Station.

enigmatic. The point is that the radio flux spectrum excludes the most natural interpretation in terms of thermal bremsstrahlung. So this component demonstrates a longlasting nonthermal process in coronal loops, like a high level of plasma turbulence and the presence of nonthermal particles.

TABLE II

Active region SD548/1980

1. loops legs, $H_{\rm long}$ =50-80 G, N-polarity, $h = 4 - 8 \cdot 10^3$ km, (Nov 19,1980)
2. loops top, $H_{\rm long}$ =15-25 G, S-polarity, $h = 75 \cdot 10^3$ km, (Nov 19,1980)
3. QT-region, $|H| = 10 - 20$ G, $h = 75 \cdot 10^3$ km, (Nov 19,1980)
4. above N-sunspot, $H = 2000$ G, N-polarity, $h = 1700$ km, (Nov18,1980) $h < 2000$ km, (Nov 19,1980)
5. above S-sunspot, $H > 3000$ G, S-polarity, $h < 3000$ km, (Nov 14,1980)
6. QT-region, $|H| = 20 - 30$ G, $h = 30 - 50 \cdot 10^3$ km, (Nov 16,1980)

3. Prominence Observations

The prominences at centimeter waves while observed off the limb have a brightness temperature comparable with its kinetic electron temperature of about 20,000 K. The position of this source corresponds to that of H_α observations. The measurements of weak polarization can be used to estimate the average longitudinal component of the magnetic field (tens of G, normally). While on the disk, prominences (filaments) are normally seen in microwaves as dark features with sizes often larger than optical ones. The origin of the above situation can be interpreted in terms of the bremsstrahlung mechanism for the generation of radio emission. The radio data supply additional and independent information on the temperature, gradients of temperature, and magnetic fields inside the prominence.

However, it is important that the filaments mark the neutral magnetic line – the line where the longitude component of the photospheric magnetic field becomes zero. Locally at this line the structure of magnetic fields becomes very complicated and even strong vertical currents have been found from the interpretation of magnetographic observations. Short cm wavelength observations reveal local bright sources with brightness temperatures up to 10^7 K and lasting for days (Akhmedov et al., 1986, 1987, 1989; Alissandrakis et al., 1993). The exceptionally high slope of the spectra leads to serious problems in the interpretation of their radiation (Korzhavin et al., 1989; Vatrushin and Korzhavin, 1989). An example of this type of peculiar sources is given in Figure 2 (source "d") and their typical parameters are summarized in Table III.

It is worth mentioning the general tendency to have brighter features in an active region near the neutral line, especially in the case of high gradients of the photospheric magnetic field. In this case the prominence become the locations of the highest brightness temperatures (at $\lambda = 3-6$ cm). Of course, strictly speaking it is not inside the prominence where the emission arises but rather in the surrounding magnetic loops of low height (of the order of

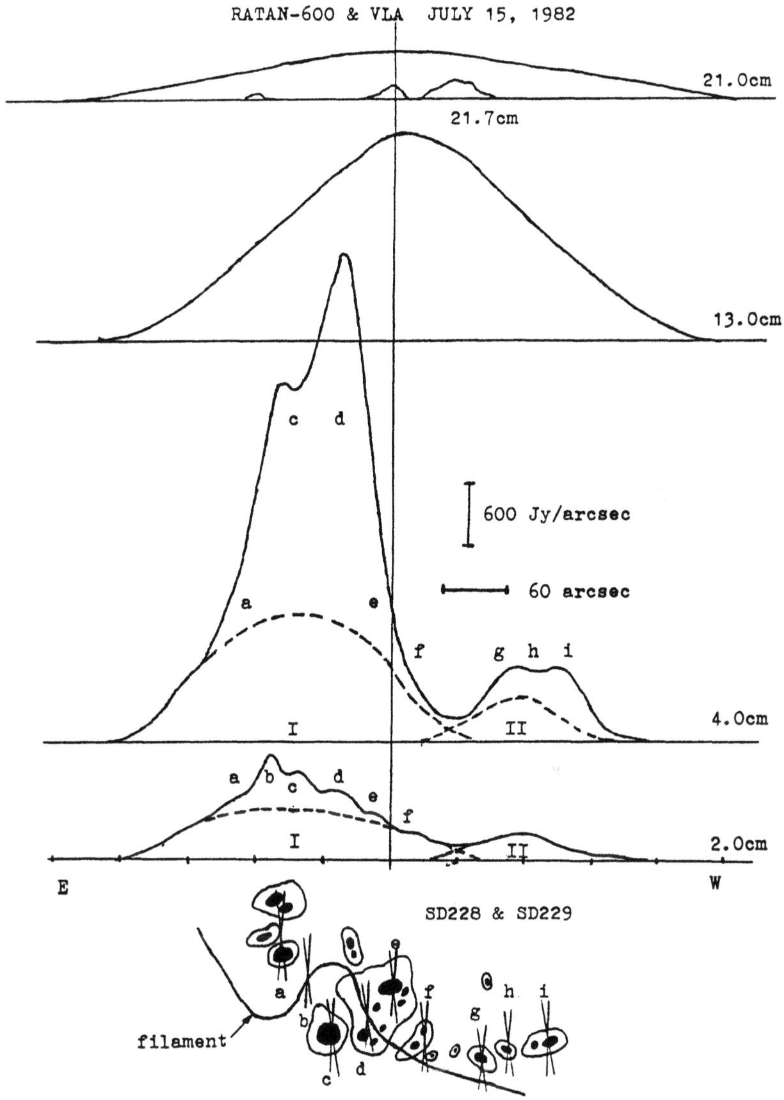

Fig. 2. Radio scans of the active region observed on July 15, 1982, near the center of the solar disk. The flux at $\lambda = 21.7$ cm from the loops observed on the VLA is shown below the upper curve (RATAN-600 scan at $\lambda = 21.0$ cm). At the bottom the optical structure of the sunspots and the H_α filament are presented. The position of the compact radio sources at microwaves from the RATAN-600 scans is shown (marked as a, b, c.....i).

15,000 km). However, their physical connection with the plasma-magnetic structures of the prominences cannot be ignored. These are certainly manifestations of local release of nonthermal energy in the solar atmosphere lasting for several days.

TABLE III

Peculiar sources

1. Peak wavelength $\lambda = 3 - 4\,\mathrm{cm}$
2. Steep spectrum index up to $n = 10$
3. Max. brightness $T_B \approx 10^7$ K
4. Life time (3 to 5 days) coincides with periods of proton flare activity
5. Height and angular size $\approx 10^4$ km
6. Location at neutral line of magnetic fields (filaments)
7. Appear simultaneously with new magnetic flux, proton flares do not destroy them

4. Conclusions

In the conclusion we would like to emphasize that the most effective analysis of the above processes and structures of coronal loops and prominences can be achieved in cooperative observations is made the combination of radio telescopes of aperture synthesis with very high spatial resolution, and instruments like RATAN-600 with good wavelength coverage. Including in the program the modern solar crosses (Nobeyama, Irkutsk) may help in studying the development of the plasma structures. X-ray mapping from space is of special interest to measure temperatures and register acceleration processes independently from radio methods .

Acknowledgements

One of the authors (A.N.Korzhavin) is very thankful to the LOC of the Second SOHO Workshop for support of this work. Part of this work was also supported by a grant of the Russian Fund of Fundamental Investigations No. 93-02-17220.

References

Akhmedov, Sh.B. et al.: 1982, *Solar Phys.* **79**, 41-58
Akhmedov, Sh.B. et al.: 1986, *Aph.J* **301, No. 1, part 1**, 460-464
Akhmedov, Sh.B. et al.: 1987, *Astrofiz.Issled.(Izv.SAO)* **25**, 105-134
Akhmedov, Sh.B. et al.: 1989, *Astrofiz.Issled.(Izv.SAO)* **28**, 111-122
Alissandrakis, C.E. et al.: 1993, *Astron.Astroph.* **270**, 509-515
Bogod, V.M.: 1992, *Astron. Nachr.* **313**, 97-100
Bogod, V.M. and Gelfreikh, G.B.: 1980, *Sol.Phys.* **67**, 29-46
Gelfreikh, G.B.: 1990, *Astron. Nachr.* **311**, 385-390
Gelfreikh, G.B.: 1992, *Workshop SOLERS22*, Donnelly, R. (ed.), Boulder, p. 196-227
Gelfreikh, G.B. et al.: 1987, *Solar Phys.* **108**, 89-97
Korzhavin, A.N. et al.: 1989, *XIII Consultation on Solar Physics: Solar Magnetic Fields and Corona*, Nauka: Novosibirsk, v.2, p. 119-124
Pariysky, Yu.N. et al.: 1976, *Astronomichesky.Zh* **53**, 1017-1025
Vatrushin, S.M. and Korzhavin A.N.: 1989, Somov, B.V. and Fomichev V.V. (eds.), *Fizika Solnechnoy Plazmy*, Nauka: Moscow, p. 100-106

MILLIMETER WAVE RADIATION SOURCES VISIBLE IN SOLAR CORONA

S. URPO and S. POHJOLAINEN
Metsähovi Radio Research Station, Espoo, Finland

and

A. KRÜGER
Astrophysical Institute, Potsdam

Abstract. Coronal millimeter wave sources (CMMS) have been observed at radio frequencies 22-90 GHz (at wavelengths 3-13 mm). The observed CMMS have been classified into three different categories according to their time scale and relation to other wavelength events. The CMMS indicate enhancement of electron density in the corona, as well as magnetic loop structures. The CMMS are evidence of dynamical processes taking place at the lower levels of the corona, propagating into the higher levels.

1. Introduction

The Metsähovi Radio Research Station has performed solar millimeter wave observations with a 14 m single dish for more than 15 years. The observations are not patrol type continuous daily measurements but 1 to 3 weeks observing sessions several times per year. The most frequent observing frequencies have been 37 GHz, 22 GHz, and 90 GHz. Measurements include solar maps and tracking data of active regions. The sensitivity of the receivers is good enough for 0.1 sfu, or 100 K in the temperature scale, resolution. The parameters of solar map measurements at the used frequencies at Metsähovi are presented in Table 1.

2. Classification of coronal mm-wave sources

The measurements have confirmed the theoretical predictions that the main part of the solar millimeter wave radiation originates in levels close to the chromosphere. The typical source height from the photosphere is 1 000 - 3 000 km. However, in rare cases the millimeter source height has been observed to be much larger, up to 100 000 km and more. A coronal radiation source related to an H_α prominence was first noted by Kundu (1972), and later 25 limb sources observed in 1979-1983 were studied in more detail by Urpo et al. (1986). A short review of earlier work and a description of the general characteristics of CMMS will be published by Krüger et al. (1993).

A reliable lower limit for the mm-wave source height can be calculated in the cases of behind-the-limb flares. In other cases the source height is estimated by telescope beam deconvolution, which is not always unambiguous. From our observations it is proposed that these high coronal sources

Space Science Reviews **70**, 199–202.

TABLE I
Parameters of solar map measurements

Frequency GHz	*Wavelength* mm	*Telescope beam size* arc min	*Estimated quiet Sun level* K
11.6	25.9	7.6	12000
22.2	13.5	4.0	9000
36.8	8.2	2.4	7800
77.1	3.9	1.2	7250
87.0	3.4	1.0	7200

can be divided into at least three different categories: events related to big energy releases, events related to post flare magnetic loops, and events in consequence of a strong active region passing the east or west limb of the Sun.

These three categories have different time scales. Millimeter coronal sources connected to big solar flares are dynamical with time scales of minutes and they are related e.g. to coronal mass ejections (CME). The second category has a time scale of few hours and they are related e.g. to long-duration soft X-ray events (LDE) and to hard X-ray (HXR) post-flare arches. The third category has a time scale of several hours. These events are related to the slowly varying component (S-component) of active regions. The highest source heights appear in the first category, the second is an intermediate case, and in the third category the source height is 5 000-20 000 km.

The proposed three categories may be connected to different magnetic structures in the solar corona. In the first category the millimeter coronal source can be related to the top of an expanding magnetic loop structure as was the case in the event of June 20, 1989, described by Graeter and Kucera (1992). The second category can be related to long living post-flare coronal arches visible in coronographic observations in H_α, see Rompolt et al. (1990). The third category can be explained by the quasi-stationary magnetic loop structures present in slowly varying active regions. Many examples of this type with source heights between 5000 and 10000 km have been measured at Metsähovi, by Urpo et al. (1987-1991).

In Figures 1-3 we present examples of coronal mm-wave sources present at Metsähovi solar maps. The off-limb source of January 25, 1991, in Figure 1, is related to a flare. In Figure 2 the mm-wave source of September 19, 1992, is related to an active prominence on the limb. The event of June 20, 1989, in Figure 3, was related to a flare as well as a coronal mass ejection, and was also recorded at X-ray wavelengths.

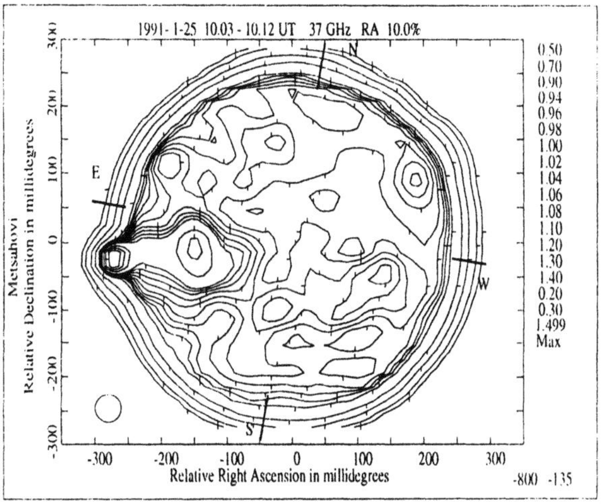

Fig. 1. A solar map measured on January 25, 1991 at 10.03-10.12 UT. The beam size at 37 GHz is 2.4 arc min, shown by the circle in the lower left corner of the plot. The maximum level recorded was 1.499 times the quiet Sun level, which corresponds to the brightness temperature of 11690 K.

3. Conclusions

We conclude that coronal mm-wave sources have been observed at frequencies 22-90 GHz (λ = 3-13 mm). The CMMS are observed to belong to three categories with different time scales: 1) Big events with duration of a few minutes, 2) Post-flare magnetic loops with duration of some hours, and 3) Strong active regions with duration of many hours. The CMMS can be related to big flares, soft X-ray long enduring events (LDE), hard X-ray postflare arches, coronal mass ejections (CME), and active prominences. The CMMS indicate the temporal enhancement of electron density in the corona, magnetic loop structures in the corona, and they can precede dynamical processes taking place at the higher levels of the corona.

References

Graeter, M., Kucera, T.A.: 1992, *Solar Phys.* **141**, 91.

Krüger, A., Hildebrandt, J., Urpo, S., and Pohjolainen, S.: 1993, *Proc. Symp. New Look at the Sun, Kofu, 1993.*

Kundu, M.R.: 1972, *Solar Phys.* **25**, 208.

Rompolt, B., Urpo, S., Pohjolainen, S., Aurass, H., Krüger, A.: 1990, *Publ. Debrecen Obs.* **7**, 206.

Urpo, S., Pohjolainen, S., and Teräsranta, H.: 1987-1991, *Helsinki University of Technology, Metsähovi Radio Research Station Report Series A* **1-7**.

Urpo, S., Krüger, A., and Hildebrandt, J.: 1986, *Astron. Astrophys.* **163**, 340.

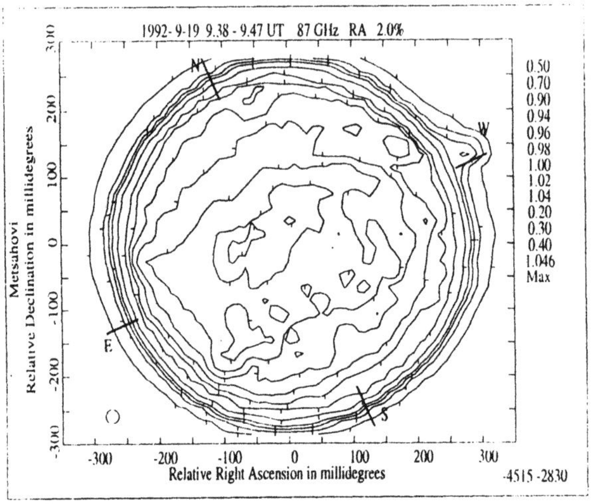

Fig. 2. A solar map measured on September 19, 1992 at 9.38-9.47 UT. The beam size at 87 GHz is 1.0 arc min. The observed brightness temperature of the off-limb source was 3600 K (level 0.5 relative to the quiet Sun).

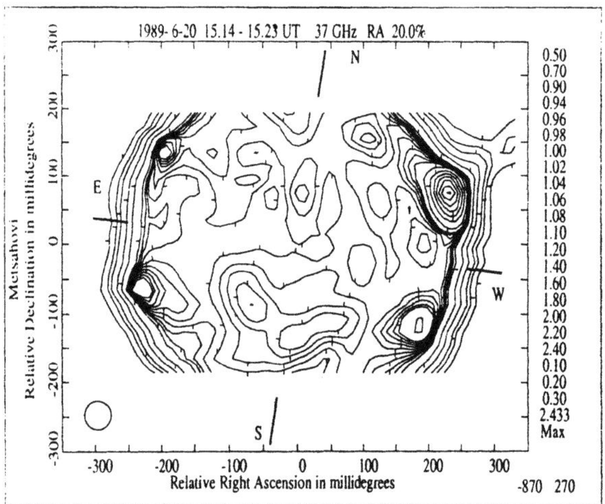

Fig. 3. A solar map measured on June 20, 1989 at 15.14-15.23 UT. The beam size and frequency as in Fig.1. The maximum level recorded was 2.433 which corresponds to the brightness temperature of 18980 K.

STATIONARY FLOWS IN CORONAL LOOPS

S. ORLANDO and S. SERIO*
Istituto ed Osservatorio Astronomico, Palermo
and
G. PERES**
Osservatorio Astrofisico di Catania

Abstract. We present a study of stationary flows in closed solar coronal loops. The hydrodynamic differential equations of plasma flow and energy balance are integrated with algorithms which achieve high reliability. We present here results on the detailed synthesis of loop emission in specific bands and lines, taking into account also non-equilibrium ionization.

Key words: Hydrodynamics - Loop Models - Solar Physics

1. Introduction

Siphon flow loop models provide an important generalization of hydrostatic models of coronal loops: a steady state flow induced in the plasma confined inside a magnetic loop by a pressure difference between the loop footpoints. The properties of these models have been studied by several authors (cf. references in Orlando, Peres and Serio 1993). Yet, for a reasonably complete model, there is not a systematic treatment of the problem nor a full exploration of the parameter space. In order to study in detail stationary siphon flows in coronal loops, we have developed a detailed model in which we take into account a wide set of physical effects including gravity, thermal conduction, radiative losses, heating, which are important for the detailed analysis of the corona and of the transition region. The model is used to tackle a variety of problems related to siphon flows. In this paper we explore also the shocked solutions and perform a detailed synthesis of loop emission in specific lines, taking into account non-equilibrium ionization.

2. Model equations

In our model of siphon flows the coronal loop is semicircular of length $2L$ and constant cross-sectional area. The model equations are

$$\rho v = j = constant \tag{1}$$

$$\rho v \frac{\partial v}{\partial s} = -\frac{\partial p}{\partial s} - \rho g_0 \cos\left(\frac{\pi s}{2L}\right) \tag{2}$$

$$\frac{1}{2}\left(5p\frac{\partial v}{\partial s} + 3v\frac{\partial p}{\partial s}\right) - \frac{\partial}{\partial s}\left(\kappa\frac{\partial T}{\partial s}\right) = E_H - \frac{p^2}{4k_B^2T^2}\Lambda(T) \tag{3}$$

$$\rho = \mu m_H n, \quad \kappa = 9.2\cdot 10^{-7}T^{5/2}, \quad p = 2nk_BT, \quad E_H = E_{H0}\, e^{-s/s_H} \tag{4}$$

* also IAIF, CNR Palermo
** present address: Istituto ed Osservatorio Astronomico, Palermo

Space Science Reviews **70**, 203–206.

where j is the mass flux, ρ the mass density of plasma, v the plasma velocity, s the field lines coordinate, p the pressure, g_0 the solar gravity, κ the thermal conductivity given by Spitzer (1962), T the temperature, E_H the volumetric input power into the solar atmosphere, k_B the Boltzmann constant, $\Lambda(T)$ the radiative losses per unit emission measure (Raymond and Smith 1977), $\mu = 1.257$ is the mean atomic mass, m_H the mass of the hydrogen atom, n the hydrogen number density, L the semilength of the loop, E_{H0} the volumetric input power at the base of the loop and s_H the heating scale length. We have chosen the footpoint temperature to be $3 \cdot 10^4$ K and we have assumed the temperature gradient to be negligible at the base of the upflowing leg. In order to achieve high reliability close to critical points, we solve the fluid equations with two independent and refined codes: the first based on a fourth-order Runge Kutta method with adaptive stepsize control and the second on a Bulirsch Stoer method (Press et al. 1986, Stoer and Bulirsch 1980). Our codes have been checked also by direct comparison of results (the difference between results obtained with the two codes are at most, confined to the 6th decimal place). The results have been compared successfully also to siphon flow models in the literature (Cargill and Priest 1980, Thomas 1988, Montesinos and Thomas 1989, Noci et al. 1989) and (setting $v = 0$) to those of hydrostatic models.

3. Lines synthesis

To synthesize line emission from siphon flow coronal loops, we have to take into account the effect of non-equilibrium ionization because the plasma in the loop moves across the steep temperature and density gradient of the transition region. The continuity equations for ionization species are

$$\frac{d}{ds}(n_i v) = n_e[n_{i+1}\alpha_{i+1} + n_{i-1}c_{i-1} - n_i(\alpha_i + c_i)] \quad (i = 1, ...N) \tag{5}$$

with $(\alpha_1 = c_0 = \alpha_{N+1} = c_N = 0)$ and where n_i is the density of the i-th ionization stage of the element, v the plasma velocity, n_e the electron number density, N the ionization states for the element analyzed, α_i and c_i respectively the collisional and dielectronic recombination coefficients and the collisional ionization coefficients (Summers 1974). As boundary conditions we assume ionization equilibrium at the base of the upflowing leg, as in Noci et al. (1989), Spadaro et al. (1990), Peres et al. (1992). The detailed synthesis of loop emission in specific bands and lines are obtained with the use of ASAP, a powerful presentation system for astrophysical plasmas developed by Reale et al. (1993). We plot in figure (1) the loop plasma emissivity vs temperature in the lines of He II (303.91 Å), Fe IX (171.07 Å), Fe XII (195.12 Å) and Fe XV (284.15 Å) in the upflowing and in the downflowing leg for a loop with $L = 10^{10}$ cm, $p_0 = 1$ $dyne$ cm^{-2}, $v_0 = 6.5 \cdot 10^4$ cm sec^{-1}. These results are of importance for the forthcoming Extreme-Ultraviolet Imaging Telescope (EIT) on board SOHO; with ASAP we can also synthesize any other important optically thin line of interest, including those observed by SUMER and CDS, and any optically thin band in the EUV/X-ray regime.

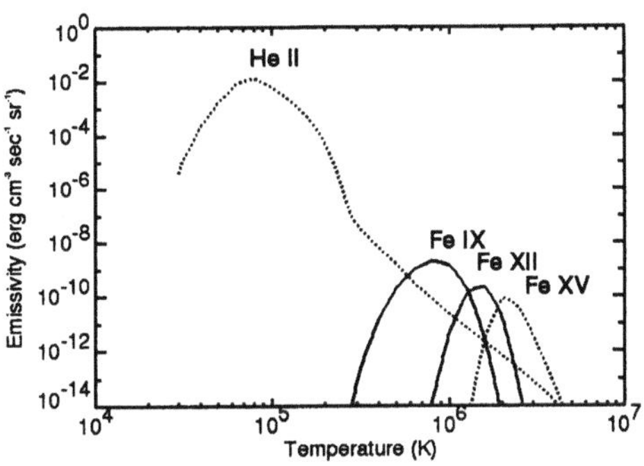

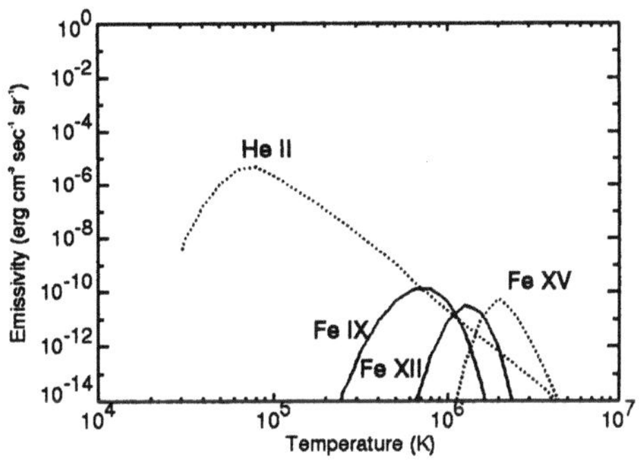

Fig. 1. Total emissivity vs. plasma temperature respectively in the upflowing leg (on the left) and in the downflowing leg (on the right)

4. Lines emission in the presence of a stationary shock

We have explored the effects that shocks can have in the structure of solutions which go across a critical point and are supersonic in the descending leg. The descending plasma flow must undergo a shock before reaching the solar surface (figure (2*a*)). The exploration of this class of solutions has showed that for fixed semilength and base pressure, the shock position depends on the heating of the loop; in addition, there is a range of heating values for which there are two alternative positions where the shock can form. The drop in velocity and density due to the shock does not alter the ionization state at the shock position because, in every case, all the densities (protons, electrons, ions) increase of the same relative amount. Yet, due to the density dependence of the emissivity, there are some differences in the emissivity profiles tied to the shock position. In particular, the emissivity in the downflowing leg changes significantly[1]: the lower is the shock position, the higher is the drop in emissivity and the higher is the emissivity in the subsonic branch (pre-shock); as a consequence, in the downflowing leg the intensity of a spectral line is higher than in a subsonic solution. As an example, for a loop with $L = 4 \cdot 10^9\ cm,\quad p_0 = 1\ dyne\ cm^{-2}$ we report in figure (2*b*), (2*c*), (2*d*) the total emissivity of the lines MgX (625.3 Å), SiXII (499.3 Å) and FeXVI (361.7 Å) vs. the spatial coordinate along the loop s/L for six shocked solutions with different heating E_H and shock point. These considerations suggest that the intensity ratio between the two loop footpoints can supply an interesting diagnostic method to verify the effective presence of coronal shocks.

5. Conclusions

We are presently studying the physics of siphon flows in loops of which we have presented here a subset of results. We are taking care both to study in detail the characteristics of physical solutions and to synthesize accurately the distribution of emission for a wide set of physical cases of interest.

[1] The differences in the upflowing leg are negligible.

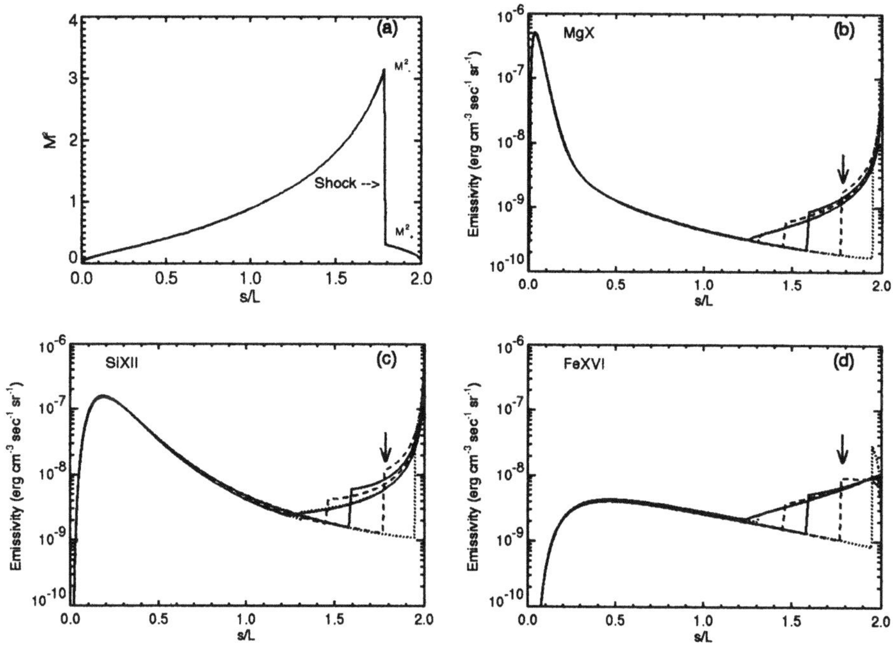

Fig. 2. (a) We report an example of shocked solution in the velocity space for a loop with $L = 4 \cdot 10^9$ *cm* and $p_0 = 1$ *dyne* cm^{-2} and $v_0 = 7.2 \cdot 10^4$ *cm* sec^{-1}. (b) The total emissivity of the lines MgX (625.3 Å), (c) SiXII (499.3 Å) and (d) FeXVI (361.7 Å) vs. the spatial coordinate along the loop s/L. The profiles marked with an arrow pertain to the solution in figure $2a$.

We acknowledge partial support from Ministero dell'Università e della Ricerca Scientifica e Tecnologica, Agenzia Spaziale Italiana and Gruppo Nazionale di Astronomia - CNR.

References

Cargill, P.J., and Priest, E.R. 1980, Solar Phys., **65**, 251.
Montesinos, B., Thomas, J.H. 1989, *Ap. J.*, **337**, 977.
Noci, G., Spadaro, D., Zappalà, R.A., Antiochos, S.K. 1989, *Ap. J.*, **338**, 1131.
Orlando, S., Peres, G., Serio, S. 1993, in preparation.
Peres, G., Spadaro, D., Noci, G. 1992, *Ap. J.*, **389**, 777.
Press, W.H., Flannery, B.P., Teukolsky, S.A., Vetterling, W.T., 1986, *'Numerical Recipes'*, (Cambridge University Press).
Raymond, J. C., Smith, B. W. 1977, *Ap. J. Suppl.*, **35**, 419.
Reale, F., Maggio, A., Ciaravella, A., Peres, G. 1993, this volume.
Spadaro, D., Zappalà, R. A., Antiochos, S. K., Lanzafame, G., Noci, G. 1990, *Ap. J.*, **362**, 370.
Spitzer, L. 1962, *'Physics of fully ionized gases'*, (Interscience tracts of physics and astronomy).
Stoer, J., Bulirsch, R., 1980, *'Introduction to numerical analysis'*, (Springer-Verlag New York Inc.).
Summers, H.P., 1974, internal memorandum 367 of Astrophysical Research Division, Appleton Laboratory, Abingdon, Oxon.
Thomas, J.H. 1988, *Ap. J.*, **333**, 407.

OBSERVATIONAL EVIDENCE FOR NON–EQUILIBRIUM IONIZATION IN THE SOLAR CORONA

D. SPADARO, P. LETO
Osservatorio Astrofisico di Catania, Italy

and

S.K. ANTIOCHOS
Naval Research Laboratory, Washington, DC

Abstract. We investigate whether temperature sensitive EUV line ratios can be used as observational signatures for the presence of non–equilibrium ionization in transition region plasma. We compute the total intensity of some EUV lines of carbon and oxygen expected from coronal loop models with a steady–state flow and which are known to have significant departures from ionization equilibrium, selecting lines whose intensity ratios are useful for deducing the electron temperature in the coronal plasma. We calculate the intensity ratios with and without the approximation of ionization equilibrium, in order to determine the effects of any deviations from equilibrium on the numerical values of the line ratios examined.

Key words: Sun – Corona – EUV lines

1. Introduction

Non-equilibrium ionization may be produced by a variety of processes in the solar corona, for example, by mass flows through the large temperature gradients of the transition region or by impulsive heating and cooling. Any deviation from equilibrium ionization would have a strong effect on the radiation from the corona and on the interpretation of solar observations; hence, it is important to determine observational signatures of non-equilibrium. We propose that temperature sensitive line ratios can be used as such signatures.

We examine here the intensity ratios of some EUV lines of carbon and oxygen. These line ratios are calculated for four coronal loop models which have a steady flow and which are known to have significant departures from ionization equilibrium.

2. Loop models and ion densities

We have adopted the simple steady–state siphon flow model formulated by Antiochos (1984). This model describes the subsonic flow of fully ionized plasma from one loop footpoint to the other driven by a pressure difference between the two chromospheric regions at the loop ends.

Noci et al. (1989) solved the fluid equations for this model, choosing the footpoint temperature to be 3×10^4 K. They computed four loop models,

Space Science Reviews **70**, 207–209.

varying the boundary conditions so as to span a wide set of values for temperature, density and velocity at the top of the loop. After determining the electron density, the flow velocity and the plasma temperature profile along the loop from the solution of the fluid equations, Noci et al. then calculated the number densities of all the ions of carbon along each loop model, with and without the approximation of ionization equilibrium. Using the same procedure, Spadaro et al. (1990) calculated the number densities of all oxygen ions for the same set of loop models. We refer the reader to Noci et al. (1989) and Spadaro et al. (1990) for a detailed description of the calculation of the loop models and ion densities.

The plasma in loop 1 can be considered in ionization equilibrium, while considerable deviations from equilibrium occur in the other loops. The densities calculated by the above authors constitute the basis of the intensity calculations performed in this work.

3. Line intensities and line ratios

We calculated the total intensities of the considered EUV spectral lines emitted by the upflowing and the downflowing legs for each loop model, using the ion densities computed both with and without the assumption of ionization equilibrium. It should be noted that coronal loops are often not spatially resolved in EUV so that only the total radiation emitted by the whole loop is observed. Therefore, in order to have an immediate comparison with these observations, we have also computed, for each loop model and for each line analyzed, the arithmetic mean between the intensity values obtained for both legs of the loop. More details on the calculations of the line intensities are reported in Spadaro, Leto & Antiochos (1993).

The line intensities so obtained have been used to calculate, both with and without the approximation of ionization equilibrium, the following line ratios: C IV I(1548.2 Å)/I(312.4 Å), O IV I(789.4 Å)/I(554.4 Å), O V I(629.7 Å)/I(172.2 Å), O VI I(1031.9 Å)/I(173.0 Å) and O VI I(1031.0 Å)/I(150.1 Å).

Note that in each case we consider the ratio of a line with lower excitation energy to a line with higher energy. The C IV ratios obtained for the upflowing and the downflowing legs of each loop, as well as for the intensities averaged between the two legs, are reported in Table 1.

As expected, the line ratio in loop 1 is negligibly affected by the approximation of ionization equilibrium. For the other loops, however, non–equilibrium effects are important, and we notice significant changes for the C IV line ratio. We find that the non–equilibrium line ratios are generally smaller than the equilibrium ones, both in the upflowing and in the downflowing leg of the loops. Therefore, the average ratios for the two legs of the loop are also smaller in non–equilibrium conditions. The behaviour of the line ratios obtained for the other examined ions is similar. We refer the reader to

TABLE I
C IV Intensity Ratios

	EQUILIB.			NON-EQUILIB.		
	up	*down*	*loop* [a]	*up*	*down*	*loop* [a]
Loop1	130.63	116.54	123.90	133.00	114.00	124.26
Loop2	130.92	71.76	111.54	118.63	45.29	108.24
Loop3	148.59	134.82	146.56	71.06	32.63	70.62
Loop4	160.93	124.32	155.03	44.26	10.07	43.85

([a]) Ratio of the intensities averaged between upflowing and downflowing leg.

Spadaro, Leto & Antiochos (1993) for a detailed discussion on the behaviour of the line ratios.

4. Conclusions

The main conclusion from our calculations is that line ratios can be significantly modified by the deviations from ionization equilibrium occurring in coronal loops with a steady-state subsonic flow from one footpoint to the other. The principal effect of non-equilibrium is a reduction of the line ratios, both in the upflowing and in the downflowing leg of the loops. This result is quite general, but depending on the temperature, density and velocity profile along the downflowing leg of the loop, it is possible to have an increase of the line ratio there due to lines forming at lower temperatures than in equilibrium.

We believe that results of our calculations are useful as predictors for future observations with high spatial resolution, such as those that will be obtained by the SOHO mission. With sufficient resolution it should be possible to observe individual legs of large coronal loops and, thereby, search for systematic differences in the line ratios. Of the ratios that we have examined, non-equilibrium effects are clearly strongest in the CIV ratio for the downflowing leg of loop 4. The deviation from the equilibrium ratio is more than an order of magnitude, and should be easily detectable by the SUMER and CDS experiments on SOHO. On the basis of our work here, we predict that the C IV 1548/312 ratio should be well below its equilibrium value of ~ 120 wherever the lines exhibit strong redshifts.

References

Antiochos, S.K. 1984, ApJ, 280, 416.
Noci, G., Spadaro, D., Zappalà, R.A., & Antiochos, S.K. 1989, ApJ, 338, 1131.
Spadaro, D., Leto, P., & Antiochos, S.K. 1993, ApJ, in press.
Spadaro, D., Zappalà, R.A., Antiochos, S.K., Lanzafame, G., & Noci, G. 1990, ApJ, 362, 370.

ASAP: A SYSTEMATIC APPROACH TO PLASMA SPECTRAL SYNTHESIS

F. REALE, A. MAGGIO and A. CIARAVELLA
Istituto e Osservatorio Astronomico di Palermo, Palazzo dei Normanni, 90134 Palermo, Italy

and

G. PERES
Osservatorio Astrofisico, Città Universitaria, 95125 Catania, Italy

Abstract. ASAP (*Analysis System for Astrophysical Plasmas*), developed at the Osservatorio Astronomico di Palermo, is a package of procedures based on IDL. It is aimed at the detailed presentation of theoretical models of astrophysical plasmas and the accurate comparison with observational data, up to fittings of specific observations. It is also useful for the prediction of specific observations, and for the simulation of the expected performances of forthcoming instruments, for instance those on board SOHO.

Key words: Hydrodynamics - Plasmas - Methods: data analysis - Ultraviolet: general - X-rays: general

1. Introduction

The group at the Osservatorio Astronomico di Palermo has been actively involved in studying astrophysical environments as diverse as solar and stellar coronae, supernova remnants, and accretion flows in galactic halos. Various models of astrophysical plasmas have been developed, including a hydrostatic model of coronal loop (Serio et al. 1981), the Palermo-Harvard hydrodynamic model (Peres et al. 1982), a hydro-code to study winds and cooling flows (Peres et al. 1989), the Palermo 2-D hydrodynamic model (Reale et al. 1990), and a siphon flow loop model (Orlando et al. 1993).

Our approach is to derive the evolution of distributions of plasma temperature, density, and velocity with these models, to compute the emission according to an appropriate radiation model, and to compare the model results with observations, usually at X-ray or UV wavelengths.

Earlier observations with spatial and spectral resolution made for instance with the Skylab S-054 and S-055 experiments (Vaiana et al. 1977; Reeves et al. 1977) have shown the value of this approach. The interpretation of data from new solar missions, like Yohkoh (Ogawara et al. 1991, Culhane et al. 1991), NIXT (Golub et al. 1990), and the forthcoming SOHO (Domingo 1988) will certainly benefit from this methodology.

2. ASAP concept

ASAP has been developed mostly to achieve presentation of model results, spectral synthesis, and direct comparison with observational data. The ASAP concept is schematically explained by the block diagram in Fig. 1. The arrows indicate the information flow needed to perform any of the ASAP tasks. Starting points for

Space Science Reviews **70**, 211–214.

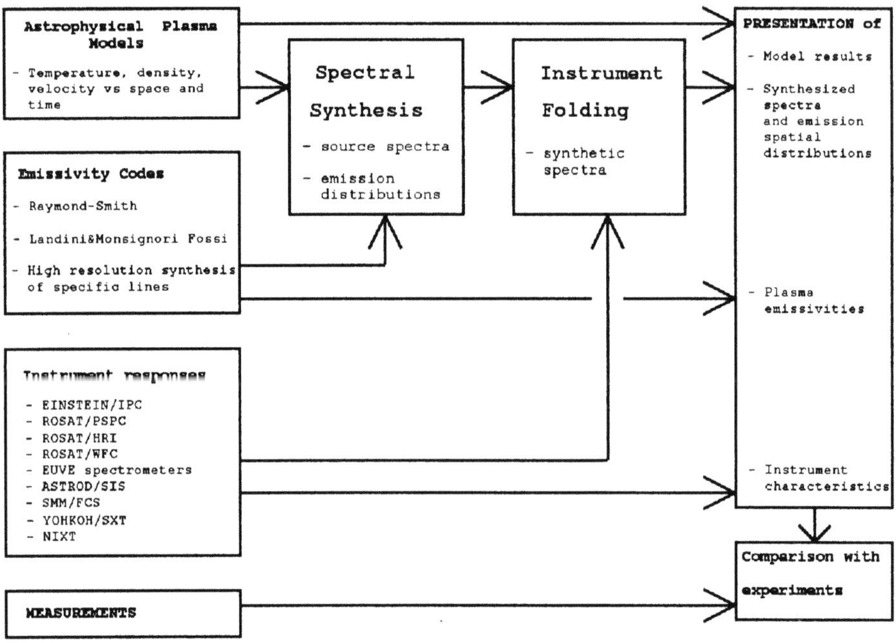

Fig. 1. Block diagram of ASAP in its present configuration. The arrows indicate the information flow needed to perform any of the ASAP tasks. Intermediate results can be scrutinized and displayed at any stage of the analysis.

the analysis are data arrays of distributions of plasma quantities computed with models, emissivity functions, and instrument characteristics. Partial results can be displayed at any stage. The final goal is to produce model results which can be directly compared with real observations from space or even laboratory experiments.

The models of emissivity of an optically thin plasma in thermal equilibrium, typically are functions of the plasma temperature, the photon energy (or wavelength), and the element abundances, and include contributions from bound-bound, free-bound, free-free, and two-photon processes.

The plasma emissivity can be computed in real time after choosing the appropriate energy band, energy resolution, temperature resolution, and plasma abundances, by running external emissivity codes through an ASAP driver – as it is the case with the Raymond thermal model (Raymond & Smith 1977; Raymond 1989) – or simply by loading a precomputed emissivity table (e.g. the emissivity tables of Landini & Monsignori Fossi, 1990). We intend to keep the emissivity tables updated, if possible, by incorporating revisions which may be released in the future, and to include new emissivity models if they will be made available. At the moment ASAP includes also emissivity tables for some X-ray lines or bands observed with specific instruments.

The synthetic spectrum is computed by folding the incident spectrum through the instrument effective area. Several effective areas are available in ASAP and others can be easily included, so that ASAP can be applied to future instruments, for instance SOHO/EIT. A direct comparison with the observations, or even with laboratory measurements, is possible after having converted the synthetic spectrum to the detector units. ASAP can take into account specific features of many detectors, such as the energy binning or the scale of the spatial pixels.

As an example we report in Fig. 2 models of coronal loop of magnetically confined hot plasma, of the kind commonly observed on the Sun with X-ray telescopes, as they would be detected in three (Fe IX, Fe XII and Fe XV) of the lines to which SOHO EIT will be sensitive. We have used the model of Serio et al. (1981) to compute the hydrostatic configuration of plasma in steady energy balance under the effect of heat conduction, radiative losses and a phenomenological heating term, and then the distribution of the emission per unit volume along the loop was obtained by means of ASAP routines.

With ASAP we already have an integrated software package which allows to present, analyze and interpret the results of models of astrophysical plasmas. For more details on this tool see Maggio et al. (1993). The approach aims also at filling the traditional gap between theory and experiment, by directly connecting the model of an astrophysical object with a system to synthesize its emission at a highly realistic level of detail, and by folding this emission with the response of existing (or planned) instruments. With ASAP we already go from the plasma fluid equations, to the determination of measurements in the focal plane of the instruments.

Acknowledgements

The authors acknowledge partial support from Ministero dell'Universitá e della Ricerca Scientifica e Tecnologica, the Italian Space Agency, and GNA-CNR.

References

Culhane J.L., et al., 1991, Sol. Phys. 136, 89
Domingo V., 1988 The SOHO Mission. ESA SP-1104, p. 49
Golub L., et al., 1990, Nature, 344, 842
Landini M., and Monsignori-Fossi B.C., 1990, A. & A. Suppl., 82, 229
Maggio A., Reale F., Ciaravella A., Peres G., 1993, A. & A., submitted
Ogawara Y., Takano T., Kato T., Kosugi T., Tsuneta S., Watanabe T., Kondo I., Uchida Y., 1991, Sol. Phys. 136, 1
Orlando S., Peres G., and Serio S., 1993, (in preparation)
Peres G., Rosner R., Serio S., Vaiana G.S., 1982, ApJ 252, 791
Peres G., Rosner R., and Bodo G., 1989, Mem. Soc. Astron. Ital., 139, 60
Raymond J.C., and Smith B.W., 1977, ApJS, 35, 419
Raymond J.C. 1989, private communication
Reale F., Peres G., Serio S., 1990, Nuovo Cimento B, 105, 1235
Reeves, E.M., Huber M.C.E., and Timothy J.C., 1977, Appl. Optics, 16, 837
Serio S., Peres G., Vaiana G.S., Golub L., Rosner R., 1981, ApJ 243, 288
Vaiana G.S., VanSpeibroeck L., Zombeck M.V., Krieger A.S., Silk J.K., and Timothy A.P., 1977, Space Sci. Instr., 3, 19

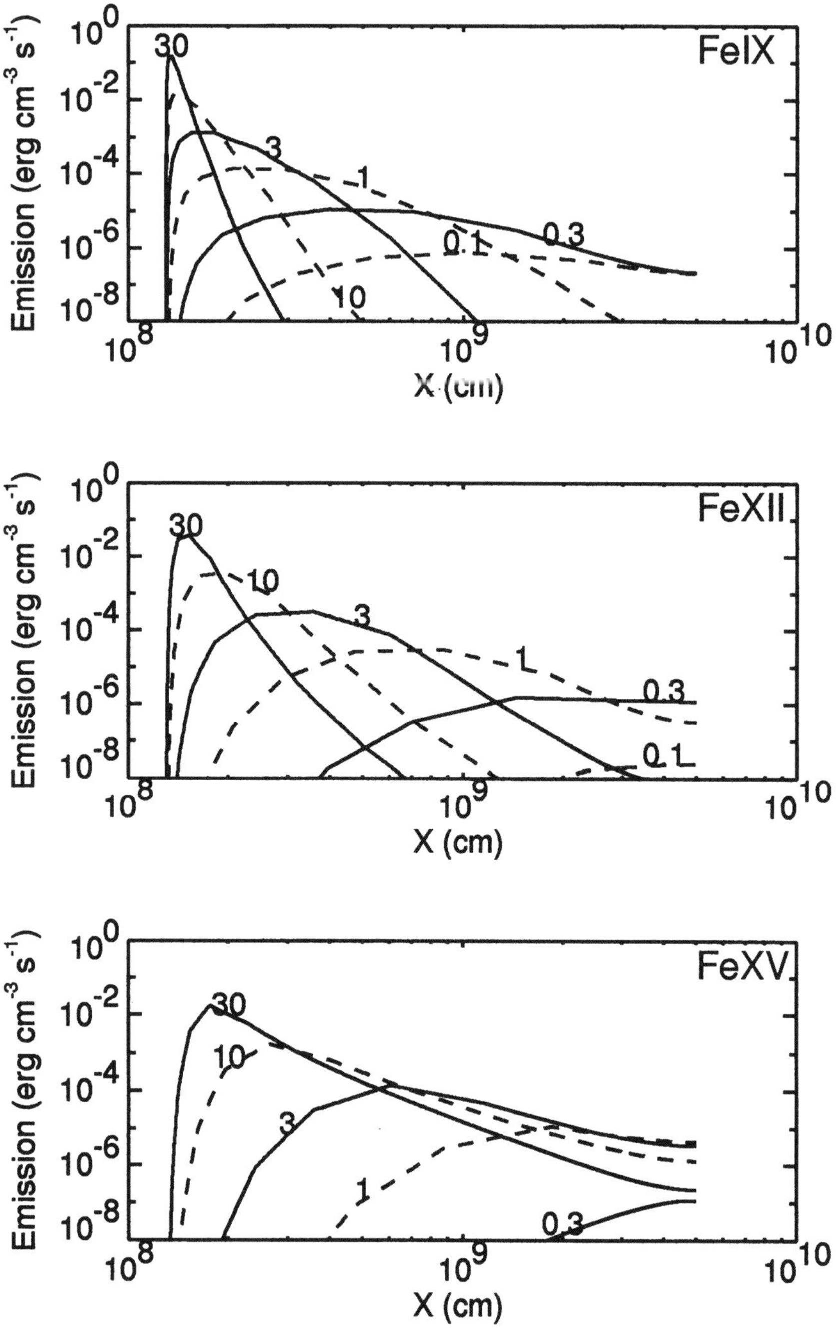

Fig. 2. Distribution of the emission in three lines (Fe IX, Fe XII, and Fe XV) to which SOHO/EIT will be sensitive for hydrostatic loops with half-length $L = 5\ 10^9$ cm and heating uniform along the loop. Only half of the loop is reported (the other one being simmetric). The labels of each curve yield the loop base pressure (dyne cm^{-2}). The model with the lowest pressure lies below the threshold in Fe XV.

ERUPTIVE PROMINENCES AS SOURCES OF MAGNETIC CLOUDS IN THE SOLAR WIND

V. BOTHMER and R. SCHWENN
Max-Planck-Institut für Aeronomie, 37189 Katlenburg-Lindau, Germany

Abstract. Large amounts of coronal material are propelled outward into interplanetary space by Coronal Mass Ejections (CMEs). Thus one might expect to find evidence for expanding flux ropes in the solar wind as well. To prove this assumption magnetic clouds were analyzed and correlated with Hα-observations of disappearing filaments. When clouds were found to be directly associated with a disappearing filament, the magnetic structure of the cloud was compared with that of the associated filament. Additionally the expansion of magnetic clouds was examined over a wide range of the heliosphere and compared with the expansion observed for erupting prominences.

Key words: Erupting Prominences – Magnetic Clouds

1. Introduction

Solar prominences, especially quiescent ones, often erupt in the shape of large twisted cylindrical flux tubes (e.g. Figure 5 in Priest, 1990). It is assumed that the helical bundles or loop-like threads of plasma observed in these prominences outline the magnetic field structure. There is a close association between Coronal Mass Ejections (CMEs) and erupting prominences. Cases have been observed where ascending prominences maintained their intrinsic structure up to a height of ca. 10 R_S (Sheeley et al., 1981).

Moving outward from the solar atmosphere into the solar wind, we find the relationship between CMEs and magnetic clouds now being well established (e.g. Burlaga, 1991). Recent results indicate that magnetic clouds can best be interpreted as large curved cylindrical magnetic flux ropes (e.g. Bothmer and Schwenn, 1994). Yet, what causes the flux rope structure remains unclear.

From the reviewed aspects, it seems plausible to assume a relationship between erupting prominences and magnetic clouds. Wilson and Hildner (1986) and Marubashi (1986) indeed found evidence for this assumption but hitherto in-situ observed clouds have not been directly correlated with Hα-observations of erupting prominences (disappearing filaments).

2. Direct Correlations of Magnetic Clouds with Disappearing Filaments

The "Catalog of Solar Filament Disappearances 1964-1980" (UAG-100, WDC-A) provides complete observations of quiescent filament disappearances (disparition brusques, DBs). Disappearance times, positions and magne-

Space Science Reviews **70**, 215–220.

tic polarities of the DBs are included. These observations and the solar wind measurements of the two Helios spacecrafts during 1975-1980 between 0.3 and 1 AU were used to prove an association of in-situ observed clouds with DBs.

Hourly averages of the Helios 1/2 plasma and IMF data were investigated for magnetic cloud events. Only clouds for which a minimum variance analysis of the magnetic field data revealed an ideal flux rope structure were selected (e.g. Bothmer and Schwenn, 1992). To avoid cases where clouds might have been caused by DBs that have gone undetected due to limb darkening or that occurred right behind the solar limbs, only events where the Helios spacecraft were within ±60° east (west) of the solar central meridian were considered. From these criteria, a set of 10 clouds was found. Assuming a disappearing filament propagates with constant speed from the sun to the spacecraft's position we examined whether DBs had occurred inside time-windows that were calculated from the minimum and maximum measured proton speeds inside each cloud. We only considered associations of clouds and DBs as confident if spacecraft and filament were separated by less than 20° in solar longitude. With these criteria five of the ten clouds were found to be uniquely associated with a DB (e.g. Figure 1). The individual events are listed in Table 1. Taking into account that the daily rate of DBs between 45° E and 45° W was only 0.1-0.4 during 1975-1980, that some DBs still might have gone undetected and that filaments of active regions have not been concerned here, the found associations can be regarded as evidence for a close relationship between clouds and erupting prominences.

3. Flux Rope Structure of Clouds and Filaments

Quiescent prominences lie above magnetic inversion lines that separate large scale solar magnetic fields of opposite polarity. The magnetic field lines that overlie a prominence axis become sheared due to the sun's differential rotation (e.g. Martens and van Ballegooijen, 1990). From the orientation and polarity of filaments, one might assume prominences to possess the four different flux rope types sketched in Figure 2, which have been found for magnetic clouds (Bothmer and Schwenn, 1994). Indeed in four of the five prominence associated cloud events, the cloud's and the filament's flux rope type were the same, with three different types being observed (Table 1).

4. Expansion of Prominences and Magnetic Clouds

Using Helios 1/2, Voyager 1/2 and Pioneer 10 hourly averages of plasma and IMF data, the radial expansion of magnetic clouds between 0.3 and 4.2 AU has been examined (Figure 3). Previous results from Voyager 1/2 observations between 2 and 4 AU have also been included (Burlaga and Behannon,

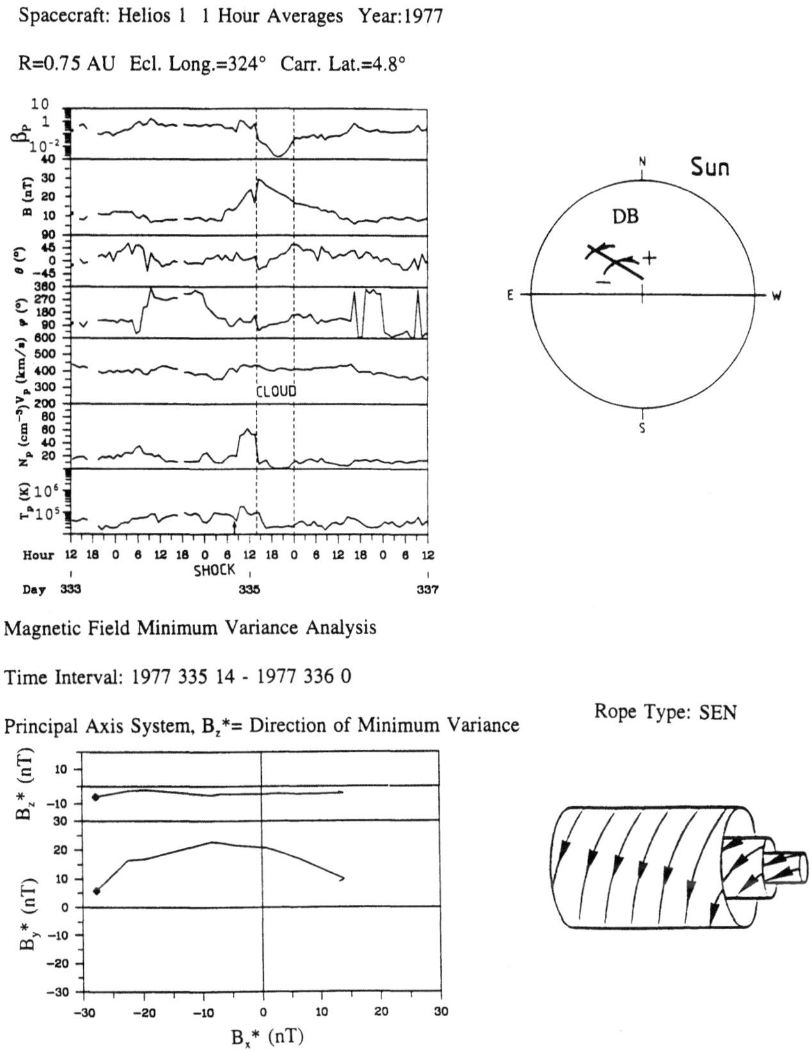

Fig. 1. A magnetic cloud observed by Helios 1 in December 1977 (dashed lines) and the associated disparition brusque (DB). The cloud was identified from the internal magnetic field rotation. Left part: One hour averages of plasma-β_P (calculated from the proton parameters), magnetic field magnitude B, elevation angle θ (90° corresponds to a northward directed IMF) and azimuth angle φ in the ecliptic plane (90° corresponds to an eastward directed IMF), proton bulk speed V_P, proton number density N_P and proton temperature T_P. Below: Results of the minimum variance analysis of the IMF data (DOY 335, 14 UT - DOY 336, 00 UT). The smooth rotation of the magnetic field in the plane of maximum variance is characteristic for magnetic clouds. Right part: Sketch of the filament's position and magnetic polarity as viewed from Helios. The lower right part shows the expected flux rope type of the filament. If the flux rope passes Helios, a south (S) to north (N) turn of the magnetic field vector should be observed, with an eastward (E) direction on the axis of the flux rope.

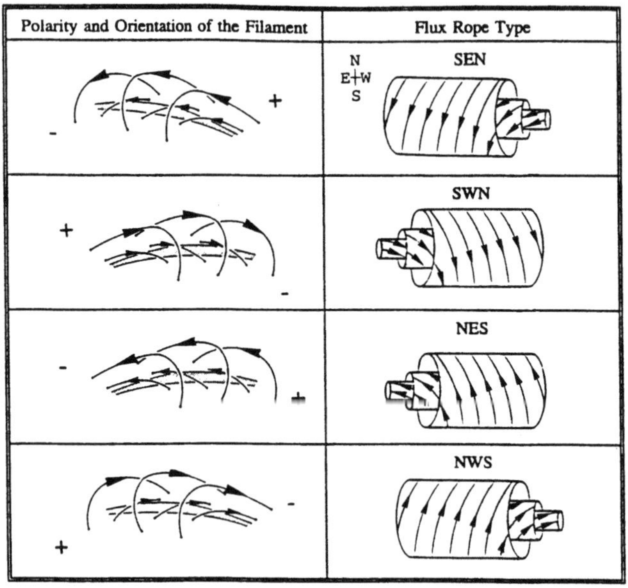

Fig. 2. Diagram sketching the magnetic field configuration of prominences and the expected magnetic flux rope types. The abbreviations have been chosen according to the directional changes of the magnetic field vector that a spacecraft would observe for the flux rope passages (for simplicity it is assumed that the flux rope's axis lies in the ecliptic plane (e.g. SEN: south (S) - north (N) turn of the magnetic field vector, being eastward (E) directed on the rope's axis; W means west).

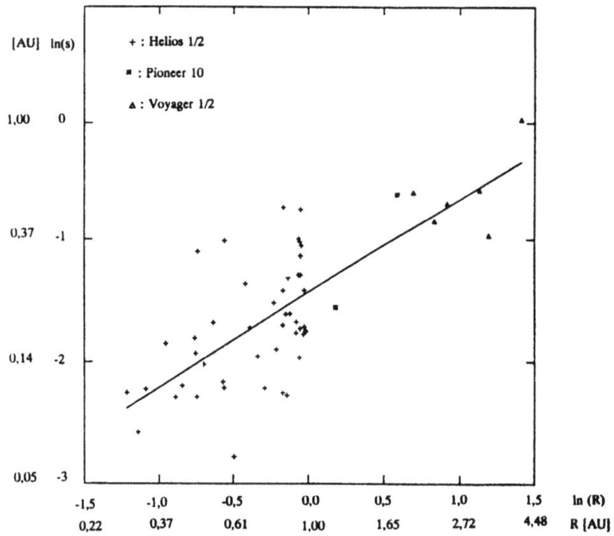

Fig. 3. Double logarithmic plot of the radial sizes, s, of magnetic clouds observed by Helios 1/2 (+), Voyager 1/2 ($\triangle$), Pioneer 10 ($*$) between 0.3 and 4.2 AU, vs. solar distance R. The straight line was obtained by linear regression: $s(R) = 0.24 \times R^{0.78 \pm 0.10} (R \text{ in AU})$.

TABLE I

Association between magnetic clouds observed with Helios 1 and DBs. The columns denote from left to right: Event number, cloud event (Year:Month:Hour), spacecraft (Helios 1/2), position of Helios, minimum and maximum measured proton bulk speeds inside the cloud, solar time-event-window calculated from V_{min} and V_{max} (Month:Day:Hour), associated DB (Month:Day:Hour, uncertainty in hours), position of the filament, polarity of the filament and expected flux rope type, in-situ observed flux rope type of the magnetic cloud.

No.	Cloud Y:M:D	SC	R [AU] Long., Lat. [°]	V_{min} V_{max} [km/s]	Time-Window M:D:H	DB M:D:H	Position [°]	Polarity; Rope Type	Cloud Type
1	75:03:04	1	0.39 W30, -7	418 507	03:03:00 03:03:07	No			
2	77:12:01	1	0.75 E36,+5	401 431	11:28:07 11:28:12	11:28:06 ±09	N60E65 N50E33	−\+ SEN	SEN
3	78:01:17	2	0.98 E09, -4	305 340	01:11:10 01:12:00	01:11:09 ±12	S37W01 S10W17	−/+ NES	NES
4	78:01:29	2	0.98 E13, -5	475 549	01:25:21 01:26:09	01:26:06 ±09	N55E25 N39E08	−\+ SEN	SEN
5	78:02:06	2	0.98 E16, -5	367 444	02:02:00 02:02:19	No			
6	78:02:16	2	0.95 E19, -6	385 491	02:11:19 02:12:17	02:11:15 ±09	S39E45 S37E26	−/+ NES	SWN
7	78:03:02	1	0.87 E55, -4	376 502	02:26:00 02:27:00	No			
8	78:04:02	2	0.61 E17, -7	420 441	03:30:12 03:30:15	03:30:03 ±12	S60E39 S45E09	+/− SWN	SWN
9	79:04:03	2	0.68 E27, -7	389 496	03:31:00 03:31:16	No			
10	79:12:31	2	0.84 E15, -1	375 442	12:27:02 12:27:16	No			

1982). The radial size, s, of a cloud was calculated as $s = <V_P> \times t$, where $<V_P>$ is the cloud's average proton speed and t its duration. The boundaries of the clouds were determined from the time intervals of the rotation of the magnetic field vector in comparison with the plasma-β slopes that were usually a minimum inside clouds. From linear regression we found that a cloud's radial size, s, increases with distance from the sun as:

$$s(R) = 0.24 \times R^{0.78\pm0.10} \qquad (R \text{ in AU}) \tag{1}$$

The average cloud size of 0.24 at 1 AU given by equation (1) is in agreement with previous cloud observations at 1 AU (Klein and Burlaga, 1982). From coronagraph observations, Mouschovias and Poland (1978) found the diameter d of an erupting prominence to increase as $d \sim R^{0.8}$, which is in agreement with equation (1). However MacQueen and Cole (1985) found d to increase rather as $R^{0.5}$. Observations from the new coronagraphs onboard of SOHO will certainly help further clarify the evolution of eruptive prominences.

5. Results and Conclusions

From a direct correlation of magnetic clouds with Hα-observations of solar disappearing filaments, we found with confidence that five of ten magnetic clouds were associated with a disparition brusque. In four of the five cases the magnetic flux rope structure of the cloud was the same as that of the associated filament. The observed similarity in the expansion of magnetic clouds and erupting prominences further indicates that magnetic clouds might result from expanding solar prominences. We conclude from these results that eruptive prominences are the most probable sources of magnetic clouds in the solar wind.

Acknowledgements

The authors would like to thank Dr. J. Cooper at NSSDC for his help in using the COHO solar wind data base.

References

Bothmer, V., R. Schwenn: 1994, 'Structure and Evolution of Magnetic Clouds', to be submitted to *Solar Physics.*

Bothmer, V., R. Schwenn: 1992, 'Magnetic Cloud Observations By The Helios Spacecraft', in *Solar Wind Seven*, ed. by E. Marsch and R. Schwenn, Pergamon, Oxford, 599.

Burlaga, L.F., K.W. Behannon: 1982, 'Magnetic Clouds: Voyager Observations between 2 and 4 AU', *Solar Phys.* **81**, 181.

Burlaga, L.F.: 1991, 'Magnetic Clouds', in *Physics of the Inner Heliosphere Vol.II*, ed. by R. Schwenn and E. Marsch, Springer Verlag, Berlin/Heidelberg, 1.

Klein, L.W., L.F. Burlaga: 1982, 'Interplanetary Magnetic Clouds At 1 AU', *JGR* **87**, 613.

MacQueen, R.M., D.M. Cole: 1985, 'Broadening Of Looplike Solar Coronal Transients', *Astrophys. J.* **299**, 526.

Martens, P.C.H., A.A. van Ballegooijen: 1990, 'Helical Flux Ropes In Solar Prominences', *AGU Geophysical Monograph* **58**, 337.

Marubashi, K.: 1986, 'Structure of the interplanetary magnetic clouds and their solar origins', *Adv. Space Res.* **6**, 335.

Mouschovias, T.C., A.I. Poland: 1978, 'Expansion and broadening of coronal loop transients: A theoretical explanation', *Astrophys. J.* **220**, 675.

Priest, E.R.: 1990, 'The Equilibrium Of Magnetic Flux Ropes', *AGU Geophysical Monograph* **58**, 1.

Sheeley, N.R. Jr., D.J. Michels, R.A. Howard, M.J. Koomen: 1981, 'The great solar eruption of May 24, 1979', *EOS* **62**, 153.

Wilson, R.M., E. Hildner: 1986, 'On the Association of Magnetic Clouds With Disappearing Filaments', *J. Geophys. Res.* **91**, 5867.

WORKING GROUP 2: LOOPS AND PROMINENCES

E.R. PRIEST
Mathematical and Computational Sciences Department, The University, ST ANDREWS, U.K., KY16 9SS

1. Introduction

We had a lively session on the subjects of prominences and loops stimulated by excellent reviews earlier in the day on Yohkoh observations by Keith Strong, the theory of loop flows by Spiro Antiochos and how SOHO can observe loop flows by Ester Antonucci. In addition an important background was provided by the contributed talks on SXT loops by E. Hiei, prominence fine structure by B. Schmieder, transition region redshifts by P. Brekke, magnetic buoyancy by B. Somov and microwaves from loops by A. Korzhavin.

Before an open discussion on the main unanswered questions about prominences and loops and how SOHO might answer them, brief summaries of the posters were given and the posters were studied during a coffee break. The key points about the posters were as follows. Ester Antonucci has developed diagnostics of flows from current sheets, both on the limb and the disc. Volker Bothmer has found that erupting prominences (and presumably their overlying coronal mass ejections) become magnetic clouds in the solar wind with a flux rope structure. Howard Garcia has deduced from PHOBOS and GOES data that compact flares lie below 50Mm in height whereas long-duration events are above 100-150 Mm with upflows in excess of 120 $km\ s^{-1}$. Viggo Hansteen has showed that transition-region redshifts are unlikely to be caused by the return of heated spicules. Ira Kim has discovered that active-region prominences have fields of 100-400G which are independent of height and of mixed polarity. Pavel Kotrc has suggested that energy transport from six-million degree SXT loops to Hα regions may be by soft X-radiation rather than thermal conduction. Natalya Kyrie has discovered an explosive disruption of a current sheet in a laboratory experiment. Giovanni Peres has developed a numerical model for siphon flow and the resulting line emission. Matt Penn has made coronal electron density measurements from the Fe XIII line in the July 1991 eclipse. Daniele Spadaro has shown that nonequilibrium ionisation due to mass flows can reduce line ratios by an order of magnitude (C IV being the most sensitive). Tatiana Stepanova has constructed a model of a coronal mass ejection as a magnetic loop that loses equilibrium when the underlying magnetic field or interior plasma temperature increases. Seppo Urpo has discovered mm radio emis-

Space Science Reviews **70**, 221–230.

sion at 22-90 GHz from CME's and flare loops at heights of 100Mm. Fabio Reale described ASAP, a systematic numerical package based on IDL. Keith Strong described the proposed Explorer mission known as TRACE.

2. Quiescent Prominences

2.1 Basic Properties

The basic properties of quiescent prominences are as follows (Tandberg-Hanssen, 1974; Priest, 1989). Their dimensions are typically 60-600Mm long, 10-100Mm high and 4-15Mm wide. Their densities, temperatures and magnetic field strengths are $10^{16} - 10^{17} m^{-3}$, 5000-8000K, 5-10G. They consist of many fine vertical threads of widths 0.3-1Mm and showing rapid UV variations in intensity and velocity (Schmieder, 1993). Prominences lie above polarity inversion lines in the photospheric magnetic field. Most quiescent prominences are of inverse polarity and are located at high latitudes with heights in excess of 30Mm (Kim, et al, 1988). The magnetic field inside the prominence is horizontal and inclined at 20 degrees to the axis. They migrate slowly towards the poles and Yohkoh SXT observations reveal an overlying arcade of coronal loops (Strong, 1993; Hiei, 1993). By comparison active-region or plage prominences have either normal or inverse polarity (Kim, 1989) and are at low latitudes with heights lower than 30Mm; their magnetic fields are 20-100G and they are smaller than quiescent prominences by about a factor of four; also their densities are in excess of $10^{17}\ cm^{-3}$.

The formation of prominences is thought to be either by condensation or injection (Antiochos and Klimchuk, 1991; Wu, et al, 1991). The classical models for prominence structure were proposed by Kippenhahn and Schluter (1957) and Kuperus and Raadu (1974). More recently a Twisted Flux Tube Model was put forward by Priest, Hood and Anzer (1989) and developed by Demoulin and Priest (1989) and Ridgeway et al (1993). The model begins with a large scale flux tube whose twist gradually increases. Eventually a dip forms in field lines near the tube summit and the prominence forms if the thermal conditions are also favourable. As the twist increases, the prominence increases in length until the twist is so large that the prominence loses equilibrium and erupts. The increase in twist is most likely caused by flux cancellation (Van Ballegooijen and Martens, 1990) and a slightly different way of forming the dip in the flux tube has been suggested by Antiochos et al (1993).

The prominence itself is highly fibril in nature (Engvold, 1976) and so both radiative transfer (Vial, 1986; Heinzel et al, 1988) and magnetohydrostatic models (Ballester and Priest, 1989) have been set up.

At the end of their lives many prominences erupt (Bothmer, 1993; Stepanova, 1993). The basic cause may well be the onset of non-equilibrium: Priest and Forbes (1990) have set up a model in which, as the twist of a prominence increases, its height increases, with the field evolving slowly through a series of equilibria: eventually, at a critical height the equilibrium ceases to exist and the prominence erupts. More details have been worked out by Forbes and Isenberg (1991), Forbes (1991), and numerical experiments on prominence eruption have been developed by Mikic et al (1989), Forbes (1992) and Steinolfson (1992).

2.2 Unanswered Questions about Prominences

There are many basic features of prominences which we do not understand at all and on which SOHO can make good progress. For example, the whole problem of formation is not understood observationally. Is it by condensation or injection? What is the evolution of loops in and over the filament channel during formation? What are the key conditions for formation? (For example, how important are cancellation and shear? Why do prominences form on some parts of polarity inversion lines and not on others?)

The whole structure of a prominence raises many questions too. The fine-scale structure is only being glimpsed at present - what is for example the filling factor and the number of fibrils along the line of sight? The global structure too is not well understood. What is the nature of the flows, the overall circulation and mass balance? If one switched off the flow it is likely that a prominence would not endure long at all. What is the structure and role of feet? If one removed the feet, a prominence may well float upward rather than fall over! What determines the polarity of an active-region prominence as inverse or normal? Are the different polarities different stages in the evolution of a prominence? How do prominences evolve as they migrate? What is the nature of the oscillations that have been observed (from 3 to 60 mins)?

Then finally the conditions for eruption or disappearance are not known at all. Can the current sheet that is thought to exist below an erupting filament (Priest and Forbes, 1990) be detected?

In attempting to answer such questions it is essential that the SOHO instruments be closely coordinated since they have been designed to complement one another and each on its own will only provide partial answers. On the disk, EIT will produce 3 arcsec images with 10 sec resolution from $6.10^4\ K$ to $3.10^6\ K$ and pressures will be deduced from emission measures or filter ratios. These will act as the environment for detailed spectroscopic measurements of temperature, density, velocity and turbulence, both from SUMER (at 1 arcsec resolution, 1 sec time resolution, $10^4 - 2.10^6\ K$ temperature range and in excess of 1 $km\ s^{-1}$ velocities) and CDS (2 arcsec, 1-10 sec, $2.10^4 - 3.10^6\ K$, $> 30\ km\ s^{-1}$). For magnetic fields we shall need to rely mainly on ground-based measurements. On the limb EIT will give images out to $1.5R_\odot$ while UVCS and LASCO images will extend to $10R_\odot$ and $30R_\odot$, respectively, and LASCO polarisation measurements will go out to $3R_\odot$.

3. Coronal Loops

3.1 Basic Properties

Many basic properties of loops were first learnt from Skylab, where six different types of loop were seen. Active-region loops have lengths 10-100 Mm, temperatures of 2.5×10^6 K and densities of 5.10^{14} – 5.10^{15} m^{-3}. Sunspot loops are similar except that they have cores with temperatures of 2.10^5 K. Quiet-region loops are cooler ($1.5-2.1 \times 10^6$ K) and less dense ($2.10^{14}-10^5$ m^{-3}), whereas interconnecting loops are 20 - 700 Mm long, have $2-3 \times 10^6$ K temperatures, a density of typically 7.10^{14} m^{-3} and last for a day. The main-phase loops in two-ribbon flares are 10-100 Mm long, have densities of 10^{17} m^{-3} and temperatures ranging between 10^4 and 4×10^6 K. Compact flare loops are shorter (5-50 Mm), hotter (4.10^7 K) and denser (10^{18} m^{-3}). Additional information has come from radio observations (e.g Kundu, 1965; Urpo, 1993; Korzhavia, 1993).

However, with its better time-resolution and lower scatter, Yohkoh's Soft X-ray Telescope (SXT) has completely transformed our view of loops, showing them to be much more dynamic (Uchida, 1992; Tsuneta, 1992; Strong, 1993). The following discoveries have been made:

1. Active regions are bright tangles of loops (Strong, 1993), which show up even better in the remarkable NIXT pictures (Golub, 1993).
2. There are many cusp loops, which are often probably reconnecting (Tsuneta, 1992; Strong, 1993);
3. 10% of the active-region plasma has temperatures of $5-6 \times 10^6$ K (Hara et al, 1992);
4. Active-region transient brightenings occur once every 3 minutes and often involve multiple loops (Shimizu et al, 1992; Hiei, 1993);
5. Active-region loops are continually expanding (Uchida, 1992);
6. X-ray jets are accelerated up to 100-1000 $km\ s^{-1}$, often in association with X-ray bright points (Shibata et al, 1992);
7. Many loops have a slowly-varying cross-section, with an expansion factor from footpoint to summit of about 1.1 (Klimchuk, 1992);
8. Twisted loops can be seen (Strong, 1993);
9. Flare loops often have brightenings at their feet or summits.

The following comments may be made in connection with (ii) and (vii). The appearance of cusp loops may be caused in two ways. The may represent steady-state magnetic configurations that are being dragged out by the solar wind, as observed as helmet streamers in white-light eclipse photographs:

in this case the magnetic field is lower on the inside of the cusp than the outside and so the pressure tends to be higher. Alternatively, cusp loops are a natural consequence of the close-down by reconnection of an open or partially open structure (Priest, 1982; Forbes and Priest, 1983). In this case, while the reconnection continues (sometimes very slowly at $1 km\ s^{-1}$ or less) energy is liberated by the reconnection process, partly in conduction fronts and partly in slow shocks which also compress the plasma flowing down from the cusp; in some cases the downflowing plasma can be slowed down by a fast-mode shock and in other cases the downflowing plasma can be compressed and cooled by radiative condensation. In other words, a range of different plasma behaviour is possible, depending on the field strength, the field shear and other parameter values.

Regarding (vii), the presence of nearly constant cross-sections in loops does not necessarily imply the presence of magnetic twist in the corona (i.e. currents), since the expansion of a loop depends critically on the boundary magnetic field profile at the solar surface. Thus, if the field profile corresponds to that of a line dipole just below the surface, there will indeed be a very strong change in cross-section for a potential field in the corona. But, if the field profile corresponds to a line dipole a long way below the surface or to a line monopole (a line current) just below the surface, then the overlying potential coronal loops will have constant cross-section.

3.2 Theory of Coronal Heating

The theory of flows along coronal loops has been well summarised by Antiochos (1993), both for steady flows driven by a spatially varying heating function and for unsteady flows driven by a time-dependent impulsive heating. Over the past ten years much work has been done on suggesting heating mechanisms.

For example, MHD waves (Califano et al, 1990, 1992) are a prime mechanism and they may be dissipated either by phase mixing (Heyvaerts and Priest, 1983; Cally, 1991; Malara et al, 1992). Alternatively, magnetic dissipation in small current sheets may be responsible (Priest, 1991). Such sheets may be created by braiding (Parker, 1972, 1990; van Ballegooijen, 1985, 1988; Mikic et al, 1988) or by shear at separatrices (Low and Wolfson, 1988) and may reveal themselves as nanoflares (Parker, 1990).

Whether the mechanism is waves or current sheets, the resulting coronal state is likely to be highly turbulent and so coronal heating by MHD turbulence has recently been investigated. Heyvaerts and Priest (1984) used a constraint of magnetic helicity injection and conservation, as did Vekstein et al (1990). A linearised version was set up by van Ballegooijen (1986) and more recently a new development has been the setting up of a self-consistent theory (Heyvaerts and Priest, 1993).

An important element of the corona is that of X-ray Bright Points, which have recently been revealed by NIXT photographs to consist of interacting loops rather than points of emission (Parnell et al, 1993). This has led to the setting up of a new Cancelling Flux Model for bright points with three phases (Priest et al, 1993). In the preinteraction phase two magnetic fragments approach one another in the photosphere. Next, in the interaction phase, fields from the fragments reconnect and produce a bright point and associated X-ray jets in the corona. Finally, in the cancellation phase photospheric reconnection allows the fragments to reconnect. During the interaction phase the reconnection point rises up from the photosphere and through the chromosphere and transition region, where it may well create an explosive event (Dere, 1993; Moses and Cook, 1993), especially if the reconnection becomes impulsive and bursty (Priest, 1986). Another point to note is that, bearing in mind that most bright points occur in the network, the same model may take place at smaller scales in the network furnaces that have been suggested by Axford (1993) to act as the source of the solar wind.

3.3 Unanswered Questions about Loops

Loops will be studied on the disc mainly by CDS, SUMER and EIT and also on the limb by UVCS and LASCO. Basic equations about the structure of coronal loops are as follows. How do loops evolve and fill with hot plasma? How do they respond to a transient heating by evaporating and condensing? Can steady siphon flows be detected? (Antiochos (1993) has considered the effect on line intensities and profiles and suggested that the transition region should be wider in upflow regions; Antonucci (1993) too has considered the signatures in terms of profile asymmetries and non-equilibrium ionisation effects.) What is the cause of transition-region redshifts? (They need to be observed as a function of position towards the limb and also as a function of temperature for hotter lines.) Can the temperature and density be measured accurately along a loop? What is the overall mass and flow balance of an active region?

Many questions arise about heating mechanisms, and in partisular what are their signatures? For wave heating, what are the key modes and damping mechanisms (phase mixing, resonant absorption or mode conversion)? Also what is the transport mechanism. Certainly the wave signatures are unlikely to be simple but rather a statistical interaction of events. For reconnection, sources in terms of EUV and radio emission have been suggested by Somov (1993) and Antonucci (1993), including line broadening. One consequence of reconnection could be jets and explosive events (Dere, 1993), and others could be active-region transient brightenings and X-ray bright points.

Other general points made during the open discussion are:

1. We need a viable FIP mechanism to explain variable abundances (does the Marsch mechanism answer the need?); it would be useful to observe

a particular area and watch the abundances change;
2. It is important to understand the transition-region redshifts;
3. We need to study the broadening of coronal lines as a function of temperature and space (for instance, the Saba-Strong (1991) measurement of $40 km\, s^{-1}$ is only at a 15 arcsec resolution);
4. There is a wide range of quite different flows in the solar atmosphere (prominence-related, surges, spicules, siphon flow, evaporation, condensation, reconnection, explosive events, jets, eruptions) and so we need to understand each detail and to relate it to the geometry and the magnetic field variations;
5. It is essential to trace each of these phenomena through the whole temperature range;
6. It is vital to have a close coordination with ground-based photospheric magnetic fields at high resolution in space and time (also with white light and Hα);
7. We need more advances in the theory in order to make more predictions;
8. Coordinated observational sequences must be set up, including ground-based and all the appropriate SOHO instruments: a blue book of observational sequences is being developed for CDS, a red book for SUMER and a yellow book for UVCS, so why not combine them and add others to form a"rainbow" book for SOHO as a whole?

Three examples were discussed in some detail. For siphon flow, it was suggested that a long-lived loop be identified with EIT and then studied with CDS and SUMER in the way detailed by Antonucci (1993). Secondly, an in-depth study of how accurate the temperature and density of a loop can be determined is required, and comparison with the loop models of, e.g., Peres (1993) would be valuable. A third example is X-ray Bright Points, which are prime examples of isolated reconnection sites (and particle accelerators, Simnett, 1993) as well as being examples of elementary heating events (Priest et al, 1993): thus a detailed study of bright points will be of fundamental interest in addition to advancing the understanding of bright points in their own right.

In conclusion, I for one felt encouraged at the end of the discussion to go away and work hard in preparation for SOHO and felt excited already at the prospect of learning much more about these basic plasma processes.

References

Antiochos, S K (1993) 'Mass flows in loops and prominences', these proceedings.

Antiochos, S K and Klimchuk, J A (1991) 'A model for the formation of solar prominences', *Astrophys. J.* **378**, 372-377.

Antiochos, S, Dahlburg, R B and Klimchuk, J A (1993) 'The magnetic field of solar prominences', *Astrophys. J*, in press.

Antonucci, E (1993) 'SOHO contribution to the understanding of mass supply and flow in the solar corona', these proceedings.

Axford, W I (1993) 'The origin of the solar wind', these proceedings.

Ballester J L and Priest, E R (1989) 'A model for the fibril structure of solar prominences', *Astron. Astrophys* **225**, 213-221.

Bothmer, V, and Schwenn, R (1993) 'Eruptive prominences as sources of magnetic clouds', these proceedings.

Brekke, P (1993) 'Observed redshifts of transition region/coronal lines', these proceedings.

Califano, F, Chiuderi, C C and Einaudi, G (1990) 'Nonresonant resistive dissipation of incompressible MHD waves', *Astrophys. J.* **365**, 757-763.

Califano, F, Chiuderi, C and Einaudi, G (1992) 'Nonresonant resistive dissipation of compressible MHD waves', *Astrophys. J.* **390**, 560-566.

Cally, P (1991) 'Phase mixing of surface waves: a new interpretation', *J. Plasma Phys.* **45**, 453-479.

Demoulin, P and Priest, E R (1989) 'A twisted flux tube model for solar prominences. II. Formation of a dip', *Astron. Astrophys.* **214**, 360-368.

Dere, K P (1993) 'Observations of mass flows in small-scale structures', these proceedings.

Engvold, O (1976) 'The fine structure of prominences ', *Solar Phys* **49**, 283-296.

Forbes, T G and Isenberg P A (1991) 'A catastrophe mechanism for coronal mass ejections', *Astrophys. J* **373**, 294-307.

Forbes, T G and Priest, E R (1983) 'A numerical experiment relevant to line-tied reconnection in two-ribbon flares', *Solar Phys.* **84**, 169-188.

Forbes, T G (1991) 'Magnetic reconnection in solar flares', *Geophys. Astrophys. Fluid Dyn.* **62**, 15-36.

Forbes, T G (1992) 'Field opening and reconnection', in *Eruptive Solar Flares* (ed Z Svestka, B Jackson and M Machado) Springer-Verlag, p79.

Golub, L (1993) in *Physics of Solar and Stellar Coronae* (ed J Linsky and S Serio), Kluwer, 71-82.

Hara, H, Tsuneta, S, Lemen, J R, Acton, L W and McTiernan, J M (1992) *Pub. Astron. Soc. Jap* **44**, L135-L140.

Heinzel, P, Gouttebroze, P and Vial, J C (1988) in *Dynamics and Structure of Solar Prominences* (ed J L Ballester and E R Priest) p71-74.

Heyvaerts J and Priest E R (1983) 'Coronal heating by phase-mixed Alfven waves', *Astron. Astrophys.* **117**, 220-234.

Heyvaerts J and Priest E R (1984) 'Coronal heating by reconnection in DC current systems. A theory based on Taylor's hypothesis', *Astron. Astrophys.* **137**, 63-78.

Heyvaerts, J and Priest, E R (1993) 'A self-consistent turbulent model for solar coronal heating', *Astrophys. J* **390**, 297-308.

Hiei, E (1993) 'Loops seen in SXT images', these proceedings.

Kim, I, Koutchmy, S, Stellmacher, G and Stepanov, A (1988) in *Role of Fine-Scale Magnetic Fields on the Structure of the Solar Atmosphere* (eds E Schroter, M Vazquez, A Wyller) Camb. Univ. Press, p289.

Kippenhahn, R and Schluter, A (1957) 'Eine theorie der solaren filamente', *Zs. Ap.*, **43**, 36-62.

Klimchuk, J, Lemen, J R, Feldman, U, Tsuneta, S and Uchida, Y (1992) 'Thickness variations along coronal loops observed by SXT on Yohkoh', *Pub. Astron. Soc. Jap.* **44**, L181-L186.

Korzhavin, A (1993) 'Coronal loops and prominences observed with RATAN 600', these proceedings.

Kundu, M (1965) *Solar Radio Astronomy*, Interscience, New York.

Kuperus, M and Raadu, M A (1974) 'The support of prominences formed in neutral sheets', *Astron. Astrophys* **31**, 189-193.

Low, B C and Wolfson, R (1988) 'Spontaneous formation of electric current sheets and the origin of solar flares', *Astrophys. J.* **324**, 574-581.

Malara, F, Velli, P, Chuideri, C, Einaudi, G (1992) 'Incompressible disturbances in nonuniform media: formation of small scales', *Astrophys. J* **396**, 297-310.

Martens, P C H and Van Ballegooijen A A(1989) 'Helical flux ropes in solar prominences' in *Physics of Magnetic Flux Ropes* (ed C T Russell, E R Priest and L C Lee), AGU Washington, p337-342.

Mikic, Z, Barnes, D C, and Schnack, D D (1988) 'Dynamical evolution of a solar coronal magnetic field arcade' *Astrophys. J.* **328**, 830-847.

Parker, E N (1972) 'Topological dissipation and the small-scale fields in turbulent gases', *Astrophys. J.* **174**, 499-510.

Parker, E N (1990) 'Tangential discontinuities and the optical analogy for stationery fields. V formal integration of the force-free equations' *Geophys. Astrophys. Fluid Dyn.* **52**, 183-210.

Parnell, C, Golub, L and Priest, E R (1993) 'The 3D structure of X-ray bright points', *Solar Phys.*, submitted.

Poedts, S, Goossens, M and Kerner, W (1990) 'On the efficiency of coronal loop heating by resonant absorption', *Astrophys. J* **360**, 279-287.

Priest, E R (1982) *Solar Magnetohydrodynamics*, Reidel, Dordrecht.

riest, E R (1986) 'Magnetic reconnection on the Sun', *Mit. Astron. Ges.* **65**, 41-51.

Priest, E R (1989) *Dynamics and Structure of Quiescent Solar Prominences*, Kluwer, Holland.

Priest, E R (1991) in *Mechanisms of Chromospheric and Coronal Heating*, (ed P. Ulmschneider, E R Priest and R Rosner) Springer-Verlag, p520-535.

Priest, E R, Hood, A W and Anzer, U (1989) 'A twisted flux tube model for solar prominences', *Astrophys.J.* **344**, 1010-1025.

Priest, E R, and Forbes, T G (1990) 'Magnetic field evolution during prominence eruptions and two-ribbon flares', *Solar Phys.* **126**, 319-350.

Priest, E R, Parnell, C E and Marten, S F (1993) 'A converging flux model of an X-ray bright point and an associated cancelling magnetic feature', *Astrophys. J.*, in press.

Ridgway, C, and Priest, E R (1993) 'Prominence Support in Helical Coronal Fields Formed by Photospheric Motions', *Solar Phys.* **146**, 277-296.

Saba, J L R and Strong, K T (1991) 'Coronal dynamics of a quiescent active region', *Astrophys. J.* **375**, 789-799.

Schmieder, B (1993) 'Hierarchy of small scales in UV prominences', these proceedings.

Shibata, K, Ishido, Y, Acton, L, Strong, K, Hirayama, T, Uchida, Y, McAllister, A, Matsumoto, R, Tsuneta, S, Shimizu, T, Hara, H, Sakurai, T, Ichimoto, K, Nishimo, Y, and Ogawara, Y (1992) 'Observations of X-ray jets with the Yohkoh soft X-ray telescope', *Pub. Astron. Soc. Japan* **44**, L173-L180.

Shimizu, T, Tsuneta, S, Acton, L W, Lemen, J R, Uchida, Y (1992) 'Transient brightenings in active regions observed by the Yohkoh soft X-ray telescope',*Pub. Astron. Soc. Japan* **44**, L147-L153.

Simnett, G (1993) 'The energy input for explosive events in the transition zone', these proceedings.

Somov, B (1993) 'Features of mass supply and flows related with coronal reconnection', these proceedings.

Steinolfson, R (1992) 'Coronal shock waves', in *Study of the Solar-Terrestrial System* ESA SP-346, p51-58.

Stepanova, T and Kosovichev, A (1993) 'Modelling of dynamic evolution of coronal loops', these proceedings.

Strong, K (1993) 'Observations of loops and prominences', these proceedings.

Tandberg-Hanssen E (1974) *Solar Prominences*, D Reidel, Dordrecht, Holland.

Tsuneta, S and Lemen, J R (1993) 'Dynamics of the solar corona observed with the Yohkoh SXT', in *Physics of Solar and Stellar Coronae* (ed J K Linsky and S Serio),Kluwer, p113-130.

Uchida, Y (1993) 'New aspects of solar coronal physics revealed by Yohkoh', in *Physics of Solar and Stellar Coronae* (ed J K Linsky and S Serio), Kluwer, p97-112.

Urpo, S, Pohjolainen, S and Kruger, A (1993) 'Millimeter wave radiation sources visible

in the solar corona', these proceedings.
Van Ballegooijen, A A (1985) 'Electric currents in the solar corona and the existence of magnetostatic equilibrium', *Astrophys. J.* **298**, 421-430.
Van Ballegooijen, A A (1986) 'Cascade of magnetic energy as a mechanism of coronal heating', *Astrophys. J.* **311**, 1001-1014.
Van Ballegooijen, A A (1988) 'Magnetic fine structure of solar coronal loops', in *Solar and Stellar Coronal Structure and Dynamics* (ed R Altrock), p115-124.
Van Ballegooijen, A A and Martens, P (1989) 'Formation and eruption of solar prominences', *Astrophys. J.* **343**, 971-984.
Vekstein, G E, Priest, E R, and Steele, C D C (1990) 'Magnetic reconnection and energy release in the solar corona by Taylor relaxation', *Solar Phys.* **131**, 297-318.
Vial, J C (1986) 'Review of diagnostics and structure', in *Coronal and Prominence Plasmas* (ed a Poland) NASA CP 2442, p89.
Wu, S T, Bao, J J, An, C H, Tandberg-Hanssen, E (1988) 'A numerical MHD simulation of prominence formation by photospheric shearing or converging motions', in *Dynamics and Structure of Solar Prominences* (ed J L Ballester and E R Priest) p85-89.

III Coronal Streamers

CORONAL STREAMERS' THEORIES

C. GÖRAN SCHULTZ
Space Research Laboratory, ERNE, University of Turku, Tykistökatu 4 A, 20520 Turku, Finland

Abstract. Some theoretical aspects of solar coronal streamers are discussed with emphasis on the current sheet and reconnection processes going on along the axis of the streamer. The dynamics of the streamer is a combination of MHD and transport, with acceleration of particles due to reconnection and leakage of plasma outwards as a "slow" solar wind as the observable results. The presence of the almost-closed magnetic bottles of streamers that can store high-energy particles for significant times provides the birdcage for solar cosmic rays, the reconnection in the sheet feeds medium-energy protons into the corona for the large-scale storage needed for certain flare models, and the build-up of excess density sets the stage for coronal mass ejections.

Key words: Corona – MHD-Sun – Sun-magnetic fields – Streamers

1. Introduction

This review is a continuation of the discussion of coronal streamers started by R. A. Kopp (1992) at the Annapolis workshop. Following the description by E. R. Priest (1982), coronal streamers are radial structures extending from heights of 0.5 to 1 solar radius up to 10 solar radii, consisting of a round base (arcade) of closed field lines surmounted by a blade of open field lines; from the end it looks like a helmet and from the side like a fan. The streamer has a density enhancement of 3 to 10 over the quiet coronal value (of some 10^{14} m^{-3} at the base of the corona). Overlying an active bipolar region one finds an active region streamer extending outwards for three or four solar radii, often narrowing to a throat or neck at two to three solar radii. Some of the thermodynamical and geometric parameters observed, and the difficulties with a stationary solution have recently been summarized by Koutchmy and Livshits (1992); a magnetostatic fit has been published by Bagenal and Gibson (1991). Helmet streamers lie above prominences: huge, almost vertical sheets of dense, cool plasma surrounded by a hotter and rarer coronal environment that may live for several months. For a general introduction to streamers, the reader is referred to Priest (1982), to Schwenn and Marsch (1990-1991) and to Hundhausen (1972). The reference list in this paper is brief.

Streamers form when the plasma in a closed-field region becomes hot enough to burst open the field and form a neutral current sheet. Alternately, they can reform by reconnection after a coronal mass ejection – a cycle, with an eruptive prominence and mass ejection as the dynamic result of quasi-steady mass build-up in a streamer (Wolfson, Conover and Illing (1987)). The magnetic structures are difficult to observe directly; some cal-

Space Science Reviews **70**, 233–239.

culations of what the Lyman-α intensity might be have been carried out by Noci, Poletto, et al (1993). The large-scale closed structures surmounted by open fields have been modelled by Pneuman and Kopp (1971) by solving numerically the MHD equations for steady, axisymmetric coronal expansion from a dipolar sphere. The resulting magnetic structure has the appearance of a large helmet streamer, with a flow that, near the equator, becomes Alfvénic around 2.5 solar radii and then sonic at 5 solar radii. The disc-like current sheet of the Pneuman-Kopp model extends out from 2.5 solar radii. The basis for our discussion is the Pneumann-Kopp model for coronal streamers. In addition, the source surface models are mentioned, and the main discussion concerns reconnection at the current sheet of streamers and the particle acceleration that might take place there.

2. Modelling streamers

The original Pneumann-Kopp model is an iterative numerical solution of the steady-state ideal MHD equations for a radial outflow from a spherical magnetically dipole surface. In his review, Kopp (1992) summarizes the time-dependent aspects of Endler's (1971) resistive treatment as follows. Starting from an empty field, one can fill with plasma until the plasma pressure forces part of the field to open up, in which case a convectively relaxed state is reached in a few relaxation times determined by the sound speed. Or, one can let an initially radial flow with a bipolar magnetic field (unbalanced $j \times B$ magnetic stresses) collapse, setting up a sheet current, which due to resistivity dissipatively relaxes – going through a series of convectively relaxed states – on a time-scale determined by the effective electrical resistivity in the diffusion region.

Once we allow time dependence, quasi-steady state solutions have been produced by Washimi, Yoshino and Ogino (1987) with cylindrical symmetry, and Linker, Van Hoven and Schnack (1990) with restricted three-dimensionality. To improve on the isothermal Pneumann-Kopp approach, a three-dimensional set of MHD and transport equations has been formulated by Cuperman, Detman et al (1992) and Cuperman, Bruma et al (1993a). The new physics comes from including transverse plasma currents. The role of boundary conditions for such simulations are discussed by Suess, Wu et al (these proceedings) as a continuation of earlier work (Wang, Wu et al (1993)); see also Suess (1992).

The almost collisionless plasma of the corona leaves some 10 orders of magnitude between what can be reasonably simulated on a computer and the real corona. The thickness of the sheet current depends on the reconnection model chosen. Joule heating inside the sheet might be larger than that supplied by thermal conduction, although elsewhere it will not be significant. Reconnection models based on electron inertia (collisionless reconnection) or turbulence (enhanced resistivity) and extrapolation are used to justify the simulation models, where numerical resistivity competes with the explicitly specified one. From the SOHO observational point of view, the global parameters in the simulations cited above are of the correct magnitude, but the flow velocities and current densities are probably locally much larger at the sheet, the key question being the pressure-driven current at the boundary of the closed region (the separatrix) and the neutral sheet.

The source surface model is an approach to fit the spectroscopically observed magnetic fields at the surface of the sun as lower boundary conditions to a force-free field structure with only radial components at an upper boundary in the corona. Aly and Seehafer (1993) give a summary of the method from the mathematical point of view; Cuperman, Bruma et al (1993b) show how to do the reconstruction in practice; Hundhausen, Hundhausen and Zweibel (1981) use the magnetostatic approach and show what the source surface boundary condition leads to in two dimensions. Special techniques have been developed to remove the 180° ambiguity in the sign of the observed transverse magnetic field.

Wolfson (1985) has presented a streamer calculation that includes volume currents in the corona as well as the current sheet. The results suggest that the current sheet is the dominant influence on the overall shape of the coronal magnetic field.

3. Partially open magnetic fields

Of particular interest for the streamer problem is the sources of free energy available. The extreme case is the low-density prominence cavity associated with coronal mass ejections. Low and Smith (1993) identify this cavity with the magnetic bubble of the static equilibrium, and come to the conclusion that the currents flowing provide sufficient excess energy for global MHD destabilization – the mass ejections. The picture by Wolfson, Conover and Illing (1987) is a quasi-static evolution of the helmet streamer as it is inflated with excess mass, but would require a stability analysis of the individual equilibria. The magnetic configurations can get rather complicated, but are usually based on the idea of a force-free field, the solution to $\nabla \times B = \alpha B$. The currents being parallel to the magnetic field, this solution is the result of minimizing the magnetic energy subject to the constraint of keeping the helicity constant (Taylor states). In cylindrical geometry with α globally constant, the solutions are Bessel functions $B = (0, J_1(\alpha r), J_0(\alpha r))$. For arbitrary boundary conditions and geometry, the problem is not necessarily well posed, nor are the solutions unique; some aspects are discussed by Cuperman, Bruma et al (1993b).

4. The current sheet

Above the helmet, the magnetic neutral line is mathematically a sheet current, which could in reality be as thin as a few ion Larmor radii. Such a current sheet is ideally MHD unstable, and tries to break up into filaments. The breakup – the tearing mode of plasma physics – involves a change in magnetic topology – reconnection, and is hence possible only through non-ideal effects – resistivity, viscosity or electron inertia. The spacing between filaments will initially be of the same order as the sheet thickness, but adjacent filaments carry current in the same direction, which means an attractive force. The merging of adjacent filaments – the coalescence instability – was found by Finn and Kaw (1977). Further studies Pritchett and Wu (1979) – have revealed that the instability has no threshold, but the merging of two current channels requires pushing magnetic flux through the X-point, the reconnection problem again, and thus very slow time-scales.

The created filament is a three-dimensional curved hoop, which by itself tries to expand ideally, subject to the constraint of being tied at the ends. An interesting observational goal would be to look for tell-tale pulsations at the starting point of the current sheet. Type III radio bursts are possibly originated by electrons travelling in these sheets.

The current sheet separating opposite magnetic flux and the "flux surface" connected to the start of the sheet – the Y-point – are the objects of reconnection studies and a possible source of high-energy particles in the solar wind. A null in the field enables breaking of the frozen-in flux condition, and dissipation of magnetic energy. Classical resistive tearing mode theories based on the original work of Furth, Killeen and Rosenbluth (1963) run into problems with the very high Reynolds numbers characteristic of coronal plasmas, leading to extremely long time-scales for changes to take place. Viscous tearing is based on the sheared flows near the current sheet. Turbulence enhances resistivity. Collisionless tearing breaks the singularity of the current sheet by the inertia of the particles themselves. The strong temperature dependence of the collisionality forces one to look at the short wavelength "resistive g" modes also.

The mechanism for breaking the constraint of ideal MHD is only half the problem: the other half is the underlying equilibrium. Spontaneous reconnection takes place on time-scales proportional to inverse fractional powers of the resistivity. Thus, the classical resistive tearing mode – breakup of the current sheet – will have an exponential temporal growth with a slow $\eta^{-3/5}$ scaling, except in a case where a current loop forces its way through a background plasma, when we get a faster $\eta^{-1/3}$ dependence. However, the tokamak fusion community will be quick to point out that the "linear" (exponential) growth is of academic interest, and the nonlinear phase has a temporal power law dependence rather that an exponential one. A new twist to the problem comes from the discovery by Dahlburg, Antiochos and Zang (1992) that the saturated quasi-steady state resulting from tearing is itself three-dimensionally ideally unstable.

Once we allow external forcing, such as attraction between two current loops, we have a magnetic configuration with an X. Sweet (1958) and Parker (1957b, 1963) have the plasma flow in slowly towards the neutral sheet (line), and squirt out through the tight part of the X at a reasonable fraction of the Alfvén speed. Petschek (1964) manages to get faster reconnection – more plasma through the X – by opening up the tight angle through the introduction of a stationary slow MHD shock wave. A number of authors have elaborated on these ideas – Vasyliunas (1975), introduced more discontinuities (Sonnerup (1970) , Yeh and Axford(1970)), or removed them (Soward and Priest(1977)). The mathematical treatment yields time-scales for the processes expressed in terms of the magnetic Reynolds number $R_m = LV_0/\eta$, where V_0 is the inflow velocity perpendicular to the neutral line, and L the length of the neutral line taken as the inner "diffusion" region of the X. For Sweet-Parker one gets time-scales $R_m^{1/2}$, and for Petschek $\log R_m$. Driven reconnection in the laboratory has been tested by Stenzel, Gekelman and Wild (1983); a careful series of numerical simulations were carried out by Biskamp (1986). The work by Biskamp has pointed out the extreme sensitivity of this problem to the imposed boundary conditions. In addition, in some of his simulations the buildup of two countercurrents just outside the diffusion region slows down the outflowing plasma to only a small fraction of the Alfvénic speed, a result present for some parameter choices in the analytic studies of Syrovatski (1976). Syrovatski has studied the possible solutions of $\nabla \cdot B = 0$; of special interest is the case where the field has been generated from the complex potential $z\sqrt{z^2 - L^2}$ with the current sheet added as an inverse hyperbolic cosine arcosh z/L. Unified summaries of the classical theory of fast reconnection have been published by Priest (1990, 1992).

For streamer simulations, there are no shortcuts. The thickness of the current sheet will decrease to a single spatial cell in the simulation long before the magnetic Reynold's number reaches values relevant for the corona. The standard approach is to make a series of simulations with decreasing sheet thickness, and attempt to extrapolate the physical behavior. One can, however, do a perturbation simulation based on a quasi-steady equilibrium.

5. Heating and acceleration of particles

The Alfvén-speed flows associated with reconnection are candidates for providing the energetic particles observed in the corona. Simnett (1993) makes a case for the actual source of energy for a large flare being globally stored protons with energies slightly below a MeV – the protons being the result of reconnection, and then trapped in large coronal loops or helmet streamers.

The presence of storage areas for energetic particles is also supported by models for solar cosmic rays. To explain the temporal behavior and energy spectra of energetic events in the MeV to GeV range, a two-dimensional diffusion in the corona – a birdcage model of particles diffusing (hopping from loop to loop) two-dimensionally until they arrive at a location that is magnetically connected to earth – seems to give the best results. Torsti and Schultz (1992) have particularly focused on the extremely energetic event of September 29, 1989 at 11:33 UT, off the west limb of the sun.

The helmet current sheet with ongoing reconnection has an associated electric field (in the fluid picture, the inflow $V \times B$ field), which (in the particle picture) accelerates particles. Individual ions, less collisional than the electrons, move on circular Larmor orbits that $E \times B$ -drift towards the current sheet, until the field is weak enough to allow acceleration in the direction of the field. Almost identical starting points can give quite different orbits even in an idealized geometry (R. Vainio and the author have done some numerical studies of particle motion in the vicinity of a Syrovatski sheet). The resulting heating of the plasma has recently been studied and summarized by Moses, Finn and Ling (1993).

6. Conclusions

Coronal streamers are the second-largest and most visible representative of a hierarchy of semi-closed magnetic structures, the label largest here reserved for the ballerina sheet itself. Source-surface and other models give streamer-like structures of solar radius size; this is consistent with what is needed for energetic particle storage. Reconnection is a source of energetic particles in the streamers, as well as the source of difficulties for streamer simulations.

Acknowledgements

Suggestions by the referee have changed the emphasis of this paper; the author thanks also for several references. The financial support of Turun yliopistosäätiö to attend the Second SOHO Workshop is gratefully acknowledged.

References

Aly, J. J. and Seehafer, N.: 1993, *Solar Phys.* **144**, 243
Bagenal, F. aand Gibson, S.: 1991, *J. Geophys. Res.* **96**, 17663
Biskamp, D.: 1986, *Phys. Fluids* **29**, 1520
Cuperman, S., Detman, T. R., Bruma, C., and Dryer, M.: 1992, *Astron. Astrophys.* **265**, 785
Cuperman, S., Bruma, C., Detman, T. R., and Dryer, M.: 1993a, *Astrophys. J.* **404**, 356
Cuperman, S., Bruma, C., Zoler, D. and Semel, M.: 1993b, *Astrophys. J.* **270**, 480
Dahlburg, R. B., Antiochos, S. K. and Zang, T. A.: 1992, *Phys. Fluids B* **4**, 3902

Endler, F.: 1971, *Ph.D. Thesis*, Göttingen University
Finn, J. M. and Kaw, P. K.: 1977, *Phys. Fluids* **20**, 72
Furth, H. P., Killeen, J. and Rosenbluth, M. N.: 1963, *Phys. Fluids* **6**, 459
Hundhausen, A. J.: 1972, *Coronal Expansion and Solar Wind*, Publisher: Springer
Hundhausen, J. R., Hundhausen, A. J. and Zweibel, E. G.: 1981, *J. Geophys. Res.* **86**, 11117
Finn, J. M. and Kaw, P. K.: 1977, *Phys. Fluids* **20**, 72
Kopp, R. A.: 1992, 'Theory of Coronal Streamer Formation', in *Proc. First SOHO Workshop*, ESA SP-348, p. 53
Koutchmy, S. and Livshits, M.: 1992, *Space Science Rev.* , 393
Linker, J. A., Van Hoven, G. and Schnack, D. D.: 1990, *Geophys. Res. Lett.* **17**, 2281
Low, B. C. and Smith, D. F.: 1993, *Astrophys. J.* **410**, 412
Moses, R. W., Finn, J. M., Ling, K. M.: 1993, *J. Geophys. Res.* **98**, 4013
Noci, G., Poletto, G., Suess, S. T., Wang, A.-H. and Wu, S. T.: 1993, *Solar Phys.* **147**, 73
Parker, E. N.: 1957a, *J. Geophys. Res.* **62**, 177
Parker, E. N.: 1957b, *J. Geophys. Res.* **62**, 509
Parker, E. N.: 1963, *Astrophys. J. Suppl.* **8**, 177
Petschek, H. E.: 1964, 'Magnetic Field Annihilation', *NASA Spec. Publ.* **SP-50**, 425
Pneuman, G. W. and Kopp, R. A.: 1971, *Solar Phys.* **18**, 258
Priest, E. R.: 1982, *Solar Magnetohydrodynamics*, Publisher: D. Reidel, Ch. 12.4.
Priest, E. R. and Krishan, V. (eds.): 1990, *Basic Plasma Processes on the Sun*, IAU, The Netherlands, 271-291
Priest, E. R. and Forbes, T. G.: 1992, *J. Geophys. Res.* **97**, 757
Pritchett, P. L. and Wu, C. C.: 1979, *Phys. Fluids* **22**, 2140
Schwenn, R., Marsch, E.: 1990,1991, *Physics of the Inner Heliosphere 1, 2*,
Simnett, G. M.: 1993, *Adv. Space Res.* **13**, (9)133
Sonnerup, B. U. Ö: 1970, *J. Plasma Physics* **4**, 161
Sonnerup, B. U. Ö: 1979, *Solar System Plasma Physics*, Vol. III, ed. L. T. Lanzerotti, C. F. Kennel and E. N. Parker, Cambridge University Press
Soward, A. M. and Priest, E. R.: 1986, *J. Plasma Physics* **35**, 333
Soward, A. M. and Priest, E. R.: 1977, *Phil. Trans. Roy. Soc. London* **A 284**, 369
Suess, S. T.: 1992, 'Modeling Solar Coronal Streamers', in *Proc. First SOHO Workshop*, ESA SP-348, p. 63
Stenzel, R. L. and Gekelman, W.: 1981, *J. Geophys. Res.* **86**, 649
Stenzel, R. L., Gekelman, W. and Wild, N.: 1983, *Phys. Fluids* **26**, 1949
Sweet, P. A.: 1958, 'The Neutral Point Theory of Solar Flares', in *Electromagnetic Phenomena in Cosmical Physics*, ed. B. Lehnert, Cambridge University Press
Syrovatskii, S. I.: 1971, *Sov. Phys. JETP* **33**, 993
Syrovatskii, S. I.: 1976, 'Neutral Current Sheets in Plasmas', in *Proc. Lebedev Physics Inst.*, ed. N. G. Basov, vol. 74, Consultants Bureau
Torsti, J. and Schultz, C. G.: 1992, 'A Double-Peaked Solar Flare – Conclusions on Particle Transport', in *Proc. First SOHO Workshop*, ESA SP-348, p. 343
Vasyliunas, V. M.: 1975, *Rev. Geophys.* **13**, 303
Wang, A.-H., Wu, S. T., Suess, S. T. and Poletto, G.: 1993, *Solar Phys.* **147**, 55
Washimi, H., Yoshino, Y. and Ogino, T.: 1987, *Geophys. Res. Lett.* **14**, 487
Wolfson, R., Conover, C. and Illing, R. M. E.: 1987, *J. Geophys. Res.* **92**, 13641
Wolfson, R.: 1985, *Astrophys. J.* **288**, 769

UNDERSTANDING SOLAR STREAMERS: THE ROLE OF SOHO

GIANNINA POLETTO
Osservatorio Astrofisico di Arcetri
Largo Fermi, 5, 50125 Firenze (ITALY)

Abstract. Streamers have been observed since far back in time, but our knowledge of their morphology and of their physical characteristics is still very limited. As a consequence, the present streamer picture is largely incomplete: because individual features are poorly known, their role in more general phenomena (like the evolution of the global corona or the solar wind mass and flow pattern) is also poorly known. In this presentation, the more relevant open problems in the understanding of streamers will be illustrated and it will be shown how new data acquired by SOHO may help us to reach a better understanding of these structures.

Key words: Sun: corona – Sun: UV radiation – Sun: solar wind

1. Introduction

Coronal streamers have been observed for a long time in white-light eclipse observations of the solar corona. More recently, coronagraphs and space experiments allowed us to observe streamers also outside of eclipses, thus providing extensive new data on these structures. Nevertheless, streamers are not well known and time dependent, rapidly evolving phenomena, like coronal transients and CMEs, have attracted much more interest than these bright quasi-stationary large features, which are always present in the corona. As a consequence, a number of problems related to streamer physics and morphology are still unclear.

The process by which streamers form, are maintained and disappear, are obscure. Streamers' three-dimensional configuration, plasma temperature, pressure and flow speed are only vaguely known and the same holds true for elemental abundances and magnetic fields. Their contribution to the solar wind is controversial, as they have not been unambiguously identified as sources of the slow solar wind.

Such a poor knowledge of most of the streamers parameters is unjustified, on the basis of the wealth of relevant information which could be derived from exhaustive analyses of these features. Streamers offer a unique opportunity to investigate the difference in physical conditions of plasma in closed vs. open magnetic structures and to acquire new data on the very region where this transition purportedly occurs, namely, at the top of the streamer "helmet". The process of global restructuring of the solar corona is obvioulsy linked to the temporal evolution of streamers: the heliosheet shape and its changes are dictated by the streamer belt and its evolution.

Space Science Reviews **70**, 241–252.

Open questions, like the previoulsy mentioned origin of the slow solar wind, cannot be answered unless streamer regions are thoroughly investigated.

SOHO experiments, supplemented by those of the Ulysses mission — now expected to operate through year the year 2001 — will provide us with the opportunity for a continued study of solar phenomena from the corona, throughout the interplanetary space, out to distances $\geq$ 1 AU. In the following, after briefly illustrating the present state of the art, I'll review a number of observations, which may be performed by SOHO experiments, that will lead to a better insight into the physics of streamers. However, it is beyond the purpose of the present review to describe specific observational sequences for SOHO experiments; complementary theoretical work, being discussed elsewhere comprehensively (Schultz, this issue), will be addressed only marginally.

2. Geometry of Streamers

So far, the geometry of streamers has been derived mainly from white light (WL) eclipse observations. These data comprise the widest range of altitudes with respect to those covered by ground-based coronagraphs — which observe only up to a limited range of heights (on the order of 2.-2.5 $R_\odot$) — and by space experiments — which observe only *beyond* some minimum height above the solar limb (on the order of 1.6 to 2.6 $R_\odot$) —, but offer only a snapshot of streamers at occasional times. At meter/decameter wavelengths, streamers can be observed and followed as they move across the disk, carried around by solar rotation — albeit with a spatial resolution of the order of a few arcminutes. The 3-D streamer geometry can be identified by means of multi-frequency radio observations (Kundu et al., 1989); however, the low spatial resolution doesn't allow for a good reconstruction of the streamer shape.

As a consequence of the limited capabilities of available techniques, two major issues are still open: the "base" of streamers has still to be clearly identified and the 3-D streamer structure is essentially unknown. SOHO will contribute to a better understanding of both issues via prolonged coordinated observations of streamers with different experiments. EIT, SUMER and CDS may observe disk and lower-corona areas, covering the lower section of streamers, over a set of wavelengths, at the same time as UVCS and LASCO perform observations at higher altitudes. Moreover, there will be the possibility of cross-checking data acquired by different instruments: for instance, CDS, expected to make disk or above limb observations up to $\approx$ 1.3 $R_\odot$, may take images in the same ion (Fe XIV) used by LASCO's inner coronagraph — which can image the solar corona down to 1.1 $R_\odot$— or in Fe XII/Fe XV, selected by EIT, as well as in OVI, which is used both by SUMER and UVCS. Thus, complementary observations from different experiments should ensure the possibility of reconstructing the structure of streamers from their lower base to their coronal extension. Obviously, because of the different field-of-view of the instruments, those with a smaller field of view should be repointed in order to cover completely the streamer area.

Coordinated streamer observations by different SOHO experiments will still lead to a 2-D streamer configuration, unless observations are performed over a time interval long enough to allow the use of tomographic techniques. So far, this technique has been scarcely used (see, e. g., Wilson, 1977 and, in the radio domain, see the new method developed by Aschwanden and Bastian, 1993, and applied, so far, only to active regions), because it is very difficult to deconvolve the time evolution from spatial variations in the 3-D feature reconstruction. A better way to derive the streamer 3-D structure is by making use of images obtained from different vantage points by different space missions: at the time of SOHO, we might be able to take advantage of a Tomography Mission (Davila, 1993), complemented by the GOES imager, TRACE, CORONAS and the next Japanese solar satellite. Although the solar community has lately expressed an interest in solar tomography from multiple-spacecraft observations, it is premature, at this stage, to rely on this possibility. A technique to derive an order-of-magnitude value for the extension of streamers along the line-of-sight via $Ly-\alpha$, $Ly-\beta$ and white light observations, has been suggested by Fineschi and Romoli (these Proceedings); detailed comparisons between streamer models and observations may also yield some information about the streamer spatial distribution (Noci et al., 1993). Uncertainties in the streamer geometry obviously cause uncertainties in the distribution of physical parameters in streamers: this topic is discussed in the next section.

3. Density and Temperature in Streamers

Traditionally, densities from eclipse polarization brightness (pB) images have been derived using the Van de Hulst method (1950), i. e., assuming an axially symmetric density distribution. Because pB is proportional to the line-of-sight integral of the density (times a scattering function), the derived densities are entirely model-dependent, as far as the distribution of density along the line-of-sight (LOS) is concerned. This kind of analysis led to the conclusion that, up to ≈ 5 $R_\odot$, the density vs. height profile, along the streamer's axis, is described by a hydrostatic law, with a "scale height" temperature T $= T_{sh} \approx$ 1-2 $\times$ $10^6 K$. Densities (N) decrease from a few units $\times$ 10^8 cm^{-3} at r $\approx$ $R_\odot$ to $N \approx 10^5$ cm^{-3} at r $\approx$ 10 $R_\odot$. However, individual features may show large departures from these values (see, e. g., Figs. 16(a-b) of Dollfus et al., 1974, and Fig. 6 of Koutchmy, 1992).

Clearly, temperatures derived by fitting the observed density vs. height profile with a hydrostatic law are upper limits to streamers temperatures, if dynamic or magnetic pressure contribute to the support of the atmosphere. Recently, Guhathakurta et al. (1992), analyzing data taken at the time of the March 17/18, 1988, eclipse, adopted the line ratio technique to derive streamer (electron) temperatures. Using ratios extremely sensitive to temperature variation (a change from 1 to 2 10^6 K implies a change in the line ratio greater than 3 orders of magnitude), these authors derived temperatures lower than T_{sh} in the three most prominent streamers (at r = 1.15 $R_\odot$). This result suggests that the hydrostatic assumption may not be representative of streamers conditions.

On the whole, today's poor knowledge of densities and temperatures in streamers is rather astounding. There is an incomplete coverage of the streamer structure (N and T are given at a few heights and, mostly, only along the streamer axis) and the hydrostatic technique used to infer temperature values is inadequate. Some problems have not been addressed at all: depending on the electron density distribution with height, proton (T_i) and electron temperatures (T_e), at some level, are decoupled and take different values. In quiet equatorial regions, Saito's density distribution (Saito, 1970) predicts $T_i \neq T_e$ at r $\geq$ 2 $R_\odot$ (Withbroe et al., 1982): depending on the distribution of physical parameters in streamers, and on streamer geometry, this effect may show up in the streamer "body", or in the region above the cusp. Thus the "scale height" temperature represents a sum over proton and electron temperatures, whose values are unknown.

SOHO experiments, making observations throughout the streamer, will provide a more complete data set. Moreover, values of densities can be obtained via alternative techniques (density sensitive line ratios), possibly coupled to estimates of filling factors and/or density irregularities. Density diagnostic via density-sensitive line ratios is to be implemented by SUMER (expected to make observations up to $\approx$ 1.5 $R_\odot$), CDS and LASCO, in lines which are sensitive to densities of the order of 10^8 cm^{-3}. The use of the S X ion (temperature of formation $\approx 1.6 \times 10^6$ K) 1213.62 to 1196.26 Å line ratio, for density diagnostic at N $\geq 5 \times 10^7$ cm^{-3} (Feldman et al., 1978, Dwivedi, 1991), has not yet been explored by the UVCS team.

Density values derived with the line-ratio technique represent an average over the line formation region and are affected by dishomogeneities in the density distribution. However, Almleaky et al. (1989) and Brown et al. (1991) have shown that in some special cases, by means of a number of line ratios, simultaneously acquired, it is possible to derive the N(T) function, also in inhomogeneous plasmas. To my knowledge, the density sensitive line ratio technique has not yet been used in streamers. Although only the lowest, higher density sections of streamers can be probed with this method, a comparison of density values derived from this and the Emission Measure technique, will provide information on filling factors within streamers, purportedly highly structured (Koutchmy, 1977, Koutchmy and Livshits,

1992).

The line ratio technique can be applied also to derive streamer temperatures, thus providing an alternative to "hydrostatic" temperatures, whenever intensities of temperature sensitive line pairs are available. T_e values above the limb, at different heights, will be derived via SUMER, CDS and LASCO observations, using this technique. UVCS has the capability of yielding an independent estimate of the electron temperature, through the profile of the electron scattered $Ly - \alpha$ component (or, alternatively, via the ratio of the resonantly scattered component of $Ly - \alpha$ to WL intensities). The electron scattered $Ly - \alpha$ component, produced by Thomson scattering of the chromospheric $Ly - \alpha$, is much weaker than the resonantly scattered component as well as much broader, due to the low mass and high thermal velocity of electrons in the corona. T_e can be derived from the width of the electron scattered $Ly - \alpha$ profile, after taking into account the geometry of features along the LOS (the scattering geometry may also have some influence on the $Ly - \alpha$ width). However, because of the weakness of the line, the observation is not easy and integration times on the order of 2 hours are needed at a distance of ≈ 2 $R_\odot$ (Hassler et al., 1992). Alternatively, in regions with no outflow velocities, the ratio of the intensity of the resonantly scattered $Ly - \alpha$ to the WL intensity depends only on the hydrogen ionization balance term — as the density dependent factors cancel out — and thus provides a means to derive electron temperatures.

The availability of different methodologies to determine electron temperatures allows us to cross-check values derived by different experiments, as well as to infer the T_e vs. height profile. Knowledge of this parameter is relevant *per se*, in regions where T_e and T_i are not coupled; at levels where the density is so high that $T_e = T_i$ can be safely assumed, it offers a simple means to infer the value of rms non-thermal velocities (by comparison with kinetic temperatures derived from the $Ly - \alpha$ resonantly scattered profile). In determining abundances, knowledge of T_e avoids the introduction of uncertainties in the ionization balance term. We will come back to these points in Section 5.

As we mentioned, all the techniques we described give average values of T, N along the LOS. Apart from tomographic techniques (see also: Altschuler and Perry, 1972, Perry and Altschuler, 1973), the only way to determine the 3-D behavior of T, N in streamers is via a detailed comparison of the observed LOS quantities with 3-D theoretical/numerical models. So far, very few attempts have been made to predict observational results on the basis of theoretical/numerical models (Bagenal and Gibson, 1991, Noci et al., 1993): moreover, being these models mostly 2-D, they do not provide any information on streamer 3-D structure. Preliminary 3-D modeling attempts have been presented by Linker and Van Hoven, 1990, Wu and Wang, 1991, Cuperman et al., 1993. It is a field surely worth further efforts.

4. Solar Wind from Streamers?

Astronaut A. Worden, from the lunar command module of Apollo 15, took, in 1971, photographic observations of a few stable streamers up to a projected distance of $\approx 50\ R_\odot$ from Sun center. Analysis of these data showed that streamers extended radially out to $\approx 35\ R_\odot$ and curved hereafter with a curvature consistent with spirals calculated assuming a constant radial velocity of $\approx 275\ km/s$. If the streamer central axis represents the path of plasma emitted from a region on the solar surface, it is obvious to conclude, from this result, that streamers cannot be the source regions of high speed solar wind streams (Wilson and McQueen, 1974).

Twenty years after this work, we have still not solved the problem of the origin of the slow solar wind. The conventional picture, after Skylab, relates high speed streams to coronal holes and slow wind streams to streamers (see e. g. Fig 2 of Hundhausen, 1977), but recent works by Wang and Sheeley (1990) and Sheeley and Wang (1991), although retaining the high speed/coronal holes connection, associate wind speed with flux tubes expansion factors in the corona and claim that the highest wind speed emanates from minimum field regions bordering holes of the same polarity. As a consequence, regions of converging field lines may be sources of high speed wind, contrary to what is traditionally envisaged.

SOHO may solve this riddle via UVCS and, possibly, LASCO experiments. UVCS will derive flow speeds, at distances of a few $R_\odot$, via the Doppler dimming technique in the $Ly-\alpha$ and OVI lines, while the LASCO team is examining the possibility to use H_α dimming observations for the same purpose. This is one of the most exciting and newest SOHO observations and much of its success depends on the yet unknown density difference between the cusp/current sheet region at the top of the closed portion of streamers and the adjacent open field regions. The Doppler dimming effect is based on the decrease in the amount of resonant scattering of chromospheric radiation which occurs in the presence of solar wind outflowing velocities. In a stationary ambient, the resonant scattering is maximum; any outflow velocity will reduce the amount of resonant scattering because of the Doppler shift of the chromospheric radiation, as seen by moving atoms. Hence, we expect Doppler dimming from open field regions flanking the streamer and our capability of determining the outflow speed from the top of streamers, depends on the density difference between this and the surrounding regions. We may also be able to determine how long does wind emanating from the cusp region mantain its identity prior to being overcome by the adjacent wind flows. It is worth pointing out that even a negative result — that is, lack of identification of the streamer originated wind — will be a valuable data, helping us to constrain densities and width of the current sheet region at the streamer top. SOHO is facing the exciting perspective of shedding light on this yet unexplored topic.

5. Further Diagnostic Capabilities from SOHO

A further issue that can be addressed by SOHO is that of element abundances in streamers. Depending on the value of the first ionization potential (FIP), elements can be divided in high FIP (ionization potential $\geq 10\ eV$) and low FIP (ionization potential $\leq 10\ eV$) elements: it is well known that in between the photosphere and the corona, because of some fractionation mechanism, the relative abundance of low to high FIP elements is enhanced (Meyer, 1991). Although the fractionation mechanism has not yet been unambiguously identified, ion-neutral separation in a magnetic field is usually invoked and different suggestions are made as to the forces (e. g., gravitational, electromagnetic) responsible for this effect (Vauclair and Meyer, 1985, Von Steiger and Geiss, 1989, Hénoux and Somov, 1992, Antiochos, 1993). A requirement common to all models is that for small scale structures: possibly, the FIP fractionation process and the coronal heating process, for which small scale events are increasingly gaining popularity, are closely related. The dependence of the Mg/O ratio on the temperature of the coronal source further hints to this relationship (Von Steiger, 1992).

A determination of element abundances in streamers is especially relevant in view of the claim that the ratio of high to low FIP elements depends on the magnetic field configuration, being highest in closed field structures and steadily decreasing to low values as we move to large scale loops and completely open fields (see Fig. 7 in Meyer, 1993). Evaluating abundances at different heights along, and across, the streamer's axis, should allow us to detect a different abundance ratio as we move from closed to open regions, if the abundance ratio is really a function of the magnetic field configuration. CDS and SUMER may determine Ne, Mg and O abundances, UVCS has the capability of determining Mg and O abundances. Uncertainties in the abundance values will be crucially reduced if, as previously mentioned, the value of T_e (and N_e) has been determined via an independent technique. It is also worth mentioning that the composition, charge state distribution, kinetic temperature and speed of Ne, Mg, O, will also be determined *in situ* by CELIAS making it possible to trace back to the Sun any abundance variation which could be measured at 1 AU.

Further issues, related to streamer physical conditions, that can be clarified by SOHO experiments, include the presence of MHD waves in a streamer and the intensity/direction of magnetic fields throughout the streamer body. Wave motions may show up in the spectral width of lines from heavy ions, which are especially sensitive to the presence of non-thermal motions, because of their large mass and small thermal width. Thus, measurements of EUV line widths offer a means of detecting waves: CDS and SUMER are expected to be able to make line-width measurements up to $\approx$ 1.5 $R_\odot$, while UVCS and LASCO will provide measurements above that height. An increase of the line width with altitude in the streamer, may be interpreted in terms of undamped waves and constant wave energy flux, requiring the velocity amplitude to increase with height. In regions where we expect $T_e = T_i$, the acquisition of the value of T_e via methods independent of the line width, allows the effect of non-thermal motions to be unambiguously identified. Alternatively, one should rely on model predictions to separate the contribution of non-thermal vs. thermal motions. Models will be necessary also to interpret LASCO magnetic field values, which will be derived via polarization measurements in Fe XIV green line. A conservative estimate suggests that, between 1.25 and 2 $R_\odot$, the magnetic field direction will be given with an uncertainty of $\pm$ 11° over a $17 \times 17''$ area: greater accuracy can be achieved by taking repeated exposures of the analyzed area. Also the measurements of density and temperatures, as Koutchmy pointed out (1992), may lead to an estimate of magnetic field values across the coronal discontinuities which are nearly always part of streamers, assuming constant pressure across the discontinuity. According to Koutchmy (1992), this indirect technique offers one of the best means of deriving magnetic field values in the intermediate corona.

6. Streamers and the Global Corona

The formation and/or disruption of streamers usually involves such a large region that is likely to affect the whole corona. However, because little is known about the process by which individual streamers form/disappear, the relation between streamer formation/disappearance and the global restructuring of the corona is a further issue which has been, so far, rarely addressed. That streamers undergo slow structural changes, during their lifetime, has been known for a long time (see, e.g., Poland, 1978); moreover, dramatic changes in their morphology have been observed at times of transients/mass ejections (Illing and Hundhausen, 1986). Hence, it appears quite likely that streamers disrupt because of CMEs blowing open the streamers magnetic fields and reform, when open field lines — possibly the *legs* at either side of the CME — eventually reconnect. However, there are cases in which streamers and CMEs behave as independent phenomena and streamers maintain intact their structure during the course of a CME (Kahler, 1991): Sime (1989) in fact argued in favor of coronal evolution driving CMEs, rather than the opposite. Clearly the issue is far from being settled; as a consequence, we don't know the mechanism by which the interplanetary magnetic field, subject to magnetic flux input by CMEs and without a clearly identifiable means to remove it, keeps roughly constant in time. The wealth of data necessary to shed light on this problem can be acquired by LASCO coronagraphs. Valuable support can be provided by X-ray or EUV experiments, like EIT: recently, Yohkoh observations provided an instance of the relationship between streamer reconfiguration and the global corona (Tsuneta, et al., 1992). This observation was anticipated on theoretical grounds by Kopp (1992), who pointed out that the elliptic MHD equations describing the behavior of coronal plasma implied the propagation of a localized disturbance throughout the whole medium. Yohkoh has also collected some evidence for the reformation of a streamer by magnetic reconnection occurring at increasingly greater heights along the neutral sheet formed by a CME (Hiei et al., 1993). However, there is no universal consensus as to how reconnection leads to the streamer formation: Kahler and Hundhausen (1992) recently suggested a new scenario for post-CME structures, which, according to these authors, consist of radial features with embedded multiple neutral sheets. Reconnection processes at these neutral sheets occurring *during* or early after a CME, would eventually lead to the formation of multiple streamers anywhere within the region of the CME. This innovative view is based on SMM coronagraph data and needs to be reconfirmed on a larger data set. SOHO may provide not only further images, but may be capable of detecting motions associated with reconnection phenomena as streamers reform.

7. Streamers and the Interplanetary Medium

The Los Alamos solar wind community, in a group of "classical" papers (see, e. g., Feldman et al., 1981; Borrini et al., 1981; Gosling et al., 1981), identified a number of signatures of streamers in the interplanetary space. According to these studies, slow wind streams, mapped back to the Sun, were associated with the streamer belt and characterized by a magnetic field reversal, high plasma density, low H^+ and He^{++} bulk velocities, low H^+ and He^{++} kinetic temperatures, low Helium abundance. We have already mentioned how, in spite of further studies that confirmed the association between streamers and slow wind (Lundstedt, 1989), the topic is still not completely settled. SOHO will provide us with the unique opportunity of simultaneous observations, by on-board experiments, of solar magnetic fields (via MDI), wind speeds close to the Sun, *in situ* measurements of particles, while complementary data will be collected by other space experiments and/or by scintillation measurements via, e.g., EISCAT (European Incoherent Scatter Radar System) and the Solar Terrestrial Laboratory in Nagoya. All these experiments will help us to get a better definition of the characteristics associated with streamers in the interplanetary space, but, besides these issues, I would like to mention a couple of recently raised questions, whose solution may depend crucially on SOHO.

Jackson, in a series of papers (Jackson, 1991, 1992, Jackson et al., 1993), has examined data from the Helios zodiacal light photometers to study the behavior of persistent elongated structures at interplanetary distances from the Sun. From his analysis, Jackson concluded that densities in these structures, usually much higher than the density of the ambient medium, either keep constant or decrease with heliocentric distance, showing, at any given time, different values along their extent. This result has been interpreted as evidence of a solar, rather than interplanetary, origin of these structures, hence identified as streamers (in the traditional interpretation they were identified as corotating interaction regions (CIR)). How and where the streamer plasma accelerates to the speed associated to these structures is, however, altogether unknown. This could be ascertained by LASCO, by tracking continuously, over many days, the position of the streamer material vs. height, as it flows outwards, with the inner coronagraphs, in the first few solar radii and with the outer C3 coronagraph, for distances up to 30 $R_\odot$. SOHO observations may help us to understand both the acceleration mechanism of the streamer plasma and the characteristics of these high density structures, as it is not clear, yet, how to discriminate between CIR and streamers (Houminer and Gallagher, 1993).

A second controversial issue is related to the presence of multiple directional discontinuities in the coronal streamer belt at sector boundary crossings in the heliosphere. Traditionally, these features have been ascribed to waves or kinks in the heliospheric current sheet. An alternative interpretation has been recently proposed by Crooker et al. (1992), who suggest that extended current sheets from a hierarchy of helmet streamers of different sizes could account for the observed directional discontinuities. Obviously, some of the difficulties in interpreting this kind of data arise because it is not easy to discriminate between phenomena originating at the Sun and phenomena occurring "en route", between the Sun and 1 AU. Detection of frequent small-scale mass outflow from small helmets over either active regions or filaments as might be revealed by LASCO and UVCS may provide the still missing evidence in favor of the Crooker et al. hypothesis. In this case we will be forced to modify our concept of the heliospheric sheet as formed by a single thin continuous surface, and adopt the view of the current sheet as a thick layer made up of a number of current sheets continuously changing, which fill the whole thickness of the streamer belt.

I would like to conclude this review reminding us that, in spite of the many brilliant perspectives that we face in the streamer research area — which I have, to some extent, described — sometimes the most brilliant results derive from unexpected data. We have to be ready for planning observing campaigns to solve today's open questions, but unanticipated observations are likely to provide us with the excitement of the unforeseen. Personally, I look forward to the post-SOHO view of streamers and I hope SOHO will bring substantial advancement in an area which has progressed, in the past years, at a somehow slow pace.

Acknowledgements

It is a pleasure to thank Roger Kopp, Giancarlo Noci and Steve Suess for many helpful discussions on this topic. This work has been partially supported by ASI (Italian Space Agency).

References

Almleaky, Y.M., Brown, J.C., Sweet, P.A.: 1989, *Astron. Astrophys.* **224**, 328
Altschuler, M.D., Perry, R.M.: 1972, *Solar Phys.* **23**, 410
Antiochos, S.: 1994, *Adv. Space Res.* , in press
Aschwanden, M.J., Bastian, T.S.: 1994, 'Symposium on New Look at the Sun' in Enome, S., ed(s)., , Kofu, September 1993, in press
Bagenal, F., Gibson, S.: 1991, *J. Geophys. Res.* **96**, 17663
Borrini, G., Gosling, J.T., Bame, S.J., Feldman, W.C., Wilcox, J.M.: 1981, *J. Geophys. Res.* **86**, 4565
Brown, J.C., Dwivedi, B.N., Almleaky, Y.M., Sweet, P.A.: 1991, *Astron. Astrophys.* **249**, 277
Crooker, N.U., Siscoe, G.L., Shodhan, S., Webb, D.F., Gosling, J.T.: 1993, *J. Geophys. Res.* **98**, 9371
Cuperman, S., Bruma, T., Detman, T., Dryer, M.: 1993, *Astrophys. J.* **404**, 356
Davila, J.M.: 1993, *Solar Tomography: Deducing the 3-D Structure of the Solar Corona* , NASA-Goddard Space Flight Center, May 14, 1993
Dollfus, A., Laffineur, M., Mouradian, Z.: 1974, *Solar Phys.* **37**, 367
Dwivedi, B. N.: 1993, *Solar Phys.* **131**, 49
Feldman, U., Doschek, G.A., Mariska, J.T., Bhatia, A.K., Mason, H.E.: 1978, *Astrophys. J.* **226**, 674
Feldman, W.C., Asbridge, J.R., Bame, S.J., Fenimore, E.E., Gosling, J.T.: 1981, *J. Geophys. Res.* **86**, 5408
Fineschi, S., Romoli, M.: 1994, these proceedings
Gosling, J.T., Borrini, G., Asbridge, J.R., Feldman, W.C., Hansen, R.T.: 1981, *J. Geophys. Res.* **86**, 5438
Guhathakurta, M., Rottman, G.J. Fisher, R.R., Orrall, F.Q., Altrock, R.C.: 1992, *Astrophys. J.* **388**, 633
Hassler, D.M., Gardner, L.D., Kohl, J.L.: 1992, in *Coronal Streamers, Coronal Loops and Solar Wind Composition*, ESA SP-348, p. 375
Hénoux, J.C., Somov, B.V.: 1992, in *Coronal Streamers, Coronal Loops and Solar Wind Composition*, ESA SP-348, p. 325
Hiei, E., Hundhausen, A.J., Sime, D.G.: 1993, *J. Geophys. Res.* , submitted
Houminer, Z., Gallagher, F.: 1993, *Solar Phys.* **145**, 359
Hundhausen, A.J.: 1977, in *Coronal Holes and High Speed Wind Streams*, ed. J.B. Zirker, p. 226
Illing, R M.E., Hundhausen, A.J.: 1986, *J. Geophys. Res.* **91**, 10951
Jackson, B.V.: 1991, *J. Geophys. Res.* **96**, 11307
Jackson, B.V.: 1992, in *Coronal Streamers, Coronal Loops and Solar Wind Composition*, ESA SP-348, p. 97

Jackson, B.V., Hick, P., Webb, D.: 1993, *Adv. Space Res.* **13**, 43
Kahler, S. : 1991, *Astrophys. J.* **378**, 398
Kahler, S., Hundhausen, A.J.: 1992, *J. Geophys. Res.* **97**, 1619
Kopp, R.A.: 1992, in *Coronal Streamers, Coronal Loops and Solar Wind Composition*, ESA SP-348, p. 53
Koutchmy, S.: 1977, *Solar Phys.* **51**, 399
Koutchmy, S., Livshits, M.: 1992, *Space Sci. Rev.* **61**, 393
Koutchmy, S.: 1992, in *Coronal Streamers, Coronal Loops and Solar Wind Composition*, ESA SP-348, p. 73
Kundu, M.R., Schmahl, E.J., Gopalswamy, N., White, S.M.: 1989, *Adv. Space Res.* **9**, 45
Linker, J.A., Van Hoven, G.: 1990, *Geophys. Res. L.* **17**, 2281
Lundstedt, H. : 1989, *Solar Phys.* **123**, 177
Meyer, J.-P.: 1991, *Adv. Space Res.* **11**, 269
Meyer, J.-P.: 1993, *Adv. Space Res.* **13**, 377
Noci, G., Poletto, G., Suess, S.T., Wang, A.-H., Wu, S.-T.: 1993, *Solar Phys.* **147**, 73
Perry, R.M., Altschuler, M.D.: 1973, *Solar Phys.* **28**, 435
Poland, A.I.: 1978, *Solar Phys.* **57**, 141
Saito, K.: 1970, *Ann. Tokyo Astr. Obs. ser. 2* **12**, 53
Schultz, G.: 1994, these proceedings
Sheeley, N.R., Jr., Wang, Y.-M.: 1991, *Solar Phys.* **131**, 165
Sime, D.G.: 1989, *J. Geophys. Res.* **94**, 151
Tsuneta, S., Hara, H., Shimizu, T., Acton, L.W., Strong, K.T., Hudson, H.S. and Ogawara, Y.: 1992, *Publ. Astron. Soc. Japan* **44**, L63
Van de Hulst, H.C.: 1950, *Bull. Astron. Inst. Neth.* **11**, 135
Vauclair, S., Meyer, J.-P.: 1985, *Proc. 19th Int. Cosmic Ray Conf.* **4**, 233
von Steiger, R.: 1992, in *Coronal Streamers, Coronal Loops and Solar Wind Composition*, ESA SP-348, p. 309
von Steiger, R., Geiss, J.: 1989, *å* **225**, 222
Wang, Y.-M., Sheeley, N.R., Jr.: 1990, *Astrophys. J.* **365**, 372
Wilson, D.C., McQueen, R.M.: 1974, *J. Geophys. Res.* **79**, 4575
Wilson, D.C.: 1977, *The Three Dimensional Solar Corona: a Coronal Streamer*, Ph.D. Thesis, University Colorado, Boulder
Withbroe, G.L., Kohl, J.L., Weiser, H., Munro, R.H.: 1982, *Space Sci. Rev.* **33**, 17
Wu, S.-T., Wang, A.-H.: 1991, *Adv. Space Res.* **11**, 195

ULTRAVIOLET SPECTROSCOPY OF THE EXTENDED SOLAR CORONA DURING THE SPARTAN 201 MISSION

J.L. KOHL, L.D. GARDNER, L. STRACHAN and D.M. HASSLER*
Harvard-Smithsonian Center for Astrophysics
Cambridge, Massachusetts 02138 U.S.A.

Abstract. The instruments on the Spartan 201 spacecraft are an Ultraviolet Coronal Spectrometer and a White Light Coronagraph. Spartan 201 was deployed by the Space Shuttle on 11 April 1993 and observed the extended solar corona for about 40 hours. The Ultraviolet Coronal Spectrometer measured the intensity and spectral line profile of HI Lyα and the intensities of OVI 103.2 and 103.7 nm. Observations were made at heliocentric heights between 1.39 and 3.5 $R_\odot$. Four coronal targets were observed, a helmet streamer at heliographic position angle 135°, the north and south polar coronal holes, and an active region above the west limb. Measurements of the HI Lyα geocorona and the solar irradiance were also made. The instrument performed as expected. Straylight suppression, spectral focus, radiometric sensitivity and background levels all appear to be satisfactory. The uv observations are aimed at determining proton temperatures and outflow velocities of hydrogen, protons and oxygen ions. Preliminary results from the north polar coronal hole observations are discussed.

Key words: Solar corona, Solar wind, UV spectroscopy

1. Introduction

The Spartan Ultraviolet Coronal Spectrometer was developed for spectroscopic studies of the expansion of the solar corona into the solar wind. The observational goal is to obtain a more complete empirical description of the extended corona than has been derived in the past from broad band polarimetry which determines electron density and the associated geometry and temporal behavior of observed coronal structures. Ultraviolet spectroscopy has the potential to provide information about the thermal and nonthermal velocities of the protons including their bulk outflow velocity and to provide direct measurements of the electron temperature. Ultraviolet spectroscopy can also provide information about the velocity distribution, bulk outflow velocity and density of minor ions in the expanding coronal plasma. It is best to combine the ultraviolet spectroscopy with visible polarimetry and in situ measurements of solar wind parameters in order to obtain as complete an empirical description as possible. In the case of Spartan, there is a companion White Light Coronagraph (Fisher and Guhathakurta, 1993) which provides polarized radiance observations that are co-registered with the spectroscopic observations. Flights at the time of the Ulysses polar passes are planned for September 1994 and 1995 when polar solar wind

* Now at High Altitude Observatory, Boulder, Colorado 80307 U.S.A.

Space Science Reviews **70**, 253–261.

parameters will be measured together with the Spartan coronagraphic observations. Later in 1995, ultraviolet spectroscopy will be carried out with the SOHO Ultraviolet Coronagraph Spectrometer (UVCS). That instrument will have higher spectral, spatial and time resolution than Spartan, and will observe a much broader spectral range.

2. Characteristics of the Spartan Ultraviolet Coronal Spectrometer

The Spartan Ultraviolet Coronal Spectrometer consists of an externally and internally occulted telescope system, a dual spectrometer, two uv detectors and electronics. There are two channels, one for observing the spectral line profile and intensity of the HI Lyα line and the other for measuring line intensities of the O VI doublet at 103.2 and 103.7 nm.

The Lyα channel has a spherical telescope mirror with a focal length of 47.5 cm, which is coated with Al + MgF_2 and an Ebert Fastie spectrometer with a discrete multi-anode detector. There are separate entrance slits for measuring the line profile and the integrated intensity. The size of the former slit corresponds to a spatial resolution element of 0.5 by 2.5 arcmin and the latter to 4.0 by 5.0 arcmin. The detector has two linear arrays, a 48 pixel array with effective spectral resolution elements of 0.025 nm, which are used for profile measurements, and a 42 pixel array with 0.2 nm elements for measuring integrated intensities. The detector has a CsI photocathode, a MgF_2 window and a 2 l s^{-1} ion pump.

The telescope mirror for the O VI channel is similar to that for Lyα except it has an irridium coating. The spectrometer has a spherical diffraction grating on a Rowland circle mount and a detector system which includes a rectangular aperture that acts as the spectrometer exit slit, and two channel electron multipliers (CEM's) which detect the light from each of the two O VI lines. The CEM's are housed in a vacuum canister with a vacuum door, and there is a 2 l s^{-1} ion pump. The CEM's are coated with KBr. The entrance slit for O VI corresponds to a spatial resolution element of 2.5 by 5.0 arcmin. For observations of the solar disk, a mechanism inserts a pinhole of about 0.01mm diameter in front of the Lyα profile slit and another in front of the O VI slit.

The field-of-view is illustrated in Figure 1. At any instant of time, the instrument observes three spatial elements corresponding to the three entrance slits. The telescope mirrors can be rotated to scan the slits parallel to a radial line that passes through the field-of-view. Heliocentric heights between about 1.5 and 3.5 $R_\odot$ can be selected. The spacecraft can be offset pointed to allow observations down to 1.39 $R_\odot$, and onto the solar disk. The spacecraft is rotated to place the field-of-view at the desired position angle.

The uv instrument, along with the White Light Coronagraph, is housed

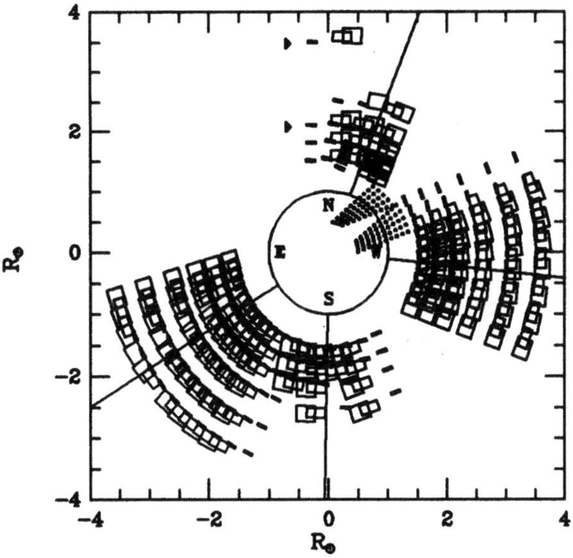

Fig. 1. Spatial elements observed by the Ultraviolet Coronal Spectrometer during the Spartan 201-1 mission are illustrated. At any instant, the instrument observes three spatial elements (e.g., the group of three rectangles at the extreme top of the figure). The smallest and largest rectangles are, respectively, for observations of the line profile and integrated line intensity of HI Lyα, and the intermediate sized one is for the intensity of OVI lines. Telescope motions are used to vary the heliocentric height and the entire spacecraft is rotated to vary the position angle.

in an evacuated cylinder with an aperture door that is opened after the Spartan is released from the shuttle. Spartan is battery powered, has no command or data links, operates autonomously executing a preprogrammed set of commands, and stores the data on magnetic tape. Four coronal targets corresponding to four primary position angles, can be selected just prior to deployment. The first is determined by the Spartan orientation at release, and the others by attitude adjusts that are keyed into the spacecraft controller about 12 hours before deployment.

3. Overview of the Observations

Spartan 201-1 was deployed from the space shuttle on 11 April 1993 and spent about 40 hours observing the extended solar corona. The four primary targets for Spartan Mission 201-1 were the north polar hole, the south polar hole, a helmet streamer at position angle 135° and an active region above the west limb. Center to limb scans (from −8.0 arcmin to the limb) of the HI Lyα profile and the OVI intensity were made at position angles between 281° and 352°. There also were measurements of the HI Lyα geocorona at five orientations between 95° and 160° from the Sun-center axis. Background measurements were made while the instrument was in the Earth's umbra and facing the Earth's surface. In order to distinguish between a light signal and background, data were accumulated with two very different areas of the telescope mirrors uncovered.

Figure 1 indicates all of the solar spatial elements observed during the mission. Concentrating on the smallest rectangles, which indicate the HI Lyα profile measurements, it can be seen that the four targets were sampled extensively, but not completely. The most complete coverage was for the helmet streamer at 135° and for the south polar hole.

The remainder of this paper will concentrate on the HI Lyα profile observations of the north polar hole. It was observed at three orientations. There were observations at 1.54, 1.83, 2.13, 2.52, 3.02, and 3.52 $R_{\odot}$. In addition, offset pointing was used to observe at 1.39, 1.49 and 1.59 $R_{\odot}$. Examples of the observations are discussed in the remainder of this paper. The observations at position 2.13 $R_{\odot}$ (height), 8.0° (position angle) and at 3.52 $R_{\odot}$, 4.0°, which are discussed in the present paper, were taken on 12 April 1993 at 18:27(UT) and 19:57 respectively.

4. Inflight Instrument Performance

The initial evaluation of the in-flight performance was based, primarily, on the observations at 3.52 $R_{\odot}$ in the north polar hole, the geocorona observation at 95° to the Sun, and the background measurements in the Earth's umbra. Inflight performance was also inferred from the preflight and postflight calibrations.

In the case of radiometric sensitivity, the laboratory calibrations indicated that the instrument retained its sensitivity during the mission.

The spectral resolution profile was checked by observing the geocoronal HI Lyα line (see Figure 2). The actual linewidth is expected to be much narrower than the instrument, and so the measured profile is indicative of the instrument's response to monochromatic light. The full width at half maximum (FWHM) is 0.027 nm.

An upper limit on the stray light levels during the mission can be estimated by comparing the shape of the HI Lyα profile on the disk (see Figure 3) to the observation of the north polar hole at 3.5 $R_{\odot}$ (see Figure 4). The latter consists of a narrow component attributable, primarily, to the geocorona, and a broader coronal component. The line shape for that observation will be discussed in more detail in section 5. Stray light is expected to be due to scattering of solar disk light in the instrument. It would therefore have the spectral line shape of the solar disk as measured with the Spartan instrument (Figure 3), which has a line width that is between that of the observed narrow and broad components of Figure 4. By comparing the two figures it is clear that the stray light level is small compared to the observed coronal intensities.

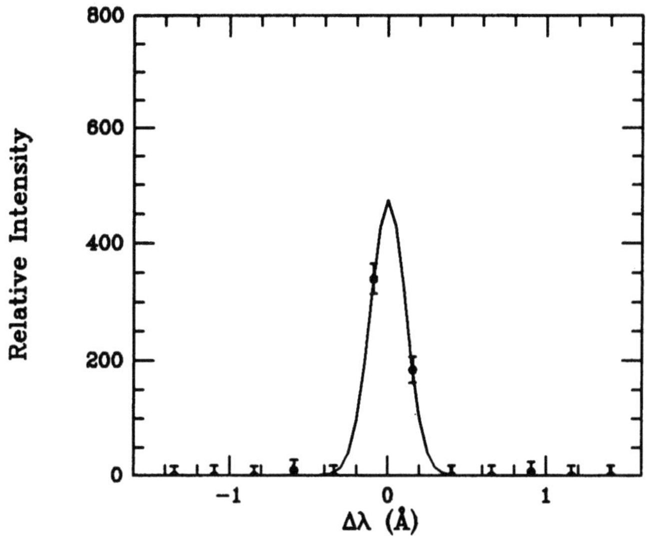

Fig. 2. The geocoronal HI Lyα line observed in the Earth's umbra. Since the geocoronal line is much narrower than the instrument resolution profile, it provides an in-flight measurement of the spectral resolution.

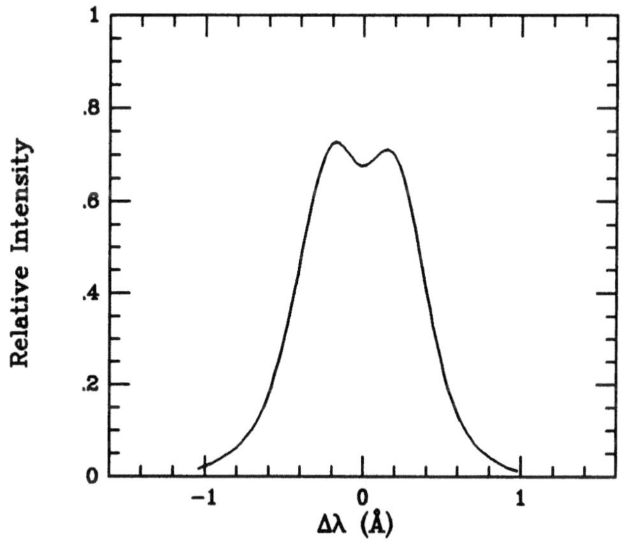

Fig. 3. Convolution of the HI Lyα line profile from the solar disk (Gouttebroze, *et al.*, 1978) and the instrument line profile. Instrument stray light will have this same line profile.

The background count rate during the mission was determined in two ways. One approach was to collect data while the instrument was in the Earth's umbra and pointing toward the Earth's surface. In that case, measurements were made with two different mirror areas exposed, a relatively

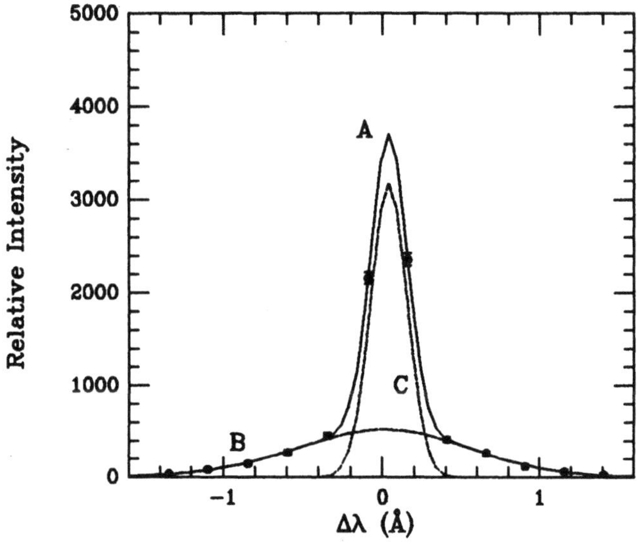

Fig. 4. The observed HI Lyα line at a position angle of 4.0° and a heliocentric height of 3.52 $R_\odot$. The solid points are the raw data, which have 1 σ statistical uncertainties of about 3 percent. The Curve A-B is a fit to the data by the two Gaussian curves indicated by B and C.

small area and a much larger area. The detector background was separated from detected photons by making use of the fact that the photon signal scaled with the unvignetted area and the background did not. The other method was based on the count rate in the far line wings. Laboratory data were used to relate the background in the central pixels to that in the line wing. It was also necessary to correct the data for cross talk using laboratory data. Both techniques indicated that the average background per pixel was about 0.013 counts per second.

5. Preliminary Results from the Observation of the North Coronal Hole

Observations by the Spartan White Light Coronagraph indicate that the north polar region is a coronal hole which is centered at position angle 0° and extends from −40° to 40°. The hole appears to have a substructure consisting of rays that extend outward along radial lines (Fisher and Guhathakurta, 1993). There may also be foreground and background structures along the line-of-sight.

The HI Lyα profiles for the observations at 3.52 and 2.13 $R_\odot$ at position angle 4.0° and 8.0°, respectively, (spatial elements indicated by arrows in Figure 1) are shown in Figures 4 and 5 respectively. The analysis presented here assumes that the line profiles only depend on the velocity distribution of

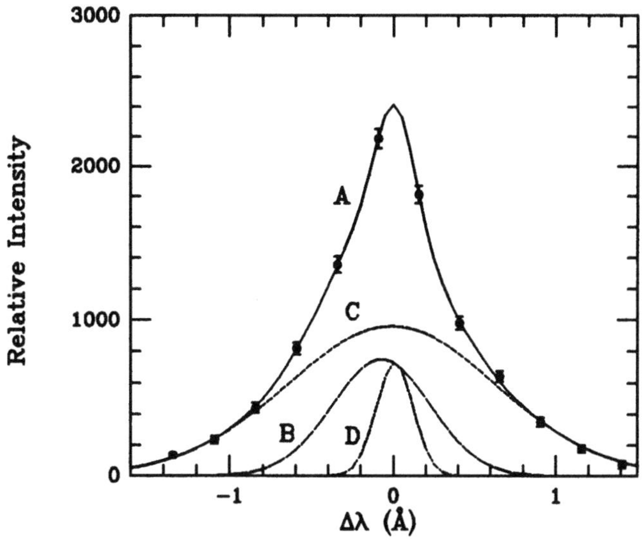

Fig. 5. The observed HI Lyα line at a position angle of 8.0° and a heliocentric height of 2.13 $R_\odot$. The solid points are the raw data, which have 1 σ statistical uncertainties of about 3 percent. The Curve A is a fit to the data by the three Gaussian curves indicated by B, C, and D.

hydrogen along the line of sight, and ignores the effect of resonant scattering on the line shapes because it is known to be small (Withbroe, *et al.*, 1982). The purpose here is to provide an indication of the data quality and some interesting characteristics of the profile shapes. Future papers will describe the observations more fully and provide a more detailed analysis.

The observation at 3.5 $R_\odot$ obviously consists of at least two components, a narrow component, which has the shape of the instrument profile, and a broader component. Most of the intensity of the narrow component is attributable to the Earth's geocorona which is formed by resonant scattering of solar HI Lyα by atomic hydrogen in the Earth's atmosphere and to absorption near the center of the coronal profile by atmospheric HI along the line-of-sight. Because the geocoronal line is very narrow compared to the instrument resolution, the shape of the geocoronal component is indistinguishable from the instrument profile. A coronal contribution to the narrow component is also a possibility.

The broad component is expected to be formed in the extended corona. It has a Doppler half width of 0.078 nm which corresponds to a hydrogen kinetic temperature of 2.2×10^6 K. Charge transfer between neutral hydrogen and the protons is expected to be rapid enough to ensure that the hydrogen and proton velocity distributions are identical at the observed heights (Withbroe, *et al.*, 1982).

The observation at 2.13 $R_\odot$ appears to consist of at least three components. As in the case of the observations at 3.5 $R_\odot$, the geocorona is expected to have a width that is essentially identical to the instrument profile. Because there is a significant geocoronal absorption of the coronal light the net geocoronal intensity is less than that observed at 3.5 $R_\odot$. The remainder of the profile cannot be fit with a simple Gaussian curve such as that to be expected for a purely thermal and spherically symmetric corona. It appears instead that the profile consists of a central portion with a Doppler half width of about 0.040 nm, which corresponds to a proton kinetic temperature of about 5.8×10^5 K and broad line wings. If you treat the line wings as a separate component, the best fit is for a Doppler half width of about 0.091 nm which corresponds to a proton kinetic temperature of about 3.0×10^6 K.

Observations at about the same position angle and at heights of 1.5 and 1.8 $R_\odot$ also exhibit enhanced line wings and a line core. The line core has a half width corresponding to a proton kinetic temperature of about 8.3×10^5 K at 1.5 $R_\odot$ and 1.03×10^6 K at 1.8 $R_\odot$.

Several possibilities for explaining the complex shapes of the observed profiles are being investigated. In addition to a deeper analysis of the present data, this involves an analysis of Yohkoh observations and several ground based observations of the north polar region.

The explanations fall into four general categories. One possibility is that one of the coronal components results from a foreground or background structure, and the other from the coronal hole itself. For example, the narrow component which has a kinetic temperature of about 8.3×10^5 K at 1.5 $R_\odot$ and 5.8×10^5 K at 2.1 $R_\odot$, may be from the coronal hole, and the broader component from a foreground or background structure. Synoptic observations by Yohkoh and ground observations of the Fe XIV green line (530.3 nm) (Esser, 1993) may provide evidence for the existence of these structures. However, any such explanation must consider the reduction in the neutral hydrogen density and hence HI Lyα intensity that would result from the ionization balance in a plasma with a high electron temperature relative to a cooler plasma.

Another explanation is that the ray structures observed by the White Light Coronagraph may be the source of either the narrow or the broad component. In this case there would be a much different proton velocity distribution in the rays than in the surrounding coronal hole plasma. However, the electron temperatures may not be so different.

It is also possible that waves propagate through the corona and across the line-of-sight. The waves would occupy some fraction of the line-of-sight leaving the rest as undisturbed corona. Transverse wave velocities could then account for the broad components. The rays are one candidate for the site of the waves.

In the above cases, there are two separate velocity distributions from two spatially isolated regions. There are also several possibilities for accounting for the observed profile shape with a single velocity distribution. For example, one could envision a coronal hole with a flow geometry that is highly non-radial with a bulk outflow velocity from the near side of the coronal hole that flows nearly directly at the observer while that from the other side flows nearly directly away. An explanation for the observed profile widths of the broad component at 2.13 $R_{\odot}$ would require outflow velocities of about 200 km s^{-1}. A difficulty in that explanation is that in order to explain the broad wings over a broad range of heights, the geometry must be such that the highly non-radial flows also occur over that same range of heights.

Another explanation is a highly non-Maxwellian velocity distribution in a predominantly single region plasma. Models of this type have been described by Scudder (1992). The general idea is that mechanisms for thermalizing the protons in the high velocity tail of the distribution are not sufficient and so the relative number of ions with high velocities becomes larger as the coronal plasma flows outward into lower density regions.

It is important to point out that the present observation is providing direct information about the hydrogen velocity distribution and probably about that of the protons. However, the electron temperature is not being measured here, although it may be possible to infer it from ionization balance modeling.

At present, all of the above explanations, and perhaps others, are in the realm of possibility. Work is currently underway to attempt to identify the most probable explanations for the observed profiles. This will require the use of a large set of coordinated ground and space based observations undertaken during the Spartan 201 mission.

Acknowledgements

This work was supported by NASA under Grant NAG5-613 to the Smithsonian Astrophysical Observatory.

References

Esser, R.: 1993, , Personal Communication

Fisher, R. and Guhathakurta, M.: 1993, , Personal Communication

Gouttebroze, P., Lemaire, P., Vial, J. C., and Artzner, G.: 1978, *Astrophysical Journal* **225**, 655

Scudder, J. D.: 1992, *Astrophysical Journal* **398**, 319

Strachan, L., *et al.*: 1993, *Astrophysical Journal* **412**, 410

Strachan, L., Gardner, L. D., Hassler, D. M. and Kohl, J. L.: 1994, *Space Science Rev.* , this issue

Withbroe, G. L., Kohl, J. L., Weiser, H., and Munro, R. H.: 1982, *Space Science Reviews* **33**, 17

PRELIMINARY RESULTS FROM SPARTAN 201: CORONAL STREAMER OBSERVATIONS

L. STRACHAN, L.D. GARDNER, D.M. HASSLER* and J.L. KOHL
Harvard-Smithsonian Center for Astrophysics
Cambridge, Massachusetts 02138 U.S.A.

Abstract. The Ultraviolet Coronal Spectrometer on Spartan 201 obtained measurements of H I Ly α line profiles and O VI λ103.2/103.7 nm integrated intensities in a helmet streamer. Similar measurements were made on the solar disk to characterize the incident radiation which is the source for the resonantly scattered component of the lines. A description of the H I Ly α observations in the streamer will be presented. A more complete physical description of the observed coronal regions will become available after the analysis of additional data from coordinated observations made by visible coronagraphs and x-ray telescopes.

Key words: Coronal Streamers, Solar Wind, UV Spectroscopy

1. Introduction

A helmet streamer on the SE limb of the Sun was observed with the Spartan 201 UV and white light coronagraphs during 11 April 1993. This paper reports on some of the preliminary results from the Spartan Ultraviolet Coronagraph Spectrometer observations. The UV coronagraph on Spartan has two separate spectrometer channels. One channel is used to provide spectral profiles and total intensities of H I Ly α (121.6 nm) and the other provides total intensities of the O VI doublet at 103.2/103.7 nm. A general overview of the Spartan UV Coronal Spectrometer flight performance and all of the UV observations is discussed by Kohl, *et al.* (1994).

The primary goal for this investigation is to derive or constrain the basic plasma parameters of the observed streamer. These parameters include the temperature and density of electrons and hydrogen, and the bulk outflow velocity of neutral hydrogen. The behavior of the protons is expected to be coupled to the neutral hydrogen through fast charge transfer processes (Withbroe *et al.*, 1982). Additionally, spectroscopic measurements of the O VI doublet line will be used to determine and constrain the abundance and outflow velocity of O^{5+} ions. The present discussion will focus on the results for the proton kinetic temperatures along the streamer axis.

2. H I Ly α Profiles

Figure 1 shows the locations in the region of the SE streamer where the 59 profile observations were made. The axis of the streamer is identified by the

* Now at the High Altitude Observatory, Boulder, CO 80307 U.S.A.

Space Science Reviews **70**, 263–266.

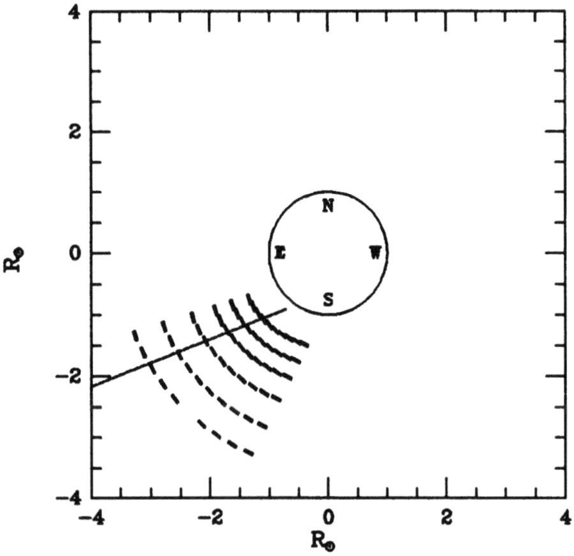

Fig. 1. The boxes show the locations where H I Ly α spectral profile measurements were obtained. The axis of the streamer is indicated by the dashed line.

dashed line in the figure. During the Spartan 201 observations, the streamer was on the East limb and its base (at 1.5 $R_\odot$) was located at a position angle of 135° from the north heliographic pole. Spatial coverage for the profile measurements included the streamer and the surrounding coronal hole and quiet regions.

Each spectral profile was obtained from a 0.5 arcmin by 2.5 arcmin region in the corona. Profiles were obtained at heliocentric heights of 1.5, 1.8, 2.1, 2.5, 3.0, and 3.5 $R_\odot$ for nine out of 10 radial scans of the Spartan UV spectrometer field of view. Nominal dwell times for the observations are 1.0, 1.0, 1.5, 3.0, 4.5, and 8.5 minutes for the heights indicated for the nine scans. The scan without an observation at $r = 3.5R_\odot$ has dwell times that are twice as long as those for the other scans.

Examples of the quality of the observed profiles are shown in Figures 2a - 2c. The profiles are sampled at 0.025 nm intervals and are plotted for the 0.4 nm band centered at 121.6 nm. The relative uncertainty in the peak is roughly 3 % (1 σ) for all profiles. A prominent feature of these profiles is that the peak intensity has been enhanced or reduced by the addition or subtraction of radiation by neutral hydrogen in the geocorona above the altitude of the Spartan orbit. This spectral feature is broader than the actual geocoronal line because of broadening by the Spartan Ly α spectrometer instrument profile which is 0.027 nm (FWHM). The geocorona is manifested as an absorption feature when the absorption of the coronal radiation is greater than the geocoronal emission (which is due to scattering

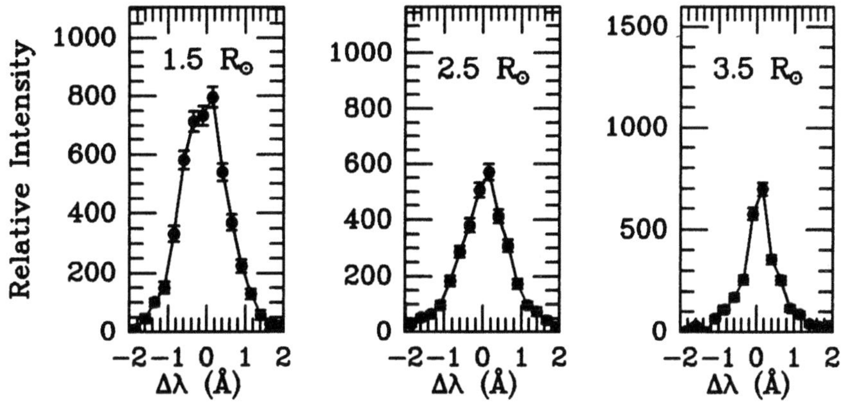

Fig. 2. H I Ly α profiles from a) 1.5, b) 2.5, and c) 3.5 $R_{\odot}$ along the streamer axis. Solid points are detector counts with $\pm 1\sigma$ uncertainties.

of radiation from the chromosphere). The net geocorona contribution to the profile can be positive or negative depending on whether the geocoronal "dayglow" is greater than or less than the total absorption in the solar coronal profile. The profile in Figure 2a is an example where there is a net absorption at the profile peak. The profiles in Figures 2b and 2c show the opposite case where there is a net emission.

To determine the true profile for the solar coronal radiation alone, each observed profile is fitted by the sum of two gaussians. A narrow gaussian which has an 1/e width that is the same as the instrument profile is normalized so that its intensity equals the net geocoronal contribution. The second gaussian is an approximation of the true coronal profile. In actuality, coronal profiles are not gaussian because of the line of sight effects of the resonant scattering process (Withbroe, *et al.* 1982) The gaussian profile fits are only used to approximate the proton rms velocity distribution along the line of sight. In a more detailed analysis, the observed profiles are to be compared with the profiles produced from a line of sight integrated model. The simple gaussian approximation for the coronal profiles is appropriate for the streamer if it is localized near the plane of the sky.

The 1/e widths of the fitted profiles can be used to derive a proton kinetic temperature which can be ascribed to both thermal and non-thermal motions of the particles. We define the relation between the coronal 1/e width $\Delta\lambda$ and the kinetic temperature T_k by $\Delta\lambda = \frac{\lambda_o}{c}\left(\frac{2kT_k}{m_p}\right)^{1/2}$, where the symbols k, c, m_p have their usual meanings and λ_o is the Ly α line center in the hydrogen rest frame. The results for the kinetic temperature as function of heliocentric distance are shown in Table I. Uncertainties (1 σ) in each temperature value is about 0.25×10^6. The kinetic temperature is approximately constant with a hint of a peak temperature at $2.1R_{\odot}$.

TABLE I
Proton kinetic temperatures corresponding to Doppler widths of the observed coronal H I Ly α profiles.

Heliocentric Height ($R_\odot$)	Kinetic Temperature (K)
1.5	2.6×10^6
1.8	2.9×10^6
2.1	3.2×10^6
2.5	3.0×10^6
3.0	2.7×10^6
3.5	2.2×10^6

3. Discussion

One explanation for the nearly constant total line widths is that the thermal and nonthermal components compensate (one decreases by the same amount as the other increases) over the interval, $1.5 \leq r \leq 3.5 R_\odot$, along the streamer axis. This possibility, while unlikely, must be investigated with appropriate models that include turbulence and nonthermal heating. It is more likely that the streamer profiles are produced from an an extended region (*e.g.*, a streamer arcade) where there are velocity outflows toward and away from the observer. The weakness in this hypothesis is that there is no large change in the profile widths when streamer configuration changes from a closed to an open (radial) flow configuration. (For the observed streamer, it appears that the field lines open up near 3 $R_\odot$.)

Future work for the data analysis involves building an empirical model of the streamer and its surrounding regions in order to test whether or not the H I Lyα intensities and profiles can be predicted. The model will include the spatial dependences for the electron and proton temperatures, densities and outflow velocities.

Acknowledgements

This work was supported by NASA under Grant NAG5-613 to the Smithsonian Astrophysical Observatory.

References

Kohl, J. L., *et al.*: 1994, *Space Sci. Rev.* , this issue
Withbroe, G. L., Kohl, J. L., Weiser, H., and Munro, R. H.: 1982, *Space Sci. Rev.* **33**, 17

SPARTAN 201 WHITE LIGHT CORONAGRAPH EXPERIMENT

R.R. FISHER and M. GUHATHAKURTA *
NASA/GSFC
Greenbelt, MD, 20771

Abstract. A white light coronagraph was launched into orbit aboard the space shuttle OV 103 (Discovery) on 7 April 1993. This device was one of two instruments included in the SPARTAN 201-1 payload, a completely autonomous sub-satellite deployed from the shuttle for a period of about 47 hours. The WLC system is an externally occulted coronagraph system which incorporates a rotating half-wave plate polarimeter, and the image data is used to infer the brightness, the polarized brightness and the degree of polarization of the white light emission from the solar corona. These data are in turn used to infer estimates of the K- and F-coronal brightnesses and density distributions. We shall present preliminary results of the electron density estimate in the coronal streamer and hole region and describe the methods employed.

Key words: Sun – Corona

1. Introduction

The SPARTAN 201 Mission has as its scientific goal the investigation of the physical conditions of the solar corona and the coronal-heliospheric interface. The large-scale features of this part of the sun's atmosphere include both helmet streamer structures and coronal holes which are associated with the generation of high speed streams of material found in the solar wind. Coronal Mass Ejection (CME) events are also detected in this region of the solar atmosphere and this type of solar activity is known to be responsible for the addition of significant mass to the steady state solar wind

The material of the corona constitutes a close approximation to a fully ionized plasma, and while it is relatively easy to perform diagnostic spectroscopy on the lower solar corona, little experimental work has been done concerning definition of the physical conditions associated with the large-scale structure of the corona, particularly those which characterize the distributions of proton density and bulk flows. To provide a comprehensive diagnostic data set for the solar corona, as viewed in the volume from 1.3 to 6.0 solar radii, the SPARTAN 201 mission was equipped with both an Ultraviolet Coronagraph Spectrometer (UVCS) and a White Light Coronagraph (WLC). The former, supplied by the Smithsonian Astrophysical Observatory of Cambridge, MA, is a spectrographic experiment which uses radiation from the Lyman-alpha line of H I (1216A) and the O VI doublet (1032A and

* affiliated to USRA

Space Science Reviews **70**, 267–272.

1035A), to build up a spatially resolved data set. The White Light Coronagraph, originally designed and fabricated at the High Altitude Observatory of Boulder, CO and now operated by the Solar Physics Branch of the Goddard Space Flight Center's Laboratory for Astronomy and Solar Physics, concentrates on the photospheric light scattered by coronal electrons.

The operational goals of the WLC experiment are the determination of the distributions of the white light brightness (B), the polarized brightness (pB) and the degree of polarization (p), all of which are viewed in the plane of the sky. From these quantities, B, pB, and p, it is then possible to obtain model dependent distributions of the three dimensional coronal electron density.

2. Observations and Analysis

2.1 White Light Coronagraph Instrument

The WLC is an externally occulted coronagraph, third in the series of the HAO orbital instruments, which incorporates a polarimeter which is sensitive to the state of linear polarization. This instrument has three features which differ from the two previous (ATM SkyLab and Solar Maximum Mission) devices. The detector used is an RCA charge coupled device (CCD) which has an operational bandpass of 500 Å FWHM centered on 4800 Å. The first external occulting disk is not circular but is serrated in an effort to control diffraction effects at this point in the instrument. Finally, the polarimeter is constructed with a fixed sheet polaroid, and a half-wave retarder is rotated during the observation sequence so that all observations are made with the polaroid held motionless with respect to the occulting disk support stem. The SP201 WLC uses a flight spare doublet objective lense 3.5 cm in clear aperture which was fabricated for the SMM Coronagraph/Polarimeter instrument. The WLC instrument is approximately 3 m in length and fits into half of the 42 cm diameter instrument carrier module provided by the SPARTAN 201 spacecraft.

2.2 Phase of the Solar Cycle

The phase of the magnetic activity cycle of the sun was such that the mission occurred after the maximum of cycle 22 (1990) at approximately the same phase (declining part of the present spot cycle) as the ATM mission of 1973. The operational planning was supported by the soft X-ray imager experiment (SXT) on the Japanese Yohkoh satellite and two groundbased instrument systems, the MK-III imaging coronameter (HAO) and the disk spectromagnetograph (NSO- GSFC). Yohkoh images indicated that at the time of the flight there existed both large-scale coronal holes near the polar regions and moderate to modest sized disk active regions were present near the west limb of the sun. Daily data from the MK-III K-coronameter indicated the presence of a large, relatively stable helmet streamer at the south east limb of the sun, as well as enhanced levels of coronal brightness over the active regions found at the west limb.

2.3 Selection of Target Positions and Data Format

On the basis of these supporting observations four large-scale coronal targets were selected: (1) the relatively bright helmet streamer seen at a position angle (PA) of approximately 133 on the SE limb, the south polar region centered at a position angle of 190, the active west limb with the WLC field of view centered near the solar equator, and the north polar region at 330 PA. The WLC has an optical configuration which off sets the occulting disk in the field of view and allows about 180 degrees of position angle to be viewed out to an altitude of 3 solar radii above the limb (Figure 1a). The extreme field of view extends to 6 solar radii opposite the occulting disk stem, and the occulting disk was selected to be 1.25 solar radii for this first mission. This is different from the previous ATM and SMM instruments which had occulting disks which were somewhat larger and restricted the inner limit of the field of view to approximately 2 solar radii. The CCD format was 256 x 320 pixels each of which subtended 22.5 arc seconds in the final image plane (Figure 1a).

The pre-launch calibration of this instrument used identical equipment for the calibration of the ATM and SMM instruments. However a re-calibration of the standard filters was performed since this experiment operates at a different central wavelength than either of the two previous experiments.

2.4 Helmet Streamer and the North Polar Hole

Here we concentrate on the analysis of the polarization sequences of the large helmet streamer on the SE limb and the north polar region. The analysis presented below is regarded as preliminary, but the results are of interest both as a comparison with previous observations of similar types of structures and in the case of the polar region for new insight into the physical conditions of the atmosphere. Photometric corrections for dark current and instrument efficiency as a function of height (v-function) have been applied, and the brightness scale used depends upon the prelaunch calibration parameters. Units are presented in terms of a fraction of the mean brightness of the solar disk at the wavelength of operation.

In Figure 1a three images have been combined to visualize the solar corona from the disk to approximately 6 $R_\odot$. The innermost image was kindly provided to us by the Yohkoh experimenter, and is a soft X-ray image of the lower corona. Next we added a moderate spatial resolution image of the inner corona, made with the MK-III imaging coronameter, and surrounding these two images is the image from the SPARTAN-201 WLC.

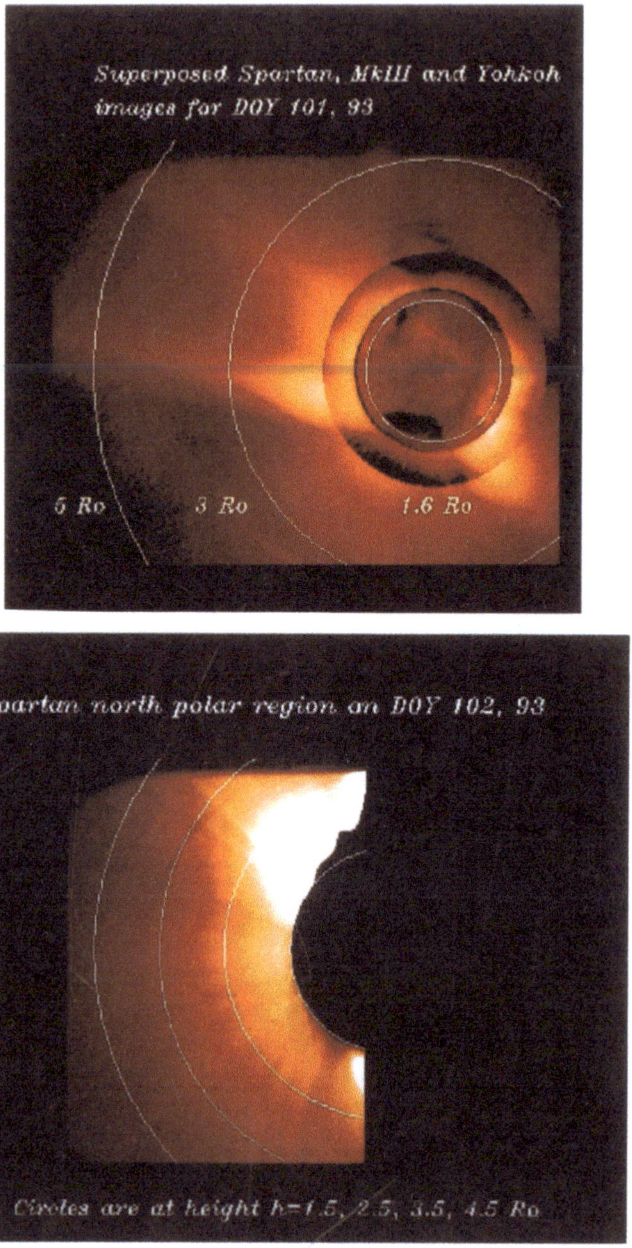

Fig. 1. a) Superposed images of Yohkoh, MkIII and Spartan 201 from sun center out to 6 $R_{\odot}$ high-lighting the south east streamer at PA 133 at 1.5 $R_{\odot}$. Here north is rotated 45° to the right from the vertical line of the frame and east is rotated 45° to the left from the vertical. b) The highly structured image of Spartan north polar region. Here north is rotated 30° counterclockwise from the equator on the left hand side.

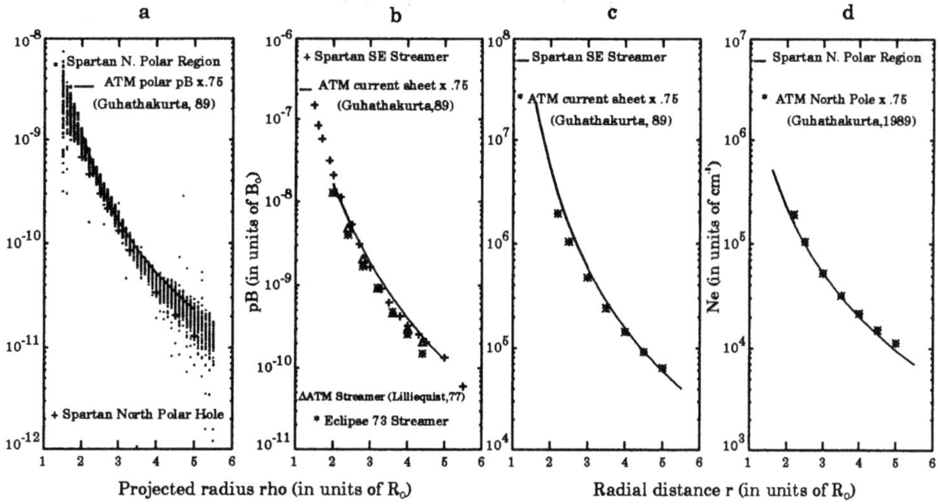

Fig. 2. a) Estimates of pB from the Spartan polar region in an angular bin of 30° with 1° resolution. The + sign represents the average pB values of the dark channels as described in text 1. b) Estimates of pB from the Spartan south east streamer and comparison with previous measurements (see section 2.4). c) Electron density model for the the south east streamer and d) average electron density for the north polar region.

The helmet streamer under investigation is a non-radial feature which is inclined somewhat toward the solar equator. The brightest portion of the streamer observed in pB at a given height is plotted (Figure 2b) as a function of height from 1.5 to 6 $R_\odot$. At 1.5 $R_\odot$ streamer axis was located at 133 PA where as at 6 $R_\odot$ the axis was at 117 PA. Over plotted on this diagram are the measurements shown by Lilliequest (1977) of a streamer observed at the 30 June 1973 total solar eclipse made by the HAO White Light Coronal Camera (WLCC) and the measurements of the same streamer from observations gathered by the WLC carried on the ATM-SkyLab. Also superposed on this diagram is the variation observed in the heliospheric current sheet (multiplied by a factor of .75) which was determined by Guhathakurta (1989) and Guhathakurta et al., 1994, from the ATM-Skylab data set. Streamers show some range of brightness and the constant pB difference between the data sets from 1973 and 1993 which are separated by roughly one full solar magnetic activity cycle, is due to differences in calibration and/or in the intrinsic variability of the solar features.

In Figure 1b we present a view of the north polar region of the sun, which is highly structured. The bright structures visualized are about 130,000 Km in width in the plane of the sky. With Spartan data set we are unable to resolve whether these structures are embedded in the coronal hole or if they are structures from the foreground and background (which originate from latitudes higher than about N50°) seen in projection over the pole. In Figure 2a we have plotted *pB* values as a function of height for the range of position angles 0-30 degrees (30° angular bin) for three different frames. Solid line represents observed ATM polar *pB* multiplied by 0.75 (Guhathakurta, 1989, Guhathakurta et al., 1994). Also shown on this diagram is the average *pB* variations observed in the two broad dark channels in the upper and lower part of Figure 1b at 320 PA and 40 PA respectively. We believe that the dark regions may be the real coronal hole and the *pB* values from these regions tend to fall at the lower boundary of the range of *pB* bars at each height and the elevated values are due to the brightest of the relatively faint rays. The radial gradient is clearly different at extended heights for the two types of structures, the bright rays and the dark lanes.

The observed values of *pB* as a function of height in the streamer and coronal hole region, combined with the assumption that these values vary with r only, permit calculation of electron density following van de Hulst (1950) technique. The density estimates yield values that are lower than those of the helmet streamers observed in 1973 by a factor of 0.75 (Figure 2c). At the present time we believe that this factor may be attributed to a recalibration of the filter system used to establish photometric scale within the facility used for setting instrument sensitivity limits. In the case of the comparison of density of the two polar regions (Figure 2d) we find two things: (1) The dark regions indicate (if they truly represent coronal hole regions) that densities obtained from SP 201 are a factor of 2 lower than those estimated from ATM-Skylab data and (2) The scale height temperature determined from these data is cooler for these types of regions, being 1.02 MK, while that estimated from the observations of the Skylab data and previous orbital coronagraph observations is around 1.3 MK. In the future we plan to attempt to analyze these data with more refined models of electron scattering from selected geometric structures.

Acknowledgements

The authors gratefully acknowledge the operational support of the YOHKOH experiment team and the Japanese Space Agency. The present work was supported in part by the High Altitude Observatory of the National Center for Atmospheric Research of Boulder, Colorado, and a grant from the NASA sub-orbital research program of the Space Physics Division.

References

Guhathakurta, M., 1989, Ph.D thesis, Univ. Denver;LASP/Univ. Colorado thesis.
Guhathakurta, M., Holzer, T.E., and MacQueen, R.M.,: 1994, *Astrophys. J.* , in press.
van de Hulst: 1950, *Bull. Astron. Inst. Neth.* **11**, 135.
Lilliequist, C., 1977, *NCAR Technical Note, NCAR/TN-128+str.*

SOFT X-RAY OBSERVATIONS OF HELMET STRUCTURES FROM SXT/YOHKOH

CHUNG-CHIEH CHENG
E.O. Hulbert Center for Space Research
Navel Research Laboratory
Washington, D.C. 20375, U.S.A.

and

SXT/YOHKOH TEAM

Abstract. We present soft X-ray observations of helmet structures in solar active regions obtained from SXT/Yohkoh. These helmet structures are observed to form in the flare decay phase and to be associated with active region loop interactions. Their morphology is similar to the much larger scale helmet streamers that appear in the outer corona as shown in optical images of solar eclipse. The observed X-ray helmet structures appear to be in quasi-equilibrium with lifetimes greater than the MHD time scale. Using the filter ratio method for the X-ray observations, we find that the cusp region has lower temperature and higher density than that in the stalk region above it. The plasma pressure in the cusp region is about the same or slightly higher than that in the stalk region.

Key words: SOLAR – X-ray – helmet structure

1. Introduction

One of the striking features in solar eclipse optical images is the helmet streamer that extends to several solar radii above the solar limb. These structures appear to be stable features that sometimes last several solar rotations (Newkirk, 1967; Koutchmy and Livshits, 1992). The steamers is produced by the complicated interaction between large-scale solar magnetic fields and the general expansion of the heated solar corona (Kopp, 1992). These features are of particular interest in the studies of the global evolution of the solar magnetic field. And the formation of the current sheet that separate the open and closed field regions is of intrinsic importance in our understanding of energy release in the stressed magnetic field. Previously, the helmet structures are only observed in the optical wavelengths either during eclipse or by coronagraph. Recent soft X-ray observations from Yohkoh have revealed a wealth of loop structures and interactions (Strong, 1993). In particular, helmet-like features with a sharply defined cusp have been seen in active regions. The morphology of these helmet-like structures is very similar to the optical helmet-streamers, but with much smaller spatial scales. In this paper, we present some X-ray observations of the active region helmet structures obtained from Yohkoh and discuss their physical properties and implications.

Space Science Reviews **70**, 273–278.

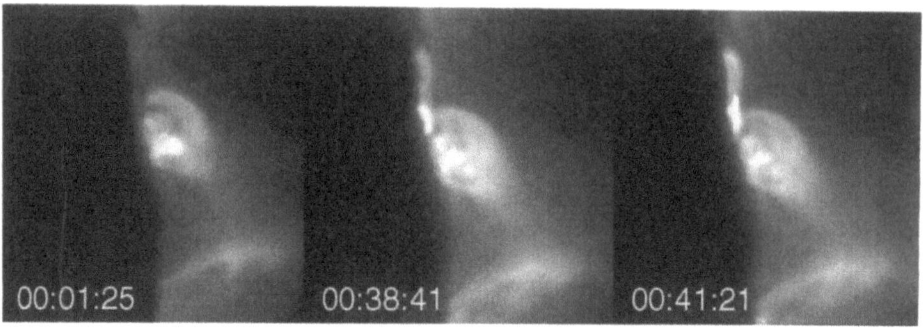

Fig. 1. Selected pre-flare X-ray images (320″ × 320″ FOV) of AR 7030 on the west limb. The images were taken with thin Al filter.

2. Helmet Structures in Flares

X-ray Helmet structures are observed in the active regions either in the decay phase of a flare or as a result of reconfiguration of magnetic fields in loop interactions. Below, we present examples of these two cases. The observations are obtained by the Soft X-ray telescope (SXT) on the Japanese solar satellite Yohkoh. The SXT can take solar images in the wavelength ranges from 3 to 60 Å. The full spatial resolution is $\approx 3''$. Depending on the observational mode, the time resolution can be as short as 0.5 sec as in flare mode (64 × 64 pixels). Active regions are generally observed in the quiet mode with a 128 × 128 pixel size and a $\approx$ 1 min time cadence. Using images taken in different filters, it is possible to derive the temperature and emission measure ([EM]) of the observed features. For a complete description of the SXT and its operational modes, see Tsuneta *et al.* (1991).

2.1 The 6 February 1992 Limb Event

2.1.1 Initial Reconfiguration

The flare occurred in NOAA AR 7030 on the west limb at 0320 UT. Many hours before the flare, the active region have been quite active. As seen in Figure 1, the region, during the pre-flare period from 0002 UT to 0041 UT, consisted of many loop systems. There were frequent transient brightening and loop enhancement. The brightenings that occurred above the limb around 0038 UT can best be described as due to loop interactions. Even at this time, there is some hint of an overall helmet-type morphology for the larger overlaying loops.

2.1.2 Explosive Energy Release

The activity on the limb continued in AR 7030 and then a flare of X-ray class M7.6 (Solar-Geophysical Data, NOAA) occurred at around 0310 UT. The

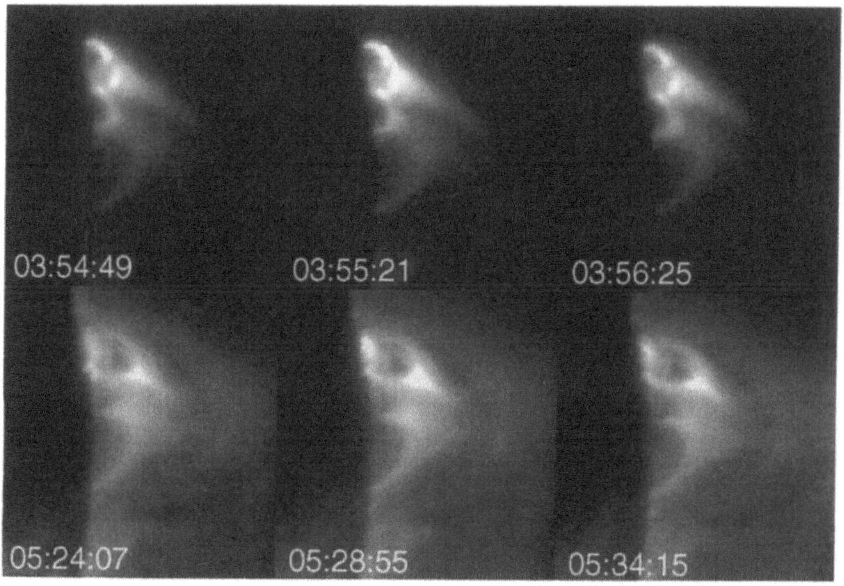

Fig. 2. Selected post-flare images (320″ × 320″ FOV) of AR 7030 on the west limb, showing the formation of multiple helmet structures. The images were taken with thin Al filter.

flare lasted a couple of hours with its X-ray intensity peaked around 0328 UT. Examination of the SXT images shows that the flare had a complicated configuration; it consisted of a bright blob and a bright loop next to it. The flare bright blob was located above the limb in the general vicinity of the brightenings at 0038 UT (Figure 1). It may not be surprising that flare energy was released in these locations where there were several intertwining loops. The energy release in the flare is itself of interest (Sterling and Shibata, 1993). For our purpose here, it suffices to note that the flare represented an explosive energy release in a highly stressed configuration, which set the stage for the helmet structure to develop in the flare decay phase.

2.1.3 Formation of the Helmet Structure

As the flare decayed after 0354 UT, the general configuration of the active region underwent drastic changes; the originally complicated morphology evolved into a much simpler helmet structure, as shown in Figure 2. These helmet structures changed gradually and were particularly striking in the images taken after 0520 UT. Two nicely delineated helmet features formed next to each other. The northern helmet structure had a clearly defined triangular cusp situated at the top of the loop with a narrow stalk extended outward. The loops in both of the helmet structures increased in height slowly with an apparent outward velocity of about 3 km $\sec^{-1}$. Since we have

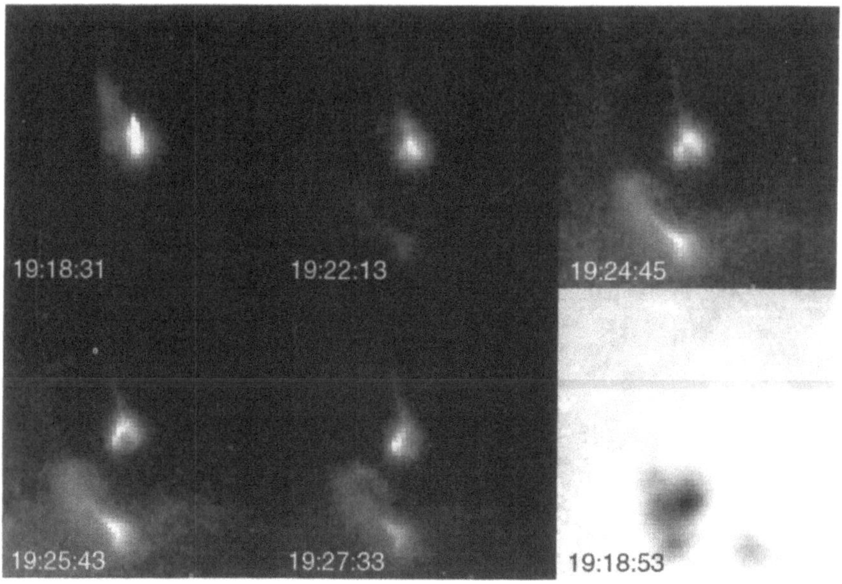

Fig. 3. X-ray images taken with thick Al filter of the 24 April 1992 flare, showing the formation of the tear-drop shaped helmet structure in the center of the FOV (160″×160″). Sunspot image was taken at 1918:53 UT.

no observations after 0620 UT, the final stage of these structures is unknown. Although the helmet structures continued to evolve gradually, they appeared to be stable with lifetimes much longer than the MHD time scales. In fact, when the active region returned to the east limb on 21 February, a flare was observed in a helmet structure whose configuration was very much similar to the one we just described (Tsuneta *et.*, 1992). The east limb helmet structure on 21 February could very well be the same structure observed on 6 February. The maintenance of the long lasting helmet structure is still unknown; magnetic reconnection could be responsible.

The temperature and [EM] can be derived from intensity ratios of X-ray images taken with different filters (Tsuneta *et al.*, 1992). For the northern helmet structure, using the thin Al and Be filter ratio, we find that the cusp region had a lower temperature of $\approx 1.4 \times 10^6$ K and a higher [EM] of $\approx 3.8 \times 10^{47}$ cm^{-47} than those in the stalk region above and the bright loop below. If we assume the same path length along the line of sight, we find that the pressure is approximately balanced along the helmet.

2.2 The 24 April 1992 Disk Event

This is another example of formation of a helmet structure in the flare decay phase. The flare occurred at N13W00 in AR 7138. In contrast to the limb event described above, the helmet structure in this flare was much smaller with a size of only $\approx 15''$. Figure 3 shows selected images of the flare. The flare was observed from 1918 UT in the flare rise phase. Due to over exposure, it is not clear what initiated the flare energy release. It appears that the flare resulted from the interaction between a preexisting larger loop and some emergent flux. The helmet structure appeared few minutes after the start of the flare and it became particularly apparent around 1925 UT. It is possible that the thin stalk might connect to some remote region. Note that the region south of the flare became bright at 1924 UT, and also showed possible helmet structure. This later region was the site of a flare some 7 hours earlier. Using filter ratio method, we find that the cusp region had lower temperature ($\approx 1.3 \times 10^6$ K) and higher [EM] ($\approx 2.8 \times 10^{47}$ cm^{-3}) than the "streamer" and the loop regions. The plasma pressure in the loop legs and the cusp were about the same but were higher than that in the stalk. Thus we expect outward mass motion along the stalk.

3. Helmet Structure from Loop Interactions

The above examples of helmet structures all involve sudden energy release in flares. Helmet structure also were observed in "quiet" active region. Figure 4 shows selected X-ray images of an active region on 3 May 1992. The dominating features is the bright helmet-like structure in the center of the field of view. At 0105 UT, a small bright loop appeared at the footpoint of a large preexisting loop. The ensuing interactions resulted in morphological change. At 0118 UT, the helmet structure became apparent. Near the end the observational period (0148 UT), the configuration has evolved into a nicely formed helmet structure with a small loop beside it.

4. Discussion

Compared with the optical helmet streamers, the X-ray helmet structures we described above have spatial scale one or two orders of magnitude less. Nevertheless, their morphological properties are strikingly similar. One important physical property the X-ray helmet structure shares with its larger optical cousin is that they both are rather stable features and represent the relaxed state of an evolving complicated magnetic field configuration. In the case of optical helmet streamer, it is the quasi-equilibrium state resulting from the interplay between the large scale magnetic field and the ever expanding corona. In the case of the active region helmet structure, it is the relaxed state of a stressed and complicated field configuration. The relaxation to the helmet structure from an original higher energy state can be caused by an explosive energy release, as in flares, or by a gradual magnetic reconnection as in the interaction between an existing loop and an

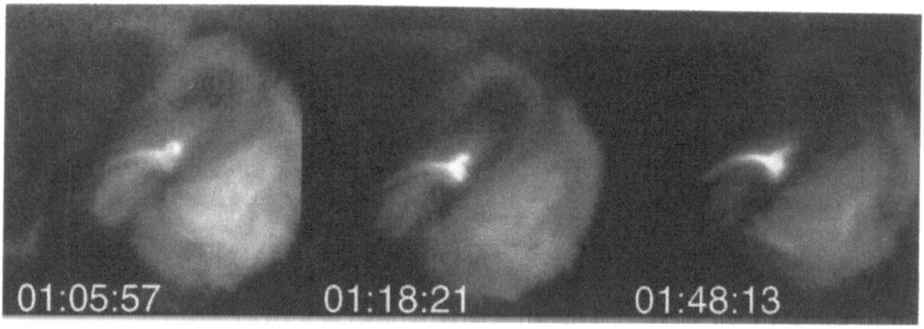

Fig. 4. Magnetic reconnection between an existing loop and an emergent flux resulted in a helmet structure in an "quiet" active region on 3 May 1992. The images were taken by SXT with thin Al filter with size of $320'' \times 320''$.

emergent flux. In either case, the helmet structure appears to be one possible quasi-equilibrium state with lower energy reached by the evolution of a complicated magnetic field-plasma configuration in the solar atmosphere.

To better define the physical properties of the helmet structure and its associated current sheet and cusp, which could be identified with a magnetic null point, we need more accurate determination of the temperature and density than that given by the filter ratio method used in this paper. Hopefully, future observations by CDS and SUMMER on SOHO will provide the necessary data. In addition, MHD numerical simulations of how an initially closed magnetic field configuration could become open and evolve into a helmet structure when energy is explosively deposited in it, are needed to sort out the physical processes involved in the formation of the helmet structures.

5. Acknowledgements

The authors sincerely thank the many people whose dedication and hard work made the Yohkoh mission a resounding success.

References

Kopp,R.: 1992, in *Pro. First SOHO Workshop*, ESA SP-348, p53.
Koutchmy, S., and Livshits, M.: 1992, *Space Sci. Rev.***61**, 393
Newkirk, G.: 1967, *Ann. Rev. Astr. Ap.***5**, 213.
Sterling,A., and Shibata, K.: 1993, private communications.
Strong, K.: 1993, in this proceedings.
Tsuneta, S. *et al.:* 1991, *Solar Phy.*, **136**, 37.

RADIO EMISSION FROM CORONAL STREAMERS

FRANCA CHIUDERI DRAGO
Dipartimento di Astronomia e Scienza dello Spazio
Università degli Studi di Firenze
Largo Enrico Fermi 5, 50125 Firenze, Italy

Abstract. Different models of coronal streamers are used to calculate the radio brightness temperature at the wavelengths of observation of the Nançays Radioheliograph. Calculation are performed assuming the location of the streamer both on the disk and at the limb. Their comparison with observations show that a satisfactory agreement with a particular model can be found in the shape and in the relative enhacement of the streamer with respect to the quiet Sun, although the absolute values of the computed brightness temperatures are much higher than the observed ones.

Key words: Sun: corona – Sun: radio emission – Sun: streamer

1. Introduction

The streamer models computed by Wang, Wu, Suess and Poletto (1993), hereafter referred to as W^2SP and a modification of one of them (Suess, private communication) hereafter referred to as SW^2P (Suess, Wang, Wu and Poletto, this issue) are used for calculating the radio brightness temperature at the radio-frequencies observed by the Nançays Radioheliograph (France) in order to allow a comparison with observations. This comparison should provide a check of the assumed parameters in the models and, in the case of disagreement, it could give the possibility to change them, in order to have, in view of the SOHO mission, a reliable streamer model. The frequencies are: 164, 236, 327 and 408 MHz corresponding to λ = 1.8, 1.3, 0.9 and 0.74 m. respectively. Calculation of the brightness temperature (T_b) are performed assuming the streamer location both at the limb and on the disk.

2. Streamer Models

The W^2SP models assume a Dipole, Quadrupole and Exapole magnetic field topology; the subsequent modifications (SW^2P) have so far been applied only to the Dipole topology.

In this poster some results obtained using the Dipole, Quadrupole and the modified Dipole ("Dipole 2") models are shown.

All the above mentioned models are two-dimensional ones: the density, temperature, velocity and magnetic field are given as a function of distance from the Sun centre, ρ , and colatitude, θ.

The three-dimensional model necessary to perform the integration of the radio transfer equation is obtained assuming a cylindrical symmetry around

Space Science Reviews **70**, 279–282.

the streamer axis.

3. Brightness Temperature and Polarization

A first calculation, performed assuming the streamer at the limb, shows that, in the assumption of a cylindrical symmetry, no models, except the Dipole-2, give an observable streamer in this position, due to the very low optical depth. Since in Dipole-2 model the density enhancement inside the streamer is coupled with a temperature decrease, the radio optical depth is greatly increased and the streamer becomes observable.

The centre to limb profiles obtained when the streamer is at its central meridian transit (centre disk for Dipoles, 34°N for the Quadrupole) are shown in Fig. 1. In this figure the quiet sun brightness temperature (dotted lines) has been obtained from the undisturbed (t = 0) model from which the corresponding streamer model was derived. In the Dipole and Quadrupole models it was necessary to multiply the quiet sun electron density by a factor f ($0.9 \leq f \leq 1$) to match the computed brightness temperature of the quiet sun, $T_b(qs)$, and that of the streamer, $T_b(s)$, at the border of this latter.

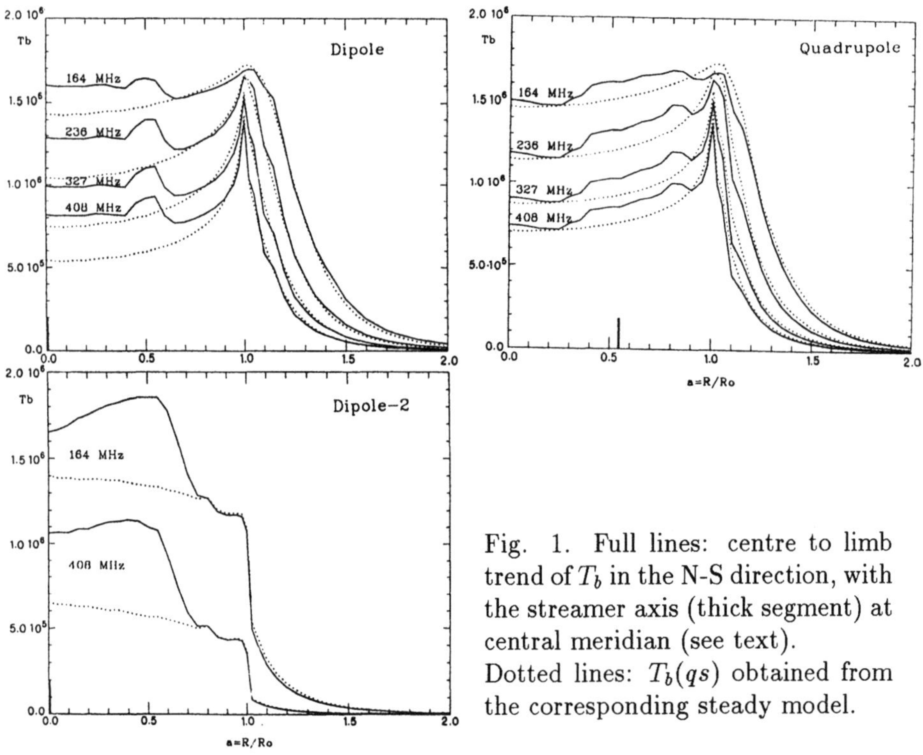

Fig. 1. Full lines: centre to limb trend of T_b in the N-S direction, with the streamer axis (thick segment) at central meridian (see text).
Dotted lines: $T_b(qs)$ obtained from the corresponding steady model.

The results obtained from the Quadrupole model, which give a quasi-constant increase of T_b over a large portion of the solar disk, reproduce quite well some aspects of the so called "Coronal Plateau" observed on Nançays radio maps by Lantos et al. (1992), who actually ascribed these features to the radio counterpart of coronal streamers. The computed radio brightness temperatures however (see Fig. 1), both of the quiet sun and of the streamer, are much larger than the observed ones. Lantos et al., in fact, find a quiet sun brightness temperature $T_b(qs) = 5.2\ 10^5$ at 164 MHz and 4.5 10^5 K at 408 MHz, while their "Plateau" temperature reaches 6.2 10^5 K and 5.3 10^5 K respectively. This large discrepancy, mostly at low frequency, must be ascribed to the large value of the coronal temperature assumed in the model: T_c= 1.8 10^6 at $\rho = 1$.

In spite of this large discrepancy in the brightness temperature, the computed difference between the streamer and quiet sun brightness temperature agrees quite well with the observed one: $\Delta T_b \simeq 10^5$ independent of frequency.

In the case of the two Dipole models on the contrary, this difference is much larger and depends on frequency; moreover the bump at the border of the streamer (Dipole) would produce, in the adopted geometry, a rather unrealistic ring around the Plateau.

Two dimensional maps showing $\Delta T_b = T_b$ - $T_b(qs)$ using the Quadrupole model are shown in Fig. 2 for the highest and the lowest considered frequencies. These maps can be compared with the observed ones shown in Fig. 13 of Lantos et al. (1992).

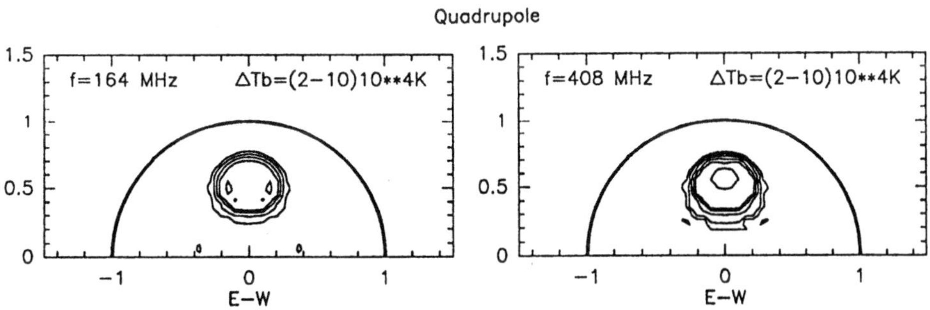

Fig. 2 Two dimensional maps of computed $\Delta T_b = T_b(s) - 4T_b(qs)$ obtained, using the Quadrupole model, with the streamer axis at zero longitude and 34°latitude.

A detailed analysis of the polarization in the streamer, computing the transfer equation separately for the ordinary (o) and the extraordinary (e) mode, has not been done. However an easy calculation shows that the maximum expected polarization cannot exceed a few percent. In fact polarized radiation can come only from optically thin regions and its maximum values, in the assumption $T_c = constant$, depends on the difference between the two optical depths τ_e and τ_o (see, for instance, Kundu, 1965):

$$P \leq \frac{\tau_e - \tau_o}{\tau_e + \tau_o} = \begin{cases} y & (Q.L.\ Propagation) \\ y^2 & (Q.T.\ Propagation) \end{cases}$$

where $y = \frac{2.8\,10^6 B(G)}{f}$ is the ratio between the Larmor and the observing frequency. Since the maximum value of the magnetic field in the streamers is B $\simeq$ 1-2 G., a polarization P $\leq$ 3% derives, in any case.

4. Conclusions

In this poster we have compared the results presented in Fig. 1 and 2, which show the aspect of a streamer on the solar disk, with the Nançays observations of the "Plateau" shown in Fig. 13 of Lantos et al. (1992). This comparison indicates that a good agreement with observations is found in the following results obtained with the Quadrupole model: the" Platform-like" structure or "Plateau", and the difference between the streamer and the quiet sun brightness temperature which turns out to be about 10^5 K, independent of frequency.

However:

- The computed brightness temperatures are larger by a factor ranging from 2.5 (164 MHz) to 1.5 (408 MHz) than the observed ones, both for the streamer and for the quiet sun. This indicates that the assumed models have a too large temperature.
- The shape of the observed plateau, although altered by the presence of A.R.'s, indicates that a cylindrical symmetry of the streamer around its axis is probably not the most suitable one. A longitudinal dimension larger than that given in latitude would probably provide better results. A larger longitudinal dimension would also provide a larger optical depth when the streamer is located at the limb, thus making it observable also in this position.

The conclusion of this comparison is that the present theoretical models of streamers should be modified, mostly for what concerns the temperature, in order to satisfy the radio observations.

References

Kundu, M.R.: 1965, *Solar Radio Astronomy*, Interscience Publishers: New York
Lantos, P., Alissandrakis, C. E., Rigaud, D.: 1992, *Solar Phys.* **137**, 225
Wang, A.H., Wu, S.T., Suess, S.T., Poletto, G.: 1993, *Solar Phys.* **147**, 55

MAGNETIC CONFIGURATION OF CORONAL STREAMERS AND THREADS

S.KOUTCHMY
Institut d'Astrophysique, CNRS, 98 Bis, Bd Arago, F-75014 Paris (France)

V.A.KOUTVITSKY, M.M.MOLODENSKY and L.S.SOLOV'IEV
IZMIRAN, Moscow Region, Troïtsk, 142092 (Russia)

and

O.KOUTCHMY
Université P. et M. Curie, Laboratoire d'Analyse Numérique,
Tour 55-65, 4, Place Jussieu, F-75252 Paris (France)

Abstract. We give a short account of the most prominent structures of the intermediate corona. Then we propose an axially symmetrical model for coronal streamers, according to which charged particles move along magnetic surfaces whose sources are electrical currents situated in the vicinity of the photosphere. The simplest current configuration is a pair of coaxial, coplanar, circular, and oppositely directed currents parallel to the photosphere. Magnetic surfaces for this current distribution exhibit a helmet-shaped separatrix and a saddle point. The temperature profile along the streamer can be predicted qualitatively if one takes into account the conservation of an adiabatic invariant in the drift theory of the charged particle motion.

1. Introduction

The structure of coronal streamers observed during a solar total eclipse (Koutchmy, 1992) poses many theoretical problems that are still unsolved :
i) the origin of the typical helmet with a cusp or onion shape,
ii) the origin of the confinement of streams at distances from the surface exceeding the solar diameter and/or in the intermediate corona,
iii) the large excess density of particles on the visible surface of the streamer, etc.

The classical approach for solving these problems is to start with a simple magnetostatic configuration around the Sun, similar to a dipole field, and to try to solve the full MHD equations, assuming a strictly radial flow presumably due to the thermally driven "wind". Then the magnetic field lines are stretched out and the neutral sheet could be a streamer, although some additional forces are needed to maintain a significant density enhancement in a neutral sheet. See Koutchmy and Livshits, (1992) for a recent review of "classical" models. Another empirical method for computing the magnetic configuration of the corona in 3D is to introduce a so-called "source surface", which considerably simplifies the computation, but leads to an artificial concept that does not have a physically plausible counterpart in the case of streamers.

Space Science Reviews **70**, 283–288.

Moreover, a modern approach for analyzing the coronal structures convincingly shows that the intermediate corona is filled with tiny and elongated structures like rays or threads as in Fig. 1. Some threads have obviously a circular cross-section that they keep over many solar radii; they extend along a direction that is never exactly radial : sometimes the deviation from the radial direction is as large as 60° (Koutchmy, 1969). Furthermore, these structures were seen outside eclipses: YOHKOH images taken in SXR show that the intermediate corona, above the inner corona made of loops, is also filled with rays, especially in the vicinity of active regions. A beautifull example is given in Shibata et al.(1992).In SXR images, time sequences show that threads are jets, although the magnetic features channeling the jet may persist. The rays often have the shape of threads rising above a loop-like structure; sometimes a cusp is well identified. YOHKOH images show more inner corona than eclipse images, but the fundamental loop-cusp-ray structure seems general. As to more quantitative aspects, little can be said from the observations of the temperature inside a range of 1 to 3 MK, but the density enhancements are at least one order of magnitude about the average background corona.

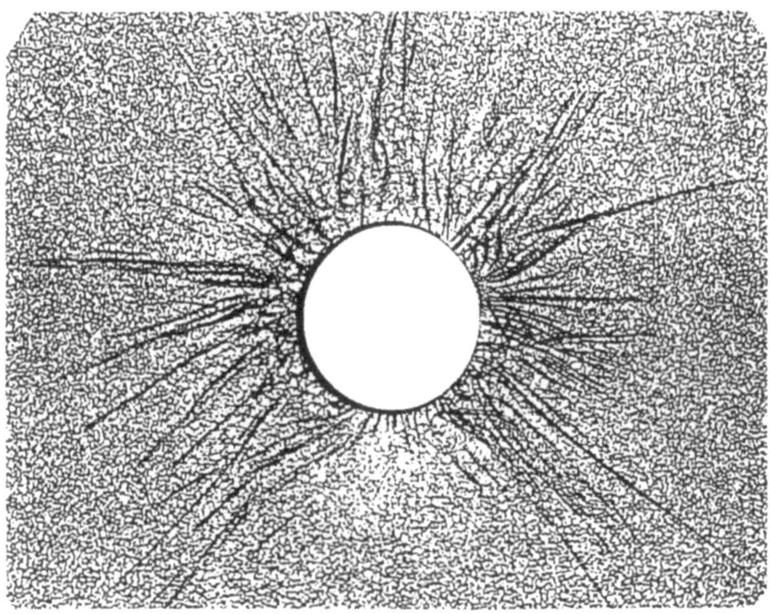

Fig. 1. Processed WL coronal eclipse image of July 31, 1990 using the "MadMax" algorithm; note the long threads.

2. Topological modelling and energy

Some of the above questions lead to the following model. We assume that charged particles from the Sun flow mainly along the separatrix magnetic surface, which has a singular X-point on the vertical axis. This magnetic surface with a sharp top (cusp) is interpreted as the visible surface of the coronal streamer. The required magnetic configuration could be due to the superposition of a circular current-carrying prominence, or facula, surrounding the spot and having a field of the opposite sign. The simplest current distribution is a pair of circular, oppositely directed, coaxial, and coplanar currents.

Let A_ρ, A_ϕ, A_z be the components of the vector potential in cylindrical coordinates (ρ, ϕ, z). The axial symmetry provides magnetic surfaces that are contour lines of $\Psi(\rho, z) = \rho\ A_\phi(\rho, z)$. Then the poloidal components of the magnetic field are $B_z = \frac{1}{\rho}\frac{\partial \Psi}{\partial \rho}$, $B_\rho = -\frac{1}{\rho}\frac{\partial \Psi}{\partial z}$, $B_\phi = 0$.

Following Landau and Lifshits (1985), the function Ψ for a pair of coils with radii a and A, and currents $-\mathcal{I}_a$ and $\mathcal{I}_A$, can be written as

$$\Psi(\rho, z) = \frac{4\ A\ \mathcal{I}_A}{C}\left[G_A(\rho, z) - \frac{a\ \mathcal{I}_a}{A\ \mathcal{I}_A}G_a(\rho, z)\right], \tag{1}$$

where

$$G_x(\rho, z) = \frac{1}{k_x}\left(\frac{\rho}{x}\right)^{1/2}\left[\left(1 - \frac{k_x^2}{2}\right)\ K(k_x) - E(k_x)\right],$$

with $k_x^2 = 4\ x\ \rho\ \left((x+\rho)^2 + z^2\right)^{-1}$, $K(k_x)$ and $E(k_x)$ full elliptic integrals of the 1st and 2nd kind respectively, and C a constant.

The magnetic field on the axis z can be described using elementary functions:

$$B_z(0, z) = \frac{2\ \pi\ A^2\ \mathcal{I}_A}{C}\left\{\frac{1}{(A^2 + z^2)^{3/2}} - \frac{a^2\ \mathcal{I}_a}{A^2\ \mathcal{I}_A}\ \frac{1}{(a^2 + z^2)^{3/2}}\right\}. \tag{2}$$

The following expansion in even powers of ρ is valid for an arbitrary axially symmetrical configuration (Morozov and Solov'iev, 1963):

$$\Psi(\rho, z) = \frac{\rho^2}{2}b(z) - \frac{\rho^4}{16}b''(z) + ... = \rho J_1(\rho\hat{b}'(z)), \tag{3}$$

where $b(z) = \hat{b}'(z) = B(0, z)$ is the field on the z-axis and $J_1(x)$ is Bessel function of the first kind.

The cross-section of a magnetic surface for the configuration (1) has a hyperbolic singular point z_s on the z-axis if the magnetic field (2) is zero at z_s, with

$$\frac{z_s^2}{A^2} = \frac{(a/A)^{2\lambda/3} - a^2/A^2}{1 - (a/A)^{2\lambda/3}} = \frac{(m_a/m_A)^{2/3} - a^2/A^2}{1 - (m_a/m_A)^{2/3}}. \tag{4}$$

Here $m_a/m_A \equiv (a/A)^\lambda$ is the ratio of the magnetic moments of the rings and $m_a/m_A = a^2\, \mathcal{I}_a/(A^2\, \mathcal{I}_A)$. The singular point can exist if $0 < \lambda < 3$.

When $\lambda \to 0$, $z_s \to \infty$ and, when $\lambda \to 3$, $z_s \to 0$. When $\lambda \to 0$, the magnetic moments and magnetic fields of both coils become equal, $m_A \to -m_a$, $B_A \to -B_a$, and the ratio of the fields at the origin tends to the third power of A/a :

$$\frac{B_a(0,0)}{B_A(0,0)} = \frac{a^2\, \mathcal{I}_a}{A^2\, \mathcal{I}_A} \left(\frac{A}{a}\right)^3 \to \left(\frac{A}{a}\right)^3. \tag{5}$$

With λ small enough, the top of a helmet corresponding to the point $\rho = 0$, $z = z_s$ occurs at an arbitrarily height above the surface of the photosphere.

The magnetic surfaces $\Psi(\rho, z) = const$ in the vicinity of the singular point $\rho = 0$, $z = z_s$ can be drawn using expansion (3). Assuming that

$$b(z) = \frac{C\, B_z(0,z)}{2\, \pi\, A^2\, \mathcal{I}_A} = \frac{1}{(A^2 + z^2)^{3/2}} - \frac{\mu}{(a^2 + z^2)^{3/2}},$$

so that $\mu = (a^2 + z_s^2)^{-3/2}(A^2 + z_s^2)^{3/2}$, we obtain, to an accuracy of ρ^4,

$$\Psi = \frac{\rho^2}{2}\left[\frac{1}{(A^2 + z^2)^{3/2}} - \frac{\mu}{(a^2 + z^2)^{3/2}}\right] - \frac{3\rho^4}{16}\left[\frac{4z^2 - A^2}{(A^2 + z^2)^{7/2}} - \frac{\mu(4z^2 - a^2)}{(a^2 + z^2)^{7/2}}\right]. \tag{6}$$

In the vicinity of $\rho = 0$ and $z = z_s$, $\Psi = P\, \rho^2\, \zeta$, with $\zeta = z - z_s$ and P a constant.

If the Larmor radii of the particles are small compared to the caracteristic size of the configuration, then the drift approximation applies (Morozov and Solov'iev, 1963). If $B_\varphi = 0$, then the drift trajectories of particles lie on the magnetic surfaces $\Psi(\rho, z) = const$. Particles drift in the azimuthal direction, and some of them are reflected from regions of strong magnetic field, which act as magnetic mirrors. Using the adiabatic invariant $I = v_\perp^2 / B = const$, the transverse and longitudinal energies of the particles are :

$$\epsilon_\perp = \epsilon_\perp^0 + \frac{mI}{2}(B - B_0), \qquad \epsilon_\parallel = \epsilon_\parallel^0 - \frac{mI}{2}(B - B_0). \tag{7}$$

When a particle moves into a strong field, its transverse energy, which determines the temperature, increases, but the longitudinal energy decreases, and vice versa. Now the dependence of the magnetic field on z has a characteristic profile and Eq. (7) provides the radial temperature dependence in the corona. The deduced behavior can be considered to be the result of the conservation of the transverse adiabatic invariant in the drift motion of charged particles.

To summarize, the coronal streamer in our model is the separatrix of an axially symmetric configuration, which arises as a result of the superposition of the magnetic fields of a facula surrounding the spot or of a filament surrounding the facula. The tendency of charged particles to escape along the separatrix magnetic surface was established by Auerbach and Boozer (1980) and by Gribkov et al.(1984).

In order to simulate an inclined streamer, while keeping the possibility of describing the system with the function Ψ, we consider also a two-dimensional problem in cartesian coordinates with $\partial\Psi/\partial y = 0$. We replace the system of circular currents by a pair of linear conductors with oppositely directed currents. Then

$$\Psi = P_y(x,z) = \frac{2}{C}\sum_i J_i ln\left[\frac{(x-x_{i1})^2-(z-z_{i1})^2}{(x-x_{i2})^2-(z-z_{i2})^2}\right]^{1/2},$$

where (x_{i1}, z_{i1}) and (x_{i2}, z_{i2}) are the coordinates of the conductors of the i-th pair.

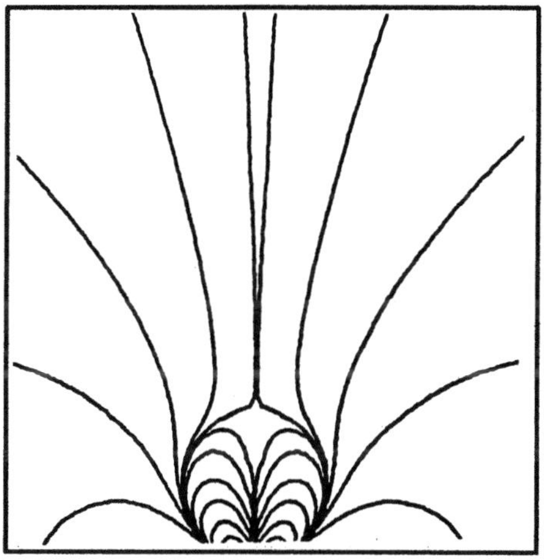

Fig.2.B–lines for $\lambda = 0.03$, $a/A = 0.1$

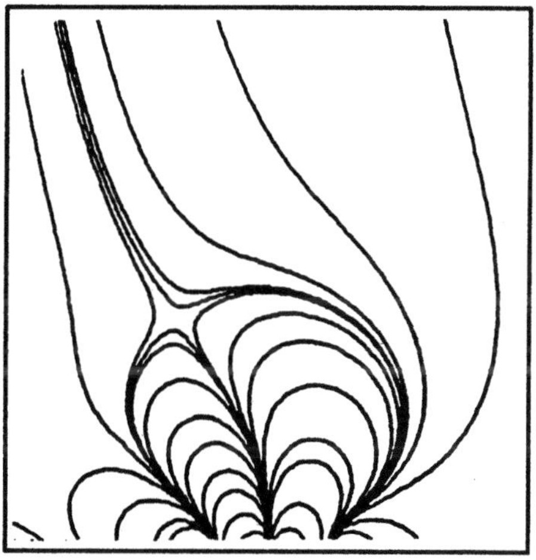

Fig.3.B–lines in the plane configuration

3. Results and Discussion

Fig. 2 shows the magnetic configurations calculated for selected parameters a/A and λ from Eq. (4). The magnetic surface cross-sections shown in Fig. 2 have the helmet shape. With increasing height, magnetic surfaces in the vicinity of the separatrix become onion-shaped (cusp). The shape of the magnetic surfaces may vary in the vertical direction.

A magnetic configuration with two saddle points in the plane $z = 0$ is possible. This configuration can be considered for modelling the coronal polar rays or plumes and possibly also corresponds to the quiet Sun field in the corona. Fig. 3 shows an inclined ray in the plane approximation. Its behaviour at infinity is determined by superposing a uniform and a weak background magnetic field $B_z^0 = const.$ The calculated and observed structures of Fig. 1 are similar. From the general appearance of helmets we can conclude that $0.001 < \lambda < 1$ in these configurations, and the best value for real structures is $\lambda \approx 0.03$. This includes straight, very long, and slightly inclined "rays".

4. Conclusion

The "separatrix model" of coronal streamers rests on the assumption of negligible influence of the proper currents on the motion of charged particles along magnetic surfaces. Magnetic surfaces are assumed to be created by external sources placed on the solar surface. The smallness of the magnetic fields does not prevent, in principle, the existence of a singular X-point on the axis where the magnetic field tends to zero. The boundaries of the helmets and cusps are mathematically isolated separatrix surfaces. The problem of the filling of the boundary with charged particles or plasma requires further work. The geometry of the magnetic field lines of the helmet configuration is analogous to that of streamlines around a sphere in a stream of fluid. In our case the non-existing lower hemisphere can be considered as a reflection of the MHD-configuration in equilibrium in the perfectly conducting plane of the photosphere. The nested meridional cross-sections of the toroidal magnetic surfaces observed inside the helmets are typical of the equilibrium MHD-configurations in TOKAMAKS inside the separatrix surface.

References

Auerbach S.P., and Boozer A.N.: 1980, *Physics of fluids* **23**, 2396

Gribkov V.M., Morozov A.I., and Pogytche O.P.: 1984, 'Plasma flow in toroidal systems with separatrix', *Fizika plazmy* **23**, 82

Koutchmy,S.: 1969, *Astrophys. Letters* **3**, 215

Koutchmy,S.: 1992, '' in , ed(s)., *Proceedings of the First SOHO Workshop*, , 73-82

Koutchmy,S., and Livshits,M.: 1992, *Space Science Rev.* **61**, 393-417

Landau L.D., and Lifshits E.M.: 1985, *Electrodynamics of continuous media*, Nauka:Moscow

Loucif M.L. and Koutchmy S.: 1989, *A.A. Suppl. Series* **77**, 45

Morozov A.I., and Solov'iev L.S.: 1963, 'Geometry of the magnetic field', *Plasma Theory* **2**,

Morozov A.I., and Solov'iev L.S.: 1963, 'Motion of charged particles in electromagnetic fields', *Plasma Theory* **2**,

Shibata, K. and the YOHKOH team: 1992, *PASJ* **44**, L173

MODELING CORONAL STREAMERS AND THEIR ERUPTION

R. S. STEINOLFSON
Department of Space Sciences, Southwest Research Institute, San Antonio, TX 78238-5166, U.S.A

Abstract. Numerical solutions of the time-dependent MHD equations are used to generate ambient coronal streamer structures in a corona characteristic of that near solar minimum. The streamers are then disrupted by slow photospheric shear motion at the base of magnetic field lines within the closed field region, which is currently believed to be responsible for producing at least some CMEs. In contrast to several other simulations of this phenomena, the polytropic index is maintained at a value of 5/3 through the addition of coronal heating. Observations are used as a guide in determining the thermodynamic structure and plasma beta in the ambient corona. For a shear speed of 2.5 km/sec, the streamer configuration evolves slowly for about 65 hours before erupting outward with the formation of a CME. The bright CME leading edge travels outward at a speed of about 240 km/sec, and the sheared field lines follow at a somewhat slower speed. A closed magnetic field region is ejected as the magnetic field lines that were opened by the CME reconnect and reform the streamer.

Key words: Sun – Corona – Streamers – Coronal Mass Ejections

1. Introduction

It is generally accepted that the rapid energy release in a flare is not the mechanism that produces the coronal mass ejection (e.g., Hundhausen, 1987; Gosling, 1993). In fact, the flare may simply be a by-product of the restructuring of the large-scale corona that occurs during the CME. One potential driver mechanism that has received some acceptance recently is slow photospheric motion that tends to shear, and thereby raise the magnetic energy level of, preexisting coronal magnetic arcades. The objective of the present study is to simulate CME generation as a result of shear motion at the coronal base of the closed field region of an ambient coronal streamer. A coronal streamer with adjacent coronal holes is taken as the initial atmosphere since observations show that many CMEs originate hear the base of, and propagate outward through, coronal streamers (e.g., Hundhausen, 1994). The fact that such a shear motion can cause the disruption of a streamer with the formation of a CME has already been demonstrated by Steinolfson (1990). Particular attention will be given to selecting parametric and input values such that the thermodynamic conditions and the plasma beta (β , the ratio of thermal pressure to magnetic pressure) in the initial atmosphere are consistent with available data. Observational consistency is achieved, in part, through the addition of a coronal heating term.

Space Science Reviews **70**, 289–294.

2. Geometry and Numerics

The two-dimensional MHD equations with an external heating term added to the energy equation are used for this study (e.g., Steinolfson, 1988). The equations are solved numerically, using an explicit differencing scheme with a high-frequency filter, in the meridional (r,Θ) plane of a spherical coordinate system. The streamer configuration is the cross-section of an equatorial streamer belt, without dipole tilt, and is most representative of the corona near solar minimum. The subsequent streamer eruption must be axisymmetric about the poles, and, consequently, the entire streamer belt is uniformly disrupted in this model. The simulation box extends from the coronal base (assumed to be at 1 solar radii, $R_\odot$) out to 10 $R_\odot$ and from the equator at 90° to the 180° pole. All physical quantities are either symmetric or antisymmetric about both the equator and the poles. A variable radial grid spacing that increases with radius allows better numerical resolution near the solar surface. The radial grid spacing varies monotonically from a minimum of 0.016 $R_\odot$ to a maximum of 0.278 $R_\odot$ with a total of 117 grid points, and the theta grid spacing is constant at 1°.

3. Initial Solar Atmosphere

The initial atmosphere is computed using a numerical relaxation procedure as used by Steinolfson et al. (1982), which is now commonly used in such simulations. The analytic solutions of Low (1986) for magnetic configurations consisting of a low-lying closed magnetic field region with over-lying open field lines provides the initial magnetic structure. The initial thermodynamics and flow speed are given by a radial hydrodynamic solution including the heating term. For a selected power law dependence of the heating on radius ($\sim r^{-0.45}$), the heating magnitude is computed so the polytropic index γ can be maintained at a constant value of 5/3. The temperature at the coronal base is set to 1.2 10^6 K, and the base density is selected to be 2.2 $10^8\,\mathrm{cm}^{-3}$. The constants for the initial magnetic configuration are such that the closed field region extends out to 2.5 $R_\odot$ at the equator, and the magnetic field strength at the coronal base at the equator is 0.87 G. The velocity vectors and the magnetic field lines in the relaxed streamer configuration are shown in Figure 1 along with contours of the plasma beta. Beta is much less than one in the open field region and becomes greater than one within the closed region and within the vertical equatorial current sheet.

4. Coronal Evolution Due to Shear Motion

A shear motion, with a maximum shear speed of 2.5 km/sec, is now applied within the closed magnetic field region of the streamer in Figure 1. The shear

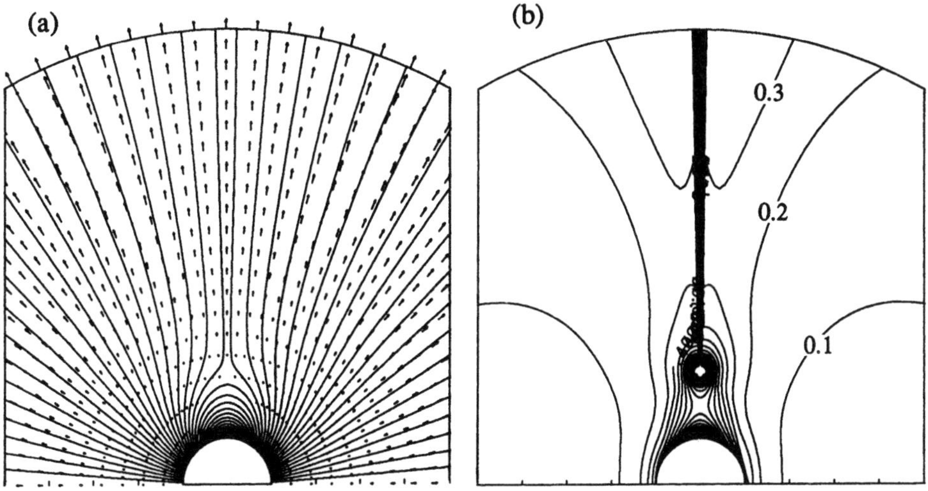

Fig. 1. Velocity vectors and magnetic field lines (a) and contours of the plasma beta (b) in the streamer configuration. The simulation is performed in the entire computational box from 1 $R_o dot$ to 10 $R_o dot$ and from 0° to 180° although the presented results are confined to a smaller region centered at the vertical equator. The velocity vector length is proportional to the flow speed at the vector tail, and the lengths are scaled to the maximum speed of 92 km/sec. The plasma beta contours range from 0.1 to 2.0 in increments of 0.1.

has a sinusoidal angular dependence such that it reverses sign at the equator and vanishes at 40° from the equator. The corona initially changes slowly in response to the shear motion as the magnetic energy increases within the closed field region. This initial phase is characterized by an evolution that is essentially force free, and the coronal evolution can be described as a series of quasi-dynamic- equilibrium states. However, eventually the increase in magnetic energy is such that the closed field portion of the streamer expands rapidly outward in an attempt to find a new dynamic equilibrium. The evolution during the eruptive stage is not force-free, and the corona is no longer in dynamic equilibrium. Magnetic field lines, velocity vectors, and density contours prior to the eruptive stage and during it are given in Figures 2(a) and 2(b). The density plotted here is the density relative to that in the streamer; i.e., $(\rho - \rho_{\rm str})/\rho_{\rm str}$. The radial trajectories along the equator of the dense CME leading edge and of the field line at the outer edge of the sheared region are shown in Figure 3, where time is referenced to the start of the shear motion.

The field lines in Figure 2(a) have clearly expanded outward from those in the initial streamer in Figure 1. The density change relative to the streamer undergoes a series of alternating compressions and expansions along the equator at this time. The large compression centered at approximately 2.1 $R_\odot$ ultimately becomes the compression that would be identified as the CME

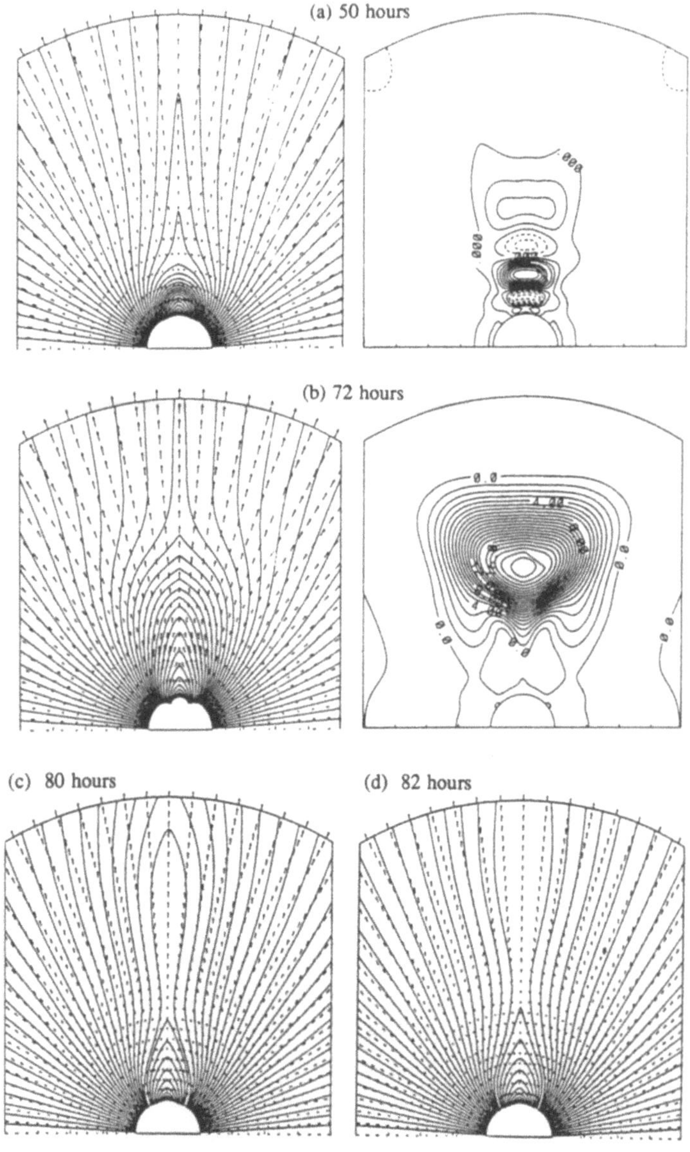

Fig. 2. Velocity vectors, magnetic field lines, and density contours prior to the ejection (a) and during it (b). The maximum flow speed at 50 hours is 113 km/sec and at 72 hours is 243 km/sec. The density contours range from -0.9 to 2.4 in increments of 0.3 at 50 hours and from 0.0 to 21.0 in increments of 1.0 at 72 hours. Velocity vectors and magnetic field lines are also shown at two times following the CME in (c) and (d). The maximum flow speed at 80 hours is 132 km/sec and at 82 hours is 120 km/sec.

in coronagraph images. The ejection of a dense blob of plasma during the eruptive (or readjustment) stage can be seen in Figure 2(b). Notice that the compression region does not conform to the shape of the magnetic field lines,

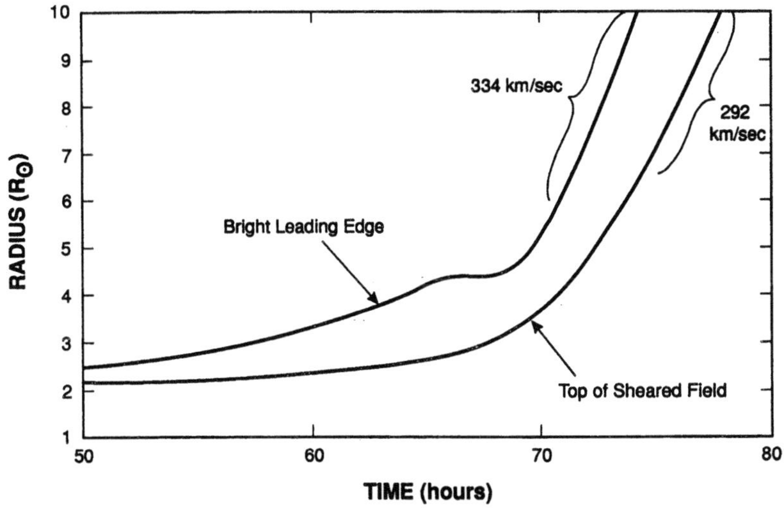

Fig. 3. The radial trajectories along the equator of the leading edge of the density compression and of the magnetic field line at the outer edge of the sheared region. The speeds are the averages over the portions of the trajectories indicated.

and that it tends to precede the outward motion of the closed field region

The average speeds along the equator of the bright leading edge and the magnetic field line at the top of the sheared region during the rapid ejection are indicated on Figure 3. The ejection speed is only slightly larger than the coronal speed ahead of it. In fact, the leading edge speed is subsonic, and there is thus no indication for the formation of an MHD shock near the leading dense edge. The outward motion of the bright leading edge essentially stops for a few hours just prior to the rapid ejection phase at approximately 70 hours in Figure 3. This type of behavior for the leading edge has been identified in observations reported on by Illing and Hundhausen (1986).

The velocity vectors and the magnetic field lines at 80 and 82 hours after the start of the shear are shown in Figure 2(c) and 2(d). The magnetic field lines are reconnecting just above the last closed field line. The location where the reconnection occurs along the equator moves outward with time, and the reconnected field lines move downward and reform the initial streamer configuration. This general behavior of the magnetic field lines after the CME is consistent with that predicted by Kopp and Pneumann (1976). The ejection of the closed magnetic field region formed by the reconnection may provide a means by which the corona can remove magnetic flux. Some mechanism for removing magnetic flux is necessary since the coronal and interplanetary magnetic field would increase indefinitely if all the field lines in CMEs remained attached to the surface (e.g., MacQueen, 1980).

5. Summary

Through the use of numerical solutions of the MHD equations, the response of an initial coronal streamer, which was also generated numerically, to slow photospheric shear motion at the coronal base of closed magnetic field lines is simulated. The energy equation includes an external heating term in order to emulate the effect of the unknown coronal heating source and to permit the use of a value for the polytropic index of 5/3. The simulated results contain several features identified in observations. The corona initially evolves slowly on a time scale characteristic of the applied shear motion. A dense blob of plasma forms near the top of the closed field region in the streamer during this slow phase and moves outward at the speed of a few km/sec. This slow stage lasts for some tens of hours during which the corona can be characterized as passing through a series of quasi-equilibrium states. Eventually the increase of magnetic energy within the closed field region becomes so large that the streamer expands rapidly outward in search of a new equilibrium. This rapid phase evolves on a time scale more typical of a characteristic speed (such as the Alfv n speed) and involves the formation of a CME and the ejection of a region of closed magnetic field that forms as the field lines reconnect following the CME.

References

Hundhausen, A.J.: 1994, in *The Many Faces of the Sun*, Eds. K. Strong, J. Saba, and B. Haisch, in press

Hundhausen, A.J.: 1993, private communication

Illing, R.M.E., and Hundhausen, A.J.: 1986, *J. Geophys. Res.* **91**, 10,951

Gosling, J.T.: 1993, *J. Geophys. Res.* **98**, 18,937

Kopp, R.A., and Pneuman, G.W.: 1976, *Solar Phys.* **50**, 85

Low, B.C.: 1986, *Astrophys. J.* **310**, 953

MacQueen, R.M.: 1980, *Phil. Trans. R. Soc. Lond. A* **297**, 605

Steinolfson, R.S., Suess, S.T., and Wu, S.T.: 1982, *J. Geophys. Res.* **255**, 730

Steinolfson, R.S.: 1988, *J. Geophys. Res.* **93**, 14,261

Steinolfson, R.S.: 1990, Max '91 Workshop #3. Max '91/SMM Solar Flares: Observations and Theory, Eds. R.M. Winglee and A.L. Kiplinger, 1990.

MAGNETOHYDRODYNAMIC SIMULATION OF A STREAMER BESIDE A REALISTIC CORONAL HOLE

S. T. SUESS
Space Science Laboratory, NASA Marshall Space Flight Center
Huntsville, Alabama 35812 U.S.A.

S. T. WU, A.-H. WANG
Center for Space Plasma and Aeronomic Research
The University of Alabama in Huntsville, Huntsville, Alabama 35899

and

G. POLETTO
Osservatorio di Arcetri, Firenze, Italy

Abstract. Existing models of coronal streamers establish their credibility and act as the initial state for transients. The models have produced satisfactory streamer simulations, but unsatisfactory coronal hole simulations. This is a consequence of the character of the models and the boundary conditions. The models all have higher densities in the magnetically open regions than occur in coronal holes (Noci, *et al.*, 1993).

Key words: Streamers – Coronal Holes – MHD Simulations

1. Introduction

An MHD numerical simulation of coronal streamers was described at the last SOHO workshop (Suess, 1992). That model used constant temperature and density at its base. The consequence was that the density in the magnetically open region was too large compared to what is expected in coronal holes. That problem is addressed here by using a varying temperature and density at the base. The reduced density in the coronal hole should improve simulation of SOHO observations of scattered light in the corona (Noci, *et al.*, 1993).

2. The Numerical Model and Results

The model is for axisymmetric, polytropic, time-dependent, ideal magnetohydrodynamic flow and is computed in a plane containing the axis of the magnetic field. Results are given between 1.0 and 7.0 $R_{\odot}$, although the solution extends to $15.0R_{\odot}$. There are 20 grid points between the pole and equator and 37 grid points, on a varying grid, from the base to 15 $R_{\odot}$. A solution is produced by a relaxation in time from an essentially arbitrary initial state. Here, the initial state is a vacuum dipole field superimposed on a Parker-type solar wind solution defined by the prescribed variation in tem-

Space Science Reviews **70**, 295–298.

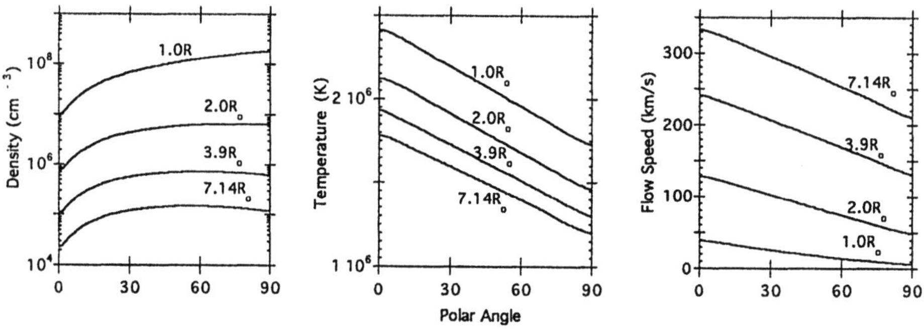

Fig. 1. The density, temperature, and flow speed at $t = 0$.

perature and density at the base. The polytropic index is 1.05. This model is described in detail by Suess (1992) and Wang, *et al.* (1992,1993).

Figure 1 shows the initial profiles ($t = 0$) of temperature, density, and flow speed versus polar angle at 1.0, 2.0, 3.9, and 7.1 $R_\odot$. The plasma β, at the coronal base, is 3.5 at the equator and 0.06 at the pole, meaning field strengths of 0.833G and 1.67G, respectively. The base densities at the equator and pole are 2×10^8 and 10^7 cm^{-3}, while the corresponding temperatures are 1.74×10^6 and 2.42×10^6 K. The values between these points vary linearly. The polar base density must be low to obtain proper densities at higher levels in the coronal hole, while the temperature must be higher in the hole than in the streamer to obtain high flow speeds in the hole. No other choice is possible in a polytropic model. However, this 'effective temperature' is not a true temperature; it reflects extended acceleration and heating of the solar wind that is known to exist in coronal holes.

Figure 2 shows the magnetic field after 20 hours of relaxation in physical time, or five Alfvén times, in comparison to the initial vacuum dipole field (left panel) and to the equivalent steady state magnetic field topology with constant base temperature and density (center and right panels). The constant temperature and density used were 1.8×10^6 degrees and 2.25×10^8 cm^{-3}, and the magnetic field strength at the equator was 1.67 G ($\beta = 1$) (Wang, *et al.*, 1993). Field line footpoints lie at 10 degree intervals. It can thus be seen that fieldlines equatorward of somewhere between 40 and 50 degrees are closed. The center and right panels show that the low density and high flow speeds that occur in the present model lead to fieldlines lying systematically poleward of the case with the constant boundary conditions. This is a natural consequence of the higher energy in the flow field.

The dramatic difference introduced by varying the temperature and the density at the base is shown in Figure 3, whose format is identical to that in Figure 1. The results for constant boundary values are shown as dashed lines. The left panel shows that at 2.0 and 3.9 $R_\odot$, the density at the center of the coronal hole is an order of magnitude less than with constant boundary

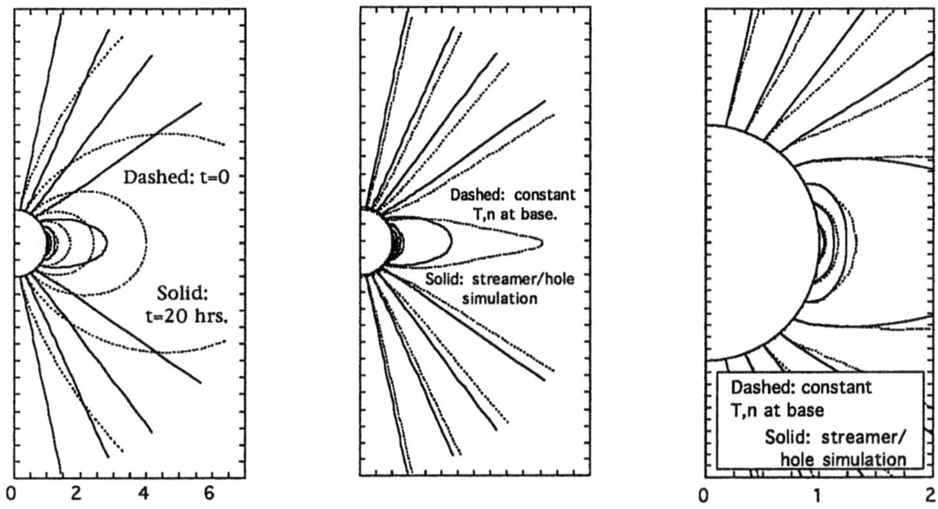

Fig. 2. Fieldline geometry for variable and constant boundary values.

values. The flow speed, in the right panel, is correspondingly much larger, reaching almost 250 km/s at the center of the hole at 3.9 $R_{\odot}$. The temperature varies little with height because of the polytropic index being 1.05.

The low density in the coronal hole is mainly due to the low density at the base of the hole - the increase in flow speed only accounts for about a factor of two reduction. The limitation is that the polytrope has a large flow speed at $1.0R_{\odot}$ at the base of the coronal hole - 40 km/s. This is an order of magnitude larger than expected at this level. The missing acceleration is not represented in the polytrope model and it is impossible to achieve both a low flow speed at the base and a high flow speed at $1.1 - 1.2R_{\odot}$. It is a limitation also inherent in thermal conduction models in the absence of energy and momentum sources low in the corona.

The overall effect of lowering the density in the coronal hole is seen using the polar gray-scale plot of the logarithm of the number density shown in Figure 4. The left panel shows the hole/streamer with overlays of temperature contours, flow speed contours, and magnetic field lines. The right panel shows a corresponding plot for constant temperature and density at the base - the case shown by the dashed lines in Figure 3. The solid contours on the left half of the panels are flow speed (Min=0, Max=350 km/s, 50 km/s contour levels). The dashed contours on the right half of the panels are temperature (Min=1×10^6, Max=2×10^6, 2×10^5 K contour levels). The solid lines on the right sides of the panels are magnetic field lines. Note that the density distribution on the right is far more spherically symmetric than on the left. Also, the temperature notch above the streamer is much narrower in the hole/streamer simulation. Finally, the density follows the contours of

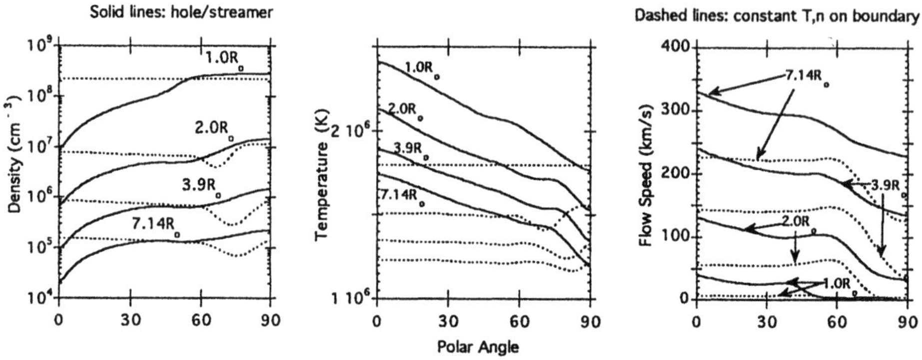

Fig. 3. The density, temperature, and flow speed at $t = 20$hrs.

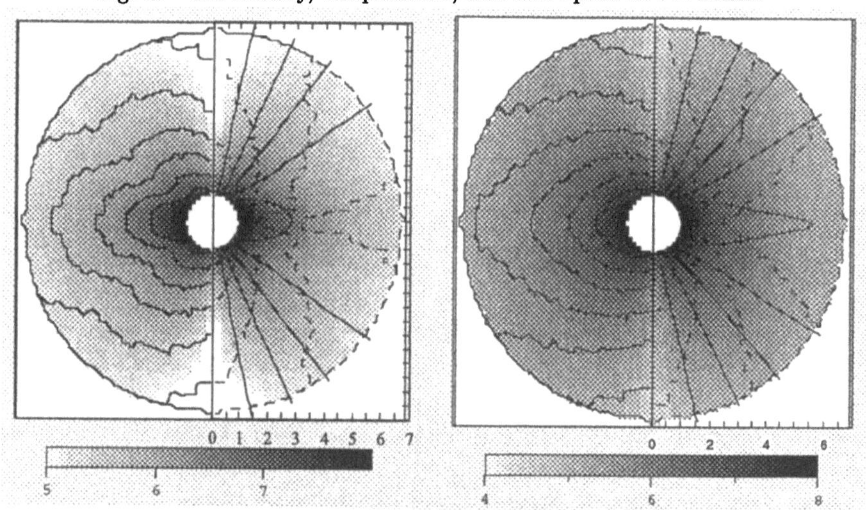

Fig. 4. Left: Coronal hole/streamer simulation. Right: Constant T,n on boundary.

velocity better than the magnetic field lines. This reflects the control flow speed exerts on density in the corona - it is not a hydrostatic medium.

References

Noci, G., Poletto, G., Suess, S. T., Wang, A.-H., and Wu, S. T.: 1993, 'Ly-α Intensity in Coronal Streamers', *Solar Phys.* **147**, 73

Parker, E. N.: 1963, *Interplanetary Dynamical Processes*, Interscience: New York

Suess, S. T.: 1992, 'Domingo, V., Poland, A., and Mariska, J.' in Modeling Solar Coronal Streamers, ed(s)., *Coronal Streamers, Coronal Loops, and Coronal and Solar Wind Composition, SP-348*, European Space Agency: Noordwijk, 63

Wang, A.-H., Wu, S. T., Suess, S. T., and Poletto, G.: 1992, 'Marsch, E., and Schwenn, R.' in A Two-Dimensional MHD Global Coronal Model: Steady-State Streamers, ed(s)., *Solar Wind Seven*, Pergamon Press: Oxford, 311

Wang, A.-H., Wu, S. T., Suess, S. T., and Poletto, G.: 1993, 'A Two-Dimensional MHD Global Coronal Model: Steady-State Streamers', *Solar Phys.* **147**, 55

MANIFESTATION OF MAGNETIC RECONNECTION IN CORONAL STREAMER CURRENT SHEETS

A.I. VERNETA, E. ANTONUCCI and D. MAROCCHI
Istituto di Fisica, University of Torino, Torino, Italy

Abstract. We investigate the possibility of observing the effects of magnetic reconnection inside a current sheet forming in a coronal streamer in the extended corona. In particular we study the possibility to observe with the UVCS of SOHO the excitation of the tearing instability in the current sheet.

Key words: UV emission lines – coronal streamers – corona of the Sun

1. Introduction

One of the general properties of helmet streamers, i.e. closed magnetic loop systems in the solar corona, is the presence of reconnecting current sheets (CS) above them (Pneuman and Kopp, 1971). Apparently, these CSs extend into the solar wind and can influence different processes in the corona and the interplanetary space.

The parameters of a streamer CS were calculated, for the first time, by Pneuman (1972), and Priest and Smith (1972). They have found that such CSs are relatively thicker (the thickness is $\delta \sim 500\,\mathrm{km}$ according to Pneuman, 1972) than the ones arising in the reconnecting high-temperature CSs associated with solar flares, where the sheet thickness can be of the order of 10 cm (Somov, 1992). The significant CS thickness together with a more moderate energetic output and a slower development of reconnection presumably makes it possible to observe directly the streamer CSs.

Here, we consider the process of excitation of tearing modes which play a fundamental role in the reconnection phenomenon (e.g., Furth et al., 1963). We find that there is the possibility to observe the excitation and suppression of the tearing instability in the extended corona with the UVCS, the ultraviolet coronagraph of the SOHO satellite.

2. Streamer Sheet Parameters

The parameters of a CS inside a coronal streamer can be estimated on the basis of the model by Pneuman (1972). These parameters, the density, the temperature and the thickness of the sheet, the electrical conductivity which determines the electric current and the radial and transverse magnetic fields in the CS, are listed in the Table I.

Here, a spherical coordinate system with the center in the Sun center is used.

Space Science Reviews **70**, 299–302.

TABLE I
Characteristic Parameters of a Coronal Streamer Current Sheet

Density	n	(cm^{-3})	$1\,10^9$
Temperature	T	(K)	$1\,10^6$
Thickness	δ	(cm)	$5\,10^7$
Conductivity	σ	(s^{-1})	$1\,10^{13}$
Radial magnetic field	B_r	(G)	0.5
Transverse magnetic field	B_θ	(G)	$3\,10^{-3}$

3. Tearing Modes and their Signatures

Now let us consider the excitation of tearing modes under such conditions. Tearing modes have a simple physical nature (e.g., Priest, 1985). Parallel electric currents flowing inside the CS tend to attract each other and, as a consequence, to tear the CS. This process implies the violation of the 'freezing in' condition in the reconnection of the magnetic field lines. It destroys the CS and forms magnetic islands.

The characteristic dimension of an island is $\lambda_f = 2\pi S^{1/4}\delta$, the time of its growth $\tau_f = \tau_R\, S^{-1/2}$ (Furth et al., 1963). Here $S = \tau_R / \tau_A$ is the magnetic Reynolds number for the CS, τ_R and τ_A are the magnetic diffusion and Alfvén times, respectively: $\tau_R = 4\pi\delta^2\sigma / c^2$, $\tau_A = \delta\,(4\pi\rho)^{1/2} / B$ ($\rho = nm_i$ is the density in the CS).

Thus assuming the CS parameters reported in Table I, we obtain for the dimension of the island $\lambda_f \sim 2\,10^{10}$ cm. When a complete disruption of the CS occurs, the thickness of the island equals the thickness of the CS. The length of the island is $l_f \sim 1\,R_\odot \sim 7\,10^{10}$ cm. The total volume is $V \sim 7\,10^{28}$ cm ^{3} and the emission measure is then $EM = n^2\delta\lambda_f\, l_f \approx 7\,10^{46}$ cm^{-3}, which is an observable quantity. The time of island formation is $\tau_f \approx 5\,10^4$ s.

Taking into account the slow plasma flow existing in the CS, we obtain that magnetic islands have to arise near the streamer and to drift outward along the CS. The drift velocity is $v \approx 4\ 10^4$ $cm\,s^{-1}$ at $r = 2\,R_\odot$ and $v \approx 4\ 10^6$ cm s^{-1} at $r = 4\,R_\odot$.

If the transverse magnetic field, which exists inside a streamer CS, is large enough the picture changes owing to the stabilization effect of this field (Somov and Verneta, 1989 and 1993). The CS manifests itself now as a single stable island with a larger volume, $5\,10^7 \times 3\,10^{11} \times 7\,10^{10}$ cm^3, and an emission measure $EM \sim 10^{48}$ cm $^{-3}$. The destruction of this island is possible when the value of the transverse field decreases. We then expect to find observational evidence not only for the existence of the tearing instability in the CS in the extended solar corona, but also for the presence of the stabilization effect of the transverse field.

TABLE II
Intensity of Line Emission from a Magnetic Island

Ion	T (K)	λ (Å)	I_1 (ph cm^{-2} s^{-1})
O VI	3.0 10^5	1032	8.4 10^5
Mg X	1.0 10^6	610	7.9 10^5
Fe XII	1.3 10^6	1242	3.4 10^4
Si XII	1.9 10^6	499	1.2 10^3

4. Expected UV Emission from a Streamer Current Sheet

We have computed the UV emission from a magnetic island formed inside the CS of a coronal streamer for the collisionally excited transitions of the O VI, Mg X, Fe XII, Si XII ions reported in Table II. We have selected UV lines which can be observed with the UVCS, the ultraviolet coronagraph of SOHO, above helmet streamers in the extended corona beyond $r \geq 2\,R_\odot$.

The CS is assumed to lie along the line-of-sight. In case of tearing instability, magnetic islands will appear as thin elongated dense features with density n $\sim 10^9$ cm^3 and about 0.7 arcsec x 4.6 arcmin wide. The emission is integrated over one solar radius along the line-of-sight.

The line intensities, I, are computed for a temperature of 10^6 K, by using the atomic data calculated by Landini and Monsignori Fossi (1990). In Table II, I_1 is the line intensity computed for a magnetic island without the stabilization effect of the transverse magnetic field; in this case, the volume is $V \sim 7\ 10^{28}$ cm^3. The quantity T in Table II represents the temperature of maximum emissivity for a given line. The UV line intensities expected from a magnetic island within a streamer CS, at 2 $R_\odot$, are at least two orders of magnitude higher than the intensities of the lines emitted by faint features such as coronal holes. The UVCS has been specifically designed to observe the weakest features in the extended corona; the detection of magnetic islands shoud be therefore well within the capabilities of the UVCS.

5. Summary and Conclusions

We have considered the possibility of observing the tearing modes inside a CS forming in a coronal streamer. The emission measure of a single magnetic island is approximately EM $\sim 7\,10^{46}$ cm^{-3}, which is an observable quantity. The island formation time is estimated to be about ten hours and its characteristic dimensions are $5\,10^{7} \times 2\,10^{10} \times 7\,10^{10}$ cm^{3}. Islands form at $r > 2\,R_{\odot}$ and drift outward with characteristic velocities $v \approx 4\,10^{4}$ cm s^{-1} and $4\,10^{6}$ cm s^{-1} at $r = 2\,R_{\odot}$ and $4\,R_{\odot}$, respectively.

The presence of a transverse magnetic field in the CS stabilizes the magnetic island. In this case the volume and emission measure of the magnetic island increase to the values V $\sim 10^{30}$ cm^{3} and EM $\sim 10^{48}$ cm^{-3}, respectively.

Acknowledgements

We thank Prof. B.V. Somov for his valuable comments. This work was supported by the Agenzia Spaziale Italiana and by the Ministero dell'Università e della Ricerca Scientifica e Tecnologica. Dr. A.I.Verneta was supported in part by the Gruppo Nazionale di Astronomia of the CNR.

References

Furth, H.P., Killeen, J., and Rosenbluth, M.N.: 1963, *Phys. Fluids* **16**, 1054

Landini, M., Monsignori Fossi, B.C.: 1990, *Astron. Astrophys. Suppl. Ser.* **82**, 229

Pneuman, G.W. and Kopp, R.A.: 1971, *Solar Phys.* **18**, 258

Pneuman, G.W.: 1972, *Solar Phys.* **23**, 223

Priest, E.R. and Smith, D.F.: 1972, *The Astrophys. J. Lett.* **12**, 25

Priest, E.R.: 1985, *Rep. Prog. Phys.* **48**, 955

Somov, B.V.: 1992, *Physical Processes in Solar Flares, Kluwer Academic Publ., Dordrecht, London*

Somov, B.V. and Verneta, A.I.: 1989, *Solar Phys.* **121**, 321

Somov, B.V. and Verneta, A.I.: 1993, *'Tearing Instability of Reconnecting Current Sheets in Space Plasmas', Space Science Reviews*, in press

CORONAL MAGNETIC FIELD EVOLUTION UNDER RECONNECTIVE RELAXATION

G.E. VEKSTEIN and E. R. PRIEST
University of St Andrews, KY16 9SS, Scotland, UK

and

R. WOLFSON
Middlebury College, Middlebury, Vermont, USA

Abstract. The nonlinear evolution of a partially open coronal magnetic configuration is considered, assuming that corona responds to photospheric footpoint motions by small-scale reconnection events that produce a relaxed lower-energy state while conserving the global magnetic helicity of the system. The results of numerical calculations for such a relaxed equilibrium show an essential role of the amount of helicity injected to the closed-field region. If photospheric perturbations are incoherent (small-scale shearing with inefficient helicity injection), the relaxed state becomes close to an initial potential field. In this case reconnective relaxation does not result in a substantial global evolution, just providing heating of the corona (Vekstein et al,1993). On the contrary, sufficient injection of the magnetic helicity can lead to a considerable restructuring of the coronal magnetic configuration, with possible change of its topology (formation of magnetic islands), and even catastrophic loss of equilibrium (Wolfson et al,1994)

Key words: Sun – Corona – Magnetic reconnection

1. Introduction

The aim of the present investigation is to obtain the response of a coronal magnetic configuration to perturbations produced by photospheric shuffling of the field-line footpoints. Since the general solution of this extremely complicated 3-D resistive MHD problem is still beyond present analytical and numerical capability, we have used a phenomenological approach first suggested for solar applications by Heyvaerts and Priest (1984) and developed further by Vekstein (1987) and Vekstein, Priest and Steele (1993).The main idea is to consider the evolution of a coronal magnetic configuration as a relaxation to a minimum magnetic energy state subject to the constraint of conservation of the total magnetic helicity (Taylor, 1974). Then the relaxed state is a linear force-free magnetic field described by

$$\nabla \times \mathbf{B} = \alpha \mathbf{B}, \qquad \alpha = \text{const} \tag{1}$$

We consider a partially open magnetic configuration of arcade type (Fig.1), assuming that the corona occupies the half-space $y > 0$ and is perturbed by shearing footpoint displacements on the photospheric plane $y = 0$.Then the shearing perturbations in the open-field regions II and III (see Figure 1) will propagate far away as Alfven waves, and so the strains will remain

Space Science Reviews **70**, 303–307.

only inside the closed-field domain I.Thus reconnective relaxation takes place only in the latter region, while in II and III the magnetic field will remain potential.

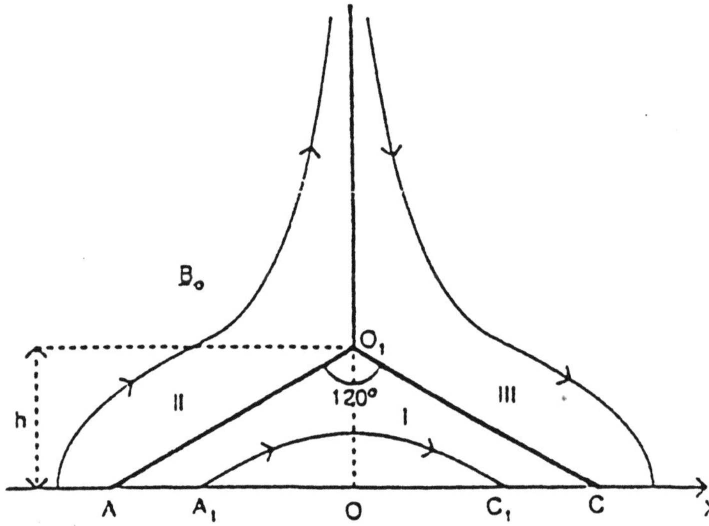

Fig. 1. A typical closed coronal arcade (region I) surrounded by open field regions (II and III).

2. Global Evolution

To model the global evolution of a partially open coronal configuration we have chosen an initial potential state of the form

$$\mathbf{B}(z) = B_y(x,y) + \mathrm{i}B_x(x,y) = \frac{\mathrm{i}(z^2+1)^{1/2}}{(z+\mathrm{i}d)^2} \tag{2}$$

where z is now the complex variable $z = x + \mathrm{i}y$. We use here dimensionless variables, so that initially the Y-type neutral point is located at $x = 0, y = 1$. A free parameter d in (2) allows to vary the strength of the magnetic field in the closed-field region I in comparison to that in the open-field domains II and III. This variation has a significant effect on the final relaxed state (see below). Helicity injection into the closed-field region and subsequent reconnective relaxation there result in the formation of a linear force-free field (1) which has to be in equilibrium with a potential magnetic field in regions II and III. These equilibrium configurations have been obtained numerically, assuming that the normal component of the magnetic field on the base remains fixed by Eq.(2), and the main results are illustrated in Figures 2-4.

The problem has been solved for boundary conditions derived from the two cases: $d = 0.8$ and $d = 3.8$ in Eq.(2). The former corresponds to a situation in which the field at the base falls off quickly beyond the closed-field region. The case $d = 3.8$, in contrast, has its field strength at the base increasing for some distance beyond the edge of the closed-field region.

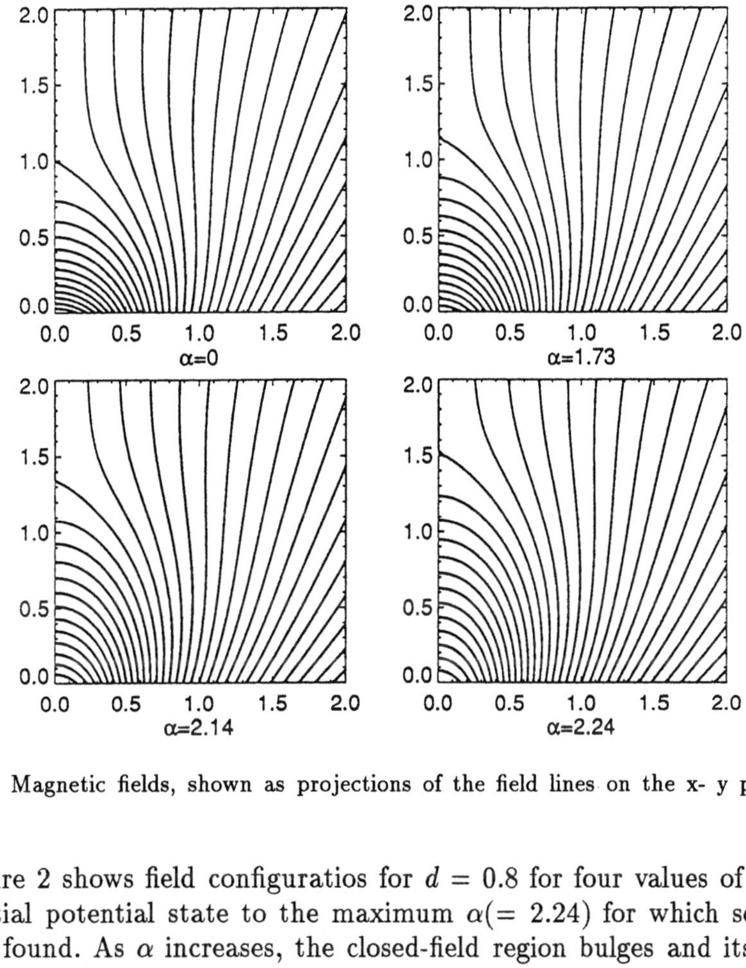

Fig. 2. Magnetic fields, shown as projections of the field lines on the x- y plane, for $d = 0.8$.

Figure 2 shows field configuratios for $d = 0.8$ for four values of α from the initial potential state to the maximum $\alpha(= 2.24)$ for which solutions can be found. As α increases, the closed-field region bulges and its height increases in response to the buildup of energy associated primarily with the increasing toroidal field component.

Figure 3 shows explicitly the increase in height of the closed-field region and the buildup of relative magnetic helicity. An important point is that evolution appears to reach a limit beyond which further equilibrium solutions seem impossible. Further increases in the injected helicity must then result either in a loss of equilibrium and perhaps eruption of the magnetic structure (Low, 1991; Priest and Forbes, 1990; Wolfson, 1993) or in a shift of the lowest-energy state actually available away from the constant-α state.

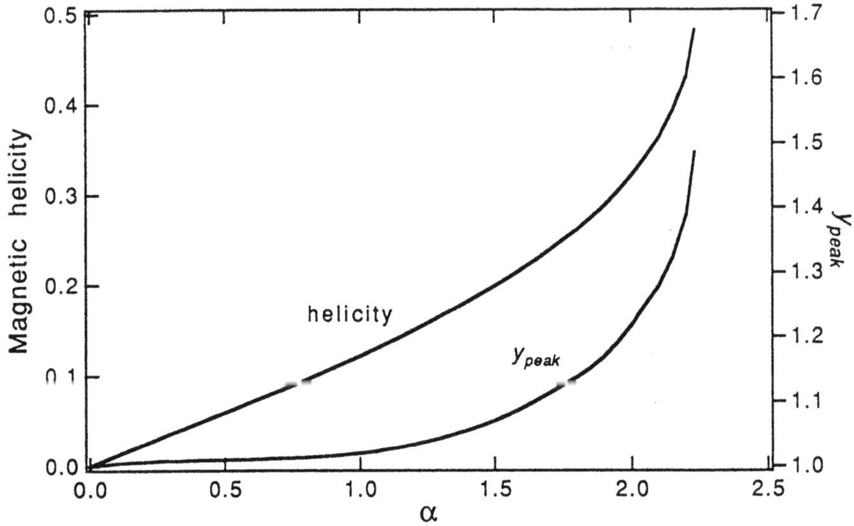

Fig. 3. Relative magnetic helicity and neutral point height as functions of α for the case $d = 0.8$.

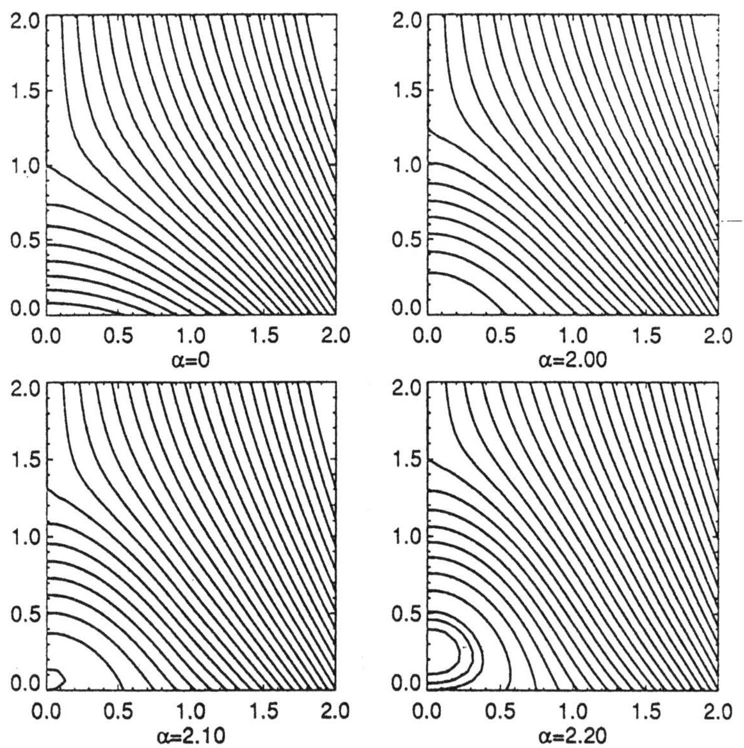

Fig. 4. Formation of magnetic islands for the case $d = 3.8$.

The same considerations apply to the case $d = 3.8$, but here we find in addition a change in magnetic topology with increasing α. As Fig.4 shows, that change consists of the formation of magnetic islands inside the closed-field region. Obviously, such a change is admissible because the steps leading from the initial potential state to the final state include non-ideal local reconnection events that can alter the overall magnetic topology. The formation of these islands, with their fields no longer tied to the photosphere, might lead to instability associated with magnetic buoyancy, and thus might provide a mechanism for eruptive events in the solar corona. In addition, these coronal magnetic islands might provide sites for prominence formation.

3. Conclusion

The results of numerical calculations for a relaxed lower-energy state in the corona show a rise and bulging of the field lines of the closed-field region with increasing injection of magnetic helicity. This evolution of a coronal configuration is similar to the slow (sub-Alfvenic) expansion of active-region magnetic fields observed recently by the Yohkoh Soft X-ray Telescope (Uchida et al,1992).

A further increase of the magnetic helicity content brings the configuration to a point where equilibrium solutions are impossible. Thus our model suggests that a catastrophic loss of magnetic equilibrium might be an initiation mechanism of coronal mass ejections.

We also find that such reconnective relaxation can result in qualitative changes in the magnetic topology, with appearance of magnetic islands within the initially line-tied magnetic field. This can change drastically the MHD stability of the system, triggering its subsequent explosive evolution.

References

Heyvaerts, J., Priest, E.R.: 1984, *A&A* **137**, 63
Low, B.C.: 1991, *Astrophys. J.* **251**, 352
Priest, E.R., Forbes, T.G.: 1990, *Solar Phys.* **126**, 319
Taylor, J.B.: 1974, *Phys. Rev. Lett.* **33**, 1139
Uchida, et al.: 1992, *Publ. Astron. Soc. Japan* **44**, L155
Vekstein, G.E.: 1987, *A&A* **182**, 324
Vekstein, G.E., Priest, E.R., Steele, C.D.C.: 1993, *Astrophys. J.* **417**, 781
Wolfson, R.: 1993, *Astrophys. J.* , in press
Wolfson, R., Vekstein, G.E., Priest, E.R.: 1994, *Astrophys. J.* , in press

WORKING GROUP 3: CORONAL STREAMERS

ROGER A. KOPP
Los Alamos National Laboratory
MS F645, Los Alamos, NM 87545 (U.S.A.)

Abstract. The working group on coronal streamers convened on the first day of the 2nd SOHO Workshop, which took place in Marciana Marina, Isola d'Elba, 27 September - 1 October 1993. Recent progress in streamer observational techniques and theoretical modeling was reported. The contribution of streamers to the mass and energy supply for the solar wind was discussed. Moreover, the importance of thin electric current sheets for determining both the gross dynamical properties of streamers and the fine-scale filamentary structure within streamers, was strongly emphasized. Potential advances to our understanding of these areas of coronal physics that could be made by the contingent of instruments aboard SOHO were pointed out.

Key words: Sun: corona – Sun: streamers – Sun: solar wind

1. Introduction

Coronal streamers are by far the most striking feature of the "quiescent" solar corona, their wispy forms having dominated the hand drawings of eclipse observers long before the advent of photography. It is fair to say that, by virtue of the sometimes-Herculean efforts of observers and theoreticians alike, there exists today an enormous wealth of knowledge regarding the morphology and evolution of streamers, the physical conditions within these structures, and the probable role of streamers within the larger context of heliospheric physics.

Unfortunately, not all of this knowledge is self-consistent, and it can certainly stand to be improved upon in many respects. Nevertheless, the reader must not be left with the mistaken impression that we are envisaging the SOHO project as starting from scratch in investigating the above topics. Rather, the goal of SOHO is to refine our perception of certain processes at work on the Sun (such as the mass and energy balance of the corona), thereby adding to the pre-existing storehouse of information. Because of their relative brightness and longevity, coronal streamers comprise a particularly attractive type of solar structure to observe and analyze for this purpose.

In addition, there are important gaps in our knowledge, and the enhanced spectroscopic capabilities of SOHO may make it possible for the first time to fill in these gaps by allowing us to ascertain the detailed properties of the coronal plasma close to the Sun.

Space Science Reviews **70**, 309–316.

2. Basic Streamer Configuration

The classical form of a coronal streamer consists of a bright dome-shaped structure in the inner corona, usually best seen at or near the solar limb at the time of a total solar eclipse; above the top of the dome, or "helmet", a region of enhanced brightness extends outward more-or-less radially, narrowing somewhat with increasing elongation from the Sun. The distance into space at which streamers can be detected seems to be limited only by the observing technique employed: see the streamer review article in this Proceedings by Poletto. It is in fact generally accepted that streamers have distinct signatures in the properties of the solar wind at, and probably well beyond, 1 A.U.

The simplest three-dimensional model that one implicitly assumes when discussing streamer physics is that of a magnetically closed bipolar region comprising the helmet, surrounded on both sides and above by a region of "open" field lines extending far out into space. The boundary between the open and closed field regions is believed to consist of a system of extremely thin sheet currents - the "magnetic separatrices".

As simple as it is, it should be stressed that this conceptual picture of a magnetized streamer structure is based more on conjecture than on fact. There are as yet no direct measurements of the magnetic field strength at coronal heights and only a few indirect inferences of the direction of the field in streamers based on coronal emission line polarization.

It is within the context of the above scenario that most of the talks and posters of the working group were presented.

3. Density and Temperature in Streamers

Except for ballpark figures, there seems to be no universal agreement on the temperature and density distribution within streamers - not to mention the values of these quantities at the streamer base. This is understandable when one considers that no two streamers are exactly alike, and there no doubt exists a natural dispersion of the physical properties of streamers as a class. Moreover, it is rare indeed that a given streamer has been analyzed extensively by two entirely different observing techniques. The Workshop presentations and discussions illustrated clearly that there is no single method of temperature determination that yields a completely unambiguous value of the kinetic (electron or ion) temperature at a fixed point within a streamer. For example, the oft-quoted "scale-height" temperatures are suspect because they are based on the assumption of dynamical stationarity, or hydrostatic equilibrium, of the coronal plasma and they ignore the possible contribution of magnetic stresses in the pressure balance. Moreover, one usually assumes a (locally) isothermal medium in approximating the observed density lapse rate by an exponential function.

Similarly, the observed intensity ratio of coronal emission lines of different ionization stages of a single element has been widely used over the years as a sensitive temperature indicator. Most recently, this technique has been applied to a large body of green- and red-line data by Guhathakurta and co-workers (this Proceedings) to infer the large-scale pattern of coronal temperature variations over a sunspot cycle. Combined with K-coronameter and photospheric magnetic field data, they were able to determine the spatial correlations of coronal density and temperature with the large-scale distribution of surface magnetic fields (see also Guhathakurta et al., 1993).

These "ionization equilibrium" temperatures implicitly assume that both of the lines used to form the intensity ratio always originate in the same coronal volume element(s) along the line of sight. Considering the large amount of fine structure likely to be present in streamers and, indeed, throughout the corona (see below), this assumption must be difficult to justify in general. A strong argument in its favor could nevertheless be made if a careful examination of similar data taken at high angular resolution were to reveal that the same fine-scale features are visible in both lines. Alternatively, to render the "identical volume" assumption less controversial, one might select line pairs arising from nearby ionization stages - or ideally even from the same stage of ionization (e.g., a resonance-line doublet in the XUV spectral region); this, however, would most likely decrease the temperature sensitivity of the technique somewhat.

The least controversial method of coronal temperature determination appears to be that of observing the spectral profile and intensity of the resonant-scattered *Lyman-α* line of neutral hydrogen. (Combined with other data, such as that from white-light coronagraphs, these observations can provide information on coronal electron densities and plasma bulk motions as well.) This method depends on only a single spectral line, whose formation mechanism is relatively simple and well-understood; moreover, that being a line of a light element, its width is fairly insensitive to turbulent motions of the coronal plasma. Of course, high in the corona there could be a large difference between the electron and proton temperatures, but this is not likely to be the case in the dense cores of streamers over the range covered by SOHO (UVCS) observations.

Interesting results using this technique were presented by the Harvard group (presentations by J. Kohl and L. Strachan). These were obtained during the April 1993 Space Shuttle deployment of the Spartan 201 instrument package containing both a white-light coronagraph and an ultraviolet coronal spectrometer. Proton kinetic temperatures were determined at a number of heights between 1.35 and 3.5 $R_\odot$, both along the axis of a large coronal streamer and in the polar coronal holes. Whereas the Spartan 201 spatial coverage was not sufficiently complete to establish a systematic trend of temperature with position in these structures, it did provide a tantalizing preview of possible things to come from the UVCS experiment on SOHO.

4. Fine-Scale Structure in Streamers

As with any other structural component of the solar atmosphere, coronal streamers exhibit a wealth of filamentary detail when observed with sufficient spatial resolution. This is not particularly new or surprising: some of the best examples of fine structure date from the early years of the present century (e.g., Campbell, 1923), when it was common practice to photograph the corona at total eclipse with very long focal-length telescopes (large plate scale). Structures of all spatial scales down to the resolution limit of the observations (typically a few arc seconds) are present on broadband eclipse photographs (Koutchmy, this Proceedings), and presumably even smaller features exist. It has long been supposed that this filamentary structure delineates the magnetic field lines imbedded in the coronal plasma. Although the brief coverage provided by eclipses does not directly reveal much evidence for plasma motions, such flows must certainly exist and these, too, should largely be aligned with the ambient magnetic field. The earliest attempt to estimate on purely theoretical grounds the effect of filamentary structure on the dynamics of the coronal plasma was made by Parker (1964).

More recently the beautiful X-ray images of the corona being returned by instruments such as NIXT and Yohkoh's SXT have sparked renewed interest in the subject of small-scale inhomogeneities, particularly within streamers. Several examples were shown during the streamer session. Perhaps the most outstanding of these were the SXT images of both bright and (relatively) dark cusp-like structures within streamer complexes (presentations by K. Strong and C.-C. Cheng). These transient brightenings may well be the long-searched-for signature of magnetic reconnection in the corona. Alternatively, they may simply reflect the plasma filling from below of a pre-existing cusp structure at the top of an arcade of loops. Detailed analyses of these fascinating observations are just now getting underway, and substantial progress in this area can be anticipated in the near future.

5. Coronal Source Regions of the Solar Wind

In the mid-1970's it became generally recognized, on the basis of kinematic mapping of the interplanetary magnetic field back to the Sun, that the large high-speed streams (HSS; $v \geq 400 km/s$) observed to recur every 29 days in the solar wind emanate primarily from long-lived coronal hole (CH) regions as the latter rotate with the Sun (Hundhausen, 1977). HSS's account for a substantial fraction of the solar wind at 1 A.U. ($\approx$ half, depending upon the number of streams present and thereby upon phase in the solar cycle); the remainder is characterized as "slow" or "background" wind ($v < 400km/s$).

Today the association of HSS's with CH's is established beyond a reasonable doubt. This scenario is presently being played out in real time by incoming solar wind data from the Ulysses mission (presentation by M. Neugebauer): as the spacecraft moves to ever-higher southern heliographic latitudes, it appears to be spending an increasing fraction of each successive solar rotation immersed in the "simpler" HSS flow pattern associated with the south polar coronal hole region.

The origin of the *slow* wind, on the other hand, is still a matter of debate, there being no obvious candidates for its source region(s) among known features on the solar surface. Some years ago it was suggested (Borrini et al., 1981) that the low-velocity, high-density, helium-poor solar wind plasma in the vicinity of the "sector boundaries" at which the interplanetary magnetic field reverses polarity abruptly, maps back to the high-density cores of coronal streamers. According to this picture, a sector boundary is nothing more than the outward extension of the current sheet postulated to lie along the symmetry axis of an underlying streamer, and the slow wind near the sector boundary originates within the body of the streamer - i.e., on the first open field lines encountered as one moves away from the streamer axis. These field lines are rooted in the solar surface at the edges of the helmet, in regions occupied by relatively high coronal density (albeit perhaps not quite as high as that within the helmet itself). Thus the slow-wind source regions would be described as being of high density at the Sun compared with, for example, a coronal hole.

On the other hand, we have heard described in the streamer session some recent work that ascribes the slow wind to the outflow of plasma near the edges of, but still within, coronal hole regions (presentation by Y.-M. Wang). On the basis of numerical simulations of coronal expansion in various types of prescribed non-r^2 geometries, in which the energy transport (like the flow itself) is assumed to be channeled along the flow tubes, Wang concludes that the (relatively) faster divergence of field lines issuing from the edges of coronal hole regions results in lower wind speeds there.

Judging from the ensuing discussion of this work, there appears to be some real confusion at present as to the slow-wind source regions. This confusion may turn out to be simply a matter of semantics; nevertheless, it should be resolved. Is the slow-wind source region near the "edge" of a (low density) coronal hole in Wang's scenario synonymous with the high density region envisaged by Borrini et al. (1981) and others to be situated at the edge of an adjacent helmet streamer? If so, should the coronal source region be considered to be part of the CH or part of the streamer? The answers to these questions ultimately revolve around a more precise operational definition of the "border" between coronal holes and streamers, as well as that of the effective boundary separating the high- and low-speed wind flows in interplanetary space.

In addition, most likely not *all* of the low-speed wind comes from these relatively narrow boundary regions between streamers and holes. The feeling that I (and I'm sure that others) have had for many years, is that the above division of the outer solar atmosphere into "streamers" and "coronal holes" is overly simplistic. A cursory glance at any high-resolution image of the X-ray corona, such as the beautiful examples from Yohkoh that were shown in this working group, reveals that there are many large regions of diffuse emission that belong neither to coronal holes nor to large-scale streamer regions. Some of these regions appear to contain faint loop-like structures, whereas others seem to be completely featureless. What is the magnetic topology of the latter? Seen with enhanced contrast and sufficient spatial resolution, they may consist of complex groupings of closed loops interspersed with "open" flux tubes extending far out into space. Such open tubes, and the plasma flowing along them, might account for much of the solar wind outside high-speed streams. However, the obvious observational difficulties associated with tracing a given parcel of low-speed wind back to a well-defined feature in one of these amorphous regions of the low corona, need not be belaboured here.

6. Theoretical Streamer Models

Early attempts to build coronal models containing streamer structures, and to predict the spatial modulation of solar wind properties by these structures, were based on the assumption of ideal magnetohydrodynamics (MHD). These proceeded by solving directly for the steady state by an iterative process (Pneuman and Kopp, 1971; Robertson, 1983). This approach is currently being pursued by Y.-M. Wang in developing a self-consistent MHD model for the solar wind outflow from CH regions, including the effects of heating and conductive energy transport. We look forward to seeing the first results from this ambitious effort at the next SOHO Workshop.

An alternative technique which seems to be more widely used nowadays is to integrate the time-dependent MHD equations with artificially enhanced resistivity, starting from an arbitrary assumed initial state and following the temporal evolution of the plasma-field system until a dynamically relaxed steady state is reached, corresponding to the assumed (invariant) boundary conditions. This "relaxation method" was first used to construct 2-D streamer models by Endler (1971).

The relaxation technique is presently being used by, among others, the Huntsville group to build composite steady state models of streamers flanked on both sides by CH regions (presentation by S. Suess). The aim is to make a direct connection to SOHO observations of the spatially resolved plasma properties (temperature, density, and bulk flow velocity) within a streamer.

To this end, G. Poletto (Arcetri) has made quantitative predictions of the intensity and shape of the hydrogen Ly-α line for these models, including the effects of Doppler dimming, for comparison with UVCS observations of the actual line profile as a function of position. Perhaps the major limitation of these simulations is that imposed by assuming polytropic flow in and around the streamer. This necessitates assigning unreasonably high temperatures to the CH field lines, in order to obtain low densities there. The next logical refinement to this model will presumably be the use of a realistic energy equation in the flow calculation. If nothing else, this should facilitate an arbitrary specification of boundary conditions at the coronal base - thereby accommodating almost any observed variation of density and temperature as one proceeds outward from a streamer to the neighboring open-field regions.

7. Streamer Evolution and the Global Corona

That streamers are intrinsically dynamical phenomena of the solar corona has been well-established by observations from Skylab, Solwind (P78-1), and SMM. Major changes are known to occur on time scales ranging from many months down to a few tens of minutes. Even the largest, most stable-appearing streamers do not persist for more than a few solar rotations (although this may be an effect of changing perspective of the streamer belt relative to the ecliptic plane). Most dramatic, of course, are instances in which a streamer is disrupted entirely by the occurrence of a nearby coronal mass ejection (CME). Many of the gradual, more subtle changes in streamer morphology manage to slip by unnoticed in day-to-day synoptic observations, being largely masked by the effects of solar rotation (Kopp, 1992). At the present time it is impossible to state with any precision what fraction of the global restructuring of the corona is caused by these slow changes, and what fraction should be ascribed to transient events (including CME's).

Again, the recent Yohkoh observations have been a boon to revealing many of the short-term changes in streamer configurations. The SXT has provided many fine examples of both the plasma filling (brightening) of pre-existing helmet shapes and the reformation of streamers following major disruptions (presentation by K. Strong). Important in themselves for depicting the various time scales on which significant changes occur, these examples and others await in-depth analysis and interpretation, which hopefully will be forthcoming in the not-too-distant future.

When it comes to modeling these transient changes in coronal streamers, we appear to be close to possessing the requisite theoretical tools. The time-dependent MHD relaxation method described in the previous section for generating steady-state streamer configurations, can in principle also be used to simulate large-scale transient phenomena in the corona, *as long as the specifics of reconnection phenomena (e.g., time scales) are not an issue*:

the inherently low resolution of existing numerical MHD codes, and the associated unrealistically large values of resistivity (small magnetic Reynolds number), have thus far precluded a physically correct treatment of magnetic dissipation in global reconnection scenarios. Notwithstanding this caveat, time-dependent simulations have been used quite successfully to illustrate the outward disruption of streamers (a convective, rather than a dissipative, process) under the action of externally generated disruptive forces in the corona (presentation by R. Steinolfson). The applications have thus far been to two-dimensional configurations, whereas in reality streamers are three-dimensional objects - probably extending in longitude not a great deal more than in latitude; thus ultimately 3-D MHD codes will have to be brought to bear on the streamer-simulation problem.

The obstacles confronting a more comprehensive modeling of streamer disruptions apparently stem more from the incomplete observational coverage of these events than from the lack of adequate computational tools. To carry out conclusive tests of our coronal-transient modeling capabilities, precise temporal and spatial knowledge of the plasma and magnetic state of streamers and their surroundings is badly needed, extending from the base of the corona (for specifying boundary conditions) out to several solar radii (for defining the initial state of an event as well as for documenting its subsequent evolution). Such thorough coverage will only be possible through a combination of coordinated observations from the SOHO instruments and correlative observations from an extensive ground observing network (e.g., high-resolution vector magnetographs).

Acknowledgements

This report was written under the auspices of the U.S.D.O.E. I am grateful for the generous travel support provided in part by the Agenzia Spaziale Italiana and in part by the Gruppo Nazionale di Astronomia of the CNR.

References

Borrini, G., Gosling, J. T., Bame, S. J., Feldman, W. C., Wilcox, J. M.: 1981, *J. Geophys. Res.* **86**, 4565

Campbell, W. W.: 1923, *PASP* **35**, 11

Guhathakurta, M., Fisher, R. R., and Altrock, R. C.: 1993, *ApJ* **414**, L145

Hundhausen, A. J.: 1977, in *Coronal Holes and High Speed Wind Streams*, Zirker, J.B., Ed., p. 225

Kopp, R. A.: 1992, in *Coronal Streamers, Coronal Loops, and Coronal and Solar Wind Composition*, ESA SP-348, p. 53

Parker, E. N.: 1964, *ApJ* **139**, 690

Pneuman, G. W., Kopp, R. A.: 1971, *Solar Phys.* **18**, 258

Robertson, B. J.: 1983, *Solar Phys.* **83**, 63

IV Coronal Holes and Solar Wind

OBSERVATIONS OF THE SOLAR WIND FROM CORONAL HOLES

M. NEUGEBAUER
Mail Stop 169-506, Jet Propulsion Laboratory, California Institute of Technology
Pasadena, CA 91109, USA

Abstract. The solar wind emanating from coronal holes (CH) constitutes a quasi-stationary flow whose properties change only slowly with the evolution of the hole itself. Some of the properties of the wind from coronal holes depend on whether the source is a large polar coronal hole or a small near-equatorial hole. The speed of polar CH flows is usually between 700 and 800 km/s, whereas the speed from the small equatorial CH flows is generally lower and can be <400 km/s. At 1 AU, the average particle and energy fluxes from polar CH are $2.5\times10^8\,cm^{-2}\,sec^{-1}$ and $2.0\,erg\,cm^{-2}\,s^{-1}$. This particle flux is significantly less than the $4x10^8\,cm^{-2}\,sec^{-1}$ observed in the slow, interstream wind, but the energy fluxes are approximately the same. Both the particle and energy fluxes from small equatorial holes are somewhat smaller than the fluxes from the large polar coronal holes.

Many of the properties of the wind from coronal holes can be explained, at least qualitatively, as being the result of the effect of the large flux of outward-propagating Alfvén waves observed in CH flows. The different ion species have roughly equal thermal speeds which are also close to the Alfvén speed. The velocity of heavy ions exceeds the proton velocity by the Alfvén speed, as if the heavy ions were surfing on the waves carried by the proton fluid.

The elemental composition of the CH wind is less fractionated, having a smaller enhancement of elements with low first-ionization potentials than the interstream wind, the wind from coronal mass ejections, or solar energetic particles. There is also evidence of fine-structure in the ratio of the gas and magnetic pressures which maps back to a scale size of roughly 1° at the Sun, similar to some of the fine structures in coronal holes such as plumes, macrospicules, and the supergranulation.

Key words: Sun – Solar Wind – Coronal Holes

1. Solar-wind Speed

The solar wind emanating from polar coronal holes forms large, quasi-stationary high-speed streams. Figure 1 shows the velocity versus time profile for 16 encounters of the Ulysses spacecraft with the high-speed stream coming from the south polar coronal hole in 1992-1993. The plot starts just after the Ulysses gravity-assist at Jupiter, in February, 1992, when the spacecraft was 5.4 AU from the Sun at a heliographic latitude of -6.4° and extends to August, 1993, at distance of 4.3 AU and −38° latitude. During the first five months after the Jupiter flyby the velocity profile was rather complex with many low-speed streams typical of solar maximum conditions at that distance from the Sun. Then, as solar activity declined and Ulysses reached slightly higher southerly latitudes, the spacecraft started to detect the flow from the southern polar coronal hole; in the stream, the speed varied from 700 to 800 km/s, while between streams the speed declined to ≈400 km/s as

Space Science Reviews **70**, 319–330.

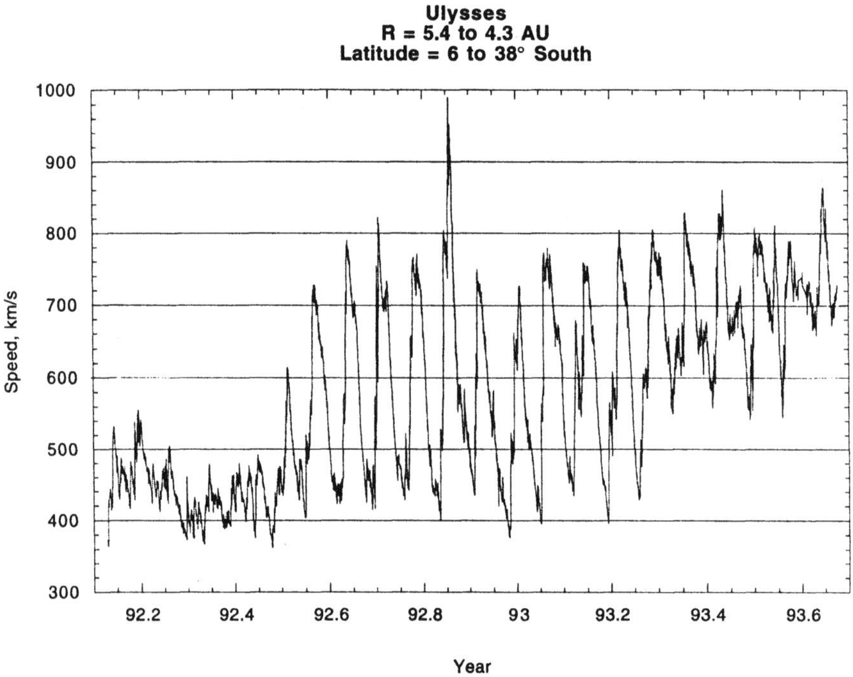

Fig. 1. Hourly averages of the solar wind speed as observed by the solar wind plasma experiment on Ulysses from February, 1992 to August, 1993 (Bame et al. 1993).

the spacecraft entered the streamer belt and crossed the heliospheric current sheet. The highest-speed peak in the Figure was caused by a coronal mass ejection, probably associated with a class 3B solar flare on November 14. The quasi-periodicity of the pattern in Figure 1 matches the solar rotation rate as viewed from the orbit of Ulysses. After May, 1993, when its latitude was approximately -29°, the spacecraft was permanently above the latitude of the streamer belt (Bame et al. 1993). There was a further change in the profile in June, 1993, after which the spacecraft was perhaps always within the coronal hole flow.

Figure 1 illustrates what we have come to think of as a typical pattern of repeated encounters with streams from polar coronal holes. But the pattern is different at other phases of the solar cycle. Figure 2 shows the velocity profile obtained near Earth by the ISEE-3 spacecraft near solar maximum in 1978-9. The x's mark the periods of flow from five relatively small, near-equatorial coronal holes whose properties were analyzed by Neugebauer and Alexander (1991). There are perhaps another five coronal hole flows in that data set which were not selected for analysis. Note that the speeds of the selected events never exceeded 700 km/s and were sometimes lower than 400 km/s. The point to be made is that the solar wind from small equatorial

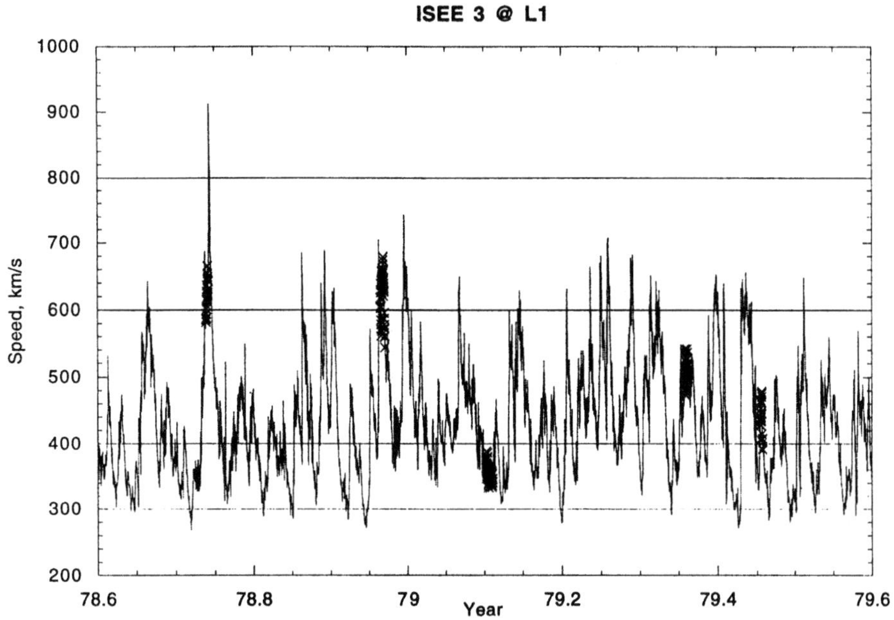

Fig. 2. Hourly averages of the solar wind speed as observed near Earth by the solar wind plasma experiment on ISEE 3 from August, 1978 to August, 1979.

coronal holes (seCH) is usually quite different from the flow from the larger polar coronal holes (pCH); in this review the properties of the two types of coronal hole flows are summarized separately.

It was discovered by the Helios observations that the latitudinal edges of the high-speed streams from pCH are very sharp (Schwenn et al. 1978), with gradients of ≈100 km/sec/deg near 0.3 AU, and probably even steeper gradients closer to the Sun. The sharp edges are not apparent in the Ulysses data in Figure 1 because of interplanetary dispersion on the trailing edges (the fastest plasma runs away from the slower plasma immediately behind it) and because of the change in profile resulting from the shocks that form on the leading edges. The velocity gradients on the edges of streams from seCH have not been studied.

2. Average Fluxes

Table I contrasts the fluxes of particles, energy, and the radial component of the magnetic field from pCH and from the slow solar wind between the fast pCH streams. Two sets of data have been used: IMP data taken at 1 AU between March, 1971 and July, 1974 (Feldman et al. 1977) and Helios data averaged between 0.3 and 1 AU between December, 1974 and May, 1976 (Marsch et al. 1984). These two non-overlapping sets of data were

TABLE I

Typical fluxes in solar-wind streams from polar coronal holes and from the slow solar wind.

		At 1 AU		At the Sun	
		Polar coronal holes	Slow wind	Polar Coronal holes	Slow wind
nv [$cm^{-2}s^{-1}$]	Helios	$2.2\pm1.1\times10^8$	$4.0\pm0.8\times10^8$	1.0×10^{13}	1.8×10^{13}
	IMP	$2.7\pm0.4\times10^8$	$3.9\pm1.5\times10^8$	1.2×10^{13}	1.8×10^{13}
Energy [$erg\,cm^{-2}\,s^{-1}$]	Helios	1.6 ± 0.4	1.7 ± 0.7	0.7×10^5	0.8×10^5
	IMP	2.2 ± 0.4	1.7 ± 0.6	1.0×10^5	0.8×10^5
B_r [G]	Helios	$3.2\pm1.5\times10^{-5}$	$2.9\pm1.9\times10^{-5}$	1.5	1.3

both obtained during periods when most of the high-speed streams could be associated with equatorward extensions of polar coronal holes. The sorting was done on the basis of solar wind speed, and so there may be a small amount of contamination of the "pCH" data with high speed streams caused by coronal mass ejections. The data were extrapolated to the solar surface by assuming radial flow — i.e., an inverse square decrease of each parameter with distance from the Sun. There is, however, good evidence for greater-than-radial expansion in the corona; therefore each of the numbers in the pCH column for the solar surface must be multiplied by an areal expansion factor, which is often modeled as being in the range of 5 to 10.

The data in Table I show that the particle flux is about a factor of 1.4-1.8 lower and the density is about a factor of 3 lower in pCH flows than in the slow solar wind. The energy fluxes, which include the energy needed to overcome solar gravity, are roughly the same, however.

The particle and energy fluxes in our sample of five seCH were 2×10^8 $cm^{-2}\,s^{-1}$ and $1.2\,erg\,cm^{-2}\,s^{-1}$, which are slightly lower than the fluxes from the pCH. The average value of B_r in seCH has not been analyzed.

3. Internal State

Figure 3 (Neugebauer 1992) gives examples of proton and electron temperatures as a function of speed for different types of solar wind flow. In this figure, quasi-stationary flow is indicated by open symbols, while transient flow from coronal mass ejections is indicated by + and x. Coronal holes are shown by squares, with the pCH and seCH having velocities above and below 650 km/s, respectively. For all the quasi-stationary solar-wind flows, the pro-

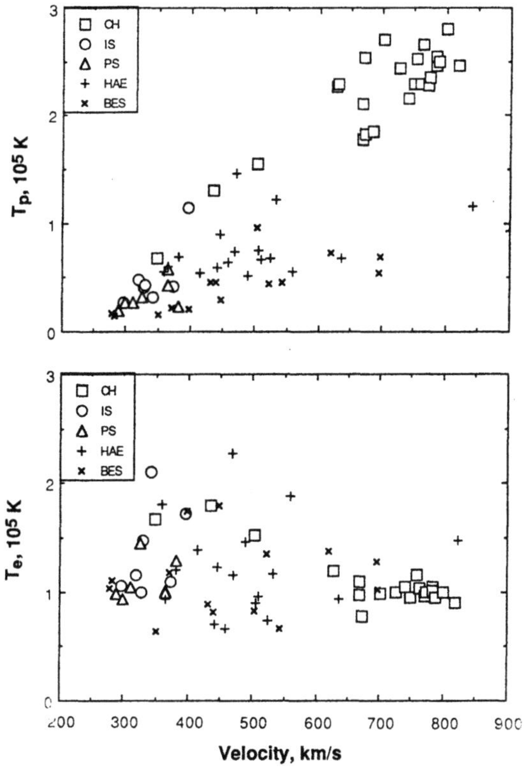

Fig. 3. Proton (top) and electron(bottom) temperatures versus solar-wind speed averaged over intervals of different types of flow (Neugebauer 1992). Open symbols denote quasi-stationary wind while + and x refer to coronal mass ejections. The notation is CH = coronal holes, IS = interstream plasma, PS = the high-density plasma sheet in which the heliospheric current sheet is embedded, HAE = helium abundance enhancements, and BES = bi-directional electron streaming.

ton temperature increases with the speed of the wind. This is in contrast to the case for transient flows, which are believed to be cooled by strong, 3-dimensional expansion en-route from the Sun to their point of observation at 1 AU. The bottom panel of Figure 3 shows, on the other hand, that the solar wind electron temperature is quite insensitive to the type of flow and that for pCH flows the electrons are substantially cooler than the protons. The electron temperature is determined principally by heat conduction from the corona, while the ion temperature is controlled by interplanetary processes, especially by waves. A simple-minded interpretation is that ion temperatures are determined by wave-particle interactions rather than by Coulomb collisions. Figure 4 gives some credence to that interpretation. It is a scatter diagram of the velocity dependence of the ratio of the proton thermal speed to the Alfvén speed observed by Ulysses at the peaks and trailing edges of pCH flows. At the highest speeds, the ratio is close to 1.

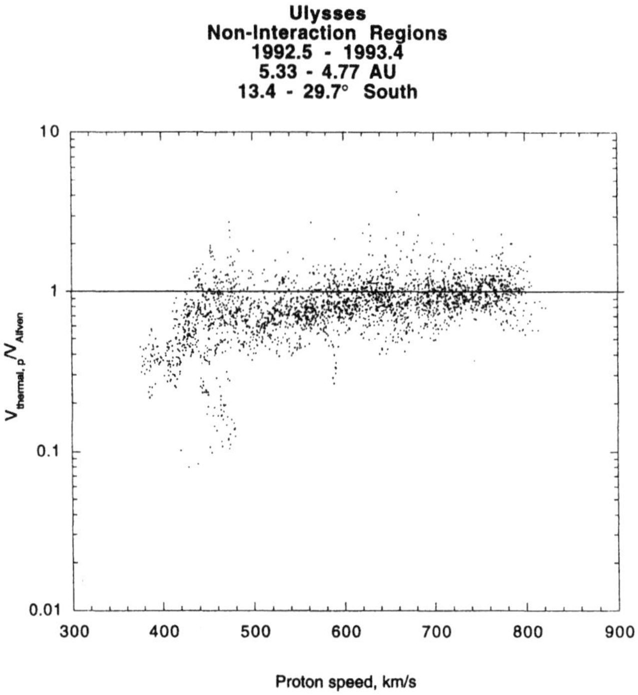

Fig. 4. Scatter plot of hourly averages of the ratio of the proton thermal speed to the Alfvén speed versus the flow speed of the wind as observed at the peaks and trailing edges of polar coronal hole streams observed by Ulysses.

At lower speeds, the ratio is less than 1, probably due to a combination of cooling due to the expansion of the gas, increased Coulomb collisions in the slower, denser gas, and a decrease in the wave flux.

4. Waves

A large flux of Alfvén waves propagates away from the Sun through the solar wind (Belcher et al. 1971). Figure 5 shows an example of the importance of these waves in the pCH flows observed by Helios 2 between 0.29 and 0.54 AU, (Denskat et al. 1983). From top to bottom, this Figure shows: (1) The fluctuations in the magnitude of the magnetic field, normalized by the average field magnitude; this parameter is low for Alfvén waves. (2) The normalized fluctuations in one of the field components, which is large for a large flux of Alfvén waves. (3) The normalized fluctuations of plasma density, which are low for Alfvén waves. (4) The solar wind speed, which is highest in the pCH, as between April 13 and April 23. (5 to 7). The correlation between velocity and field for each of the three vector components, with the normalization and the sign of the field vector taken such that correlations

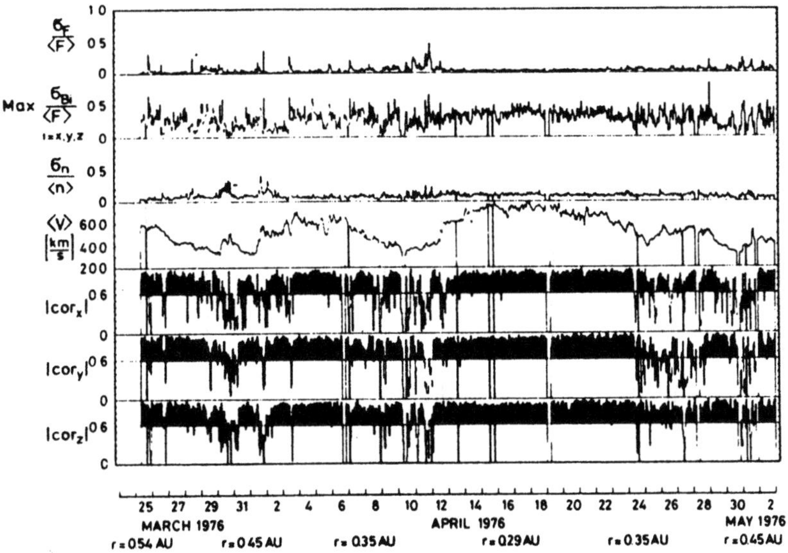

Fig. 5. Plots of various parameters indicative of the presence of Alfvén waves as observed by Helios 2 (Denskat and Neubauer 1983). See the text for further explanation.

of +1.0, 0, and −1.0 correspond to a pure flux of Alfvén waves propagating away from the Sun, to random fluctuations, and to an inward propagating wave flux, respectively. Correlations greater than +0.6 have been shaded. The relationship between high speeds, in this case associated with pCH flows, and strong fluxes of outward propagating Alfvén waves, often with $|\delta \mathbf{B}|/B \approx 0.5$, is striking. Similar results are observed for seCH (Neugebauer 1992).

In his talk, Axford (these Proceedings) emphasized another indicator of the presence and importance of waves in the solar wind from coronal holes, namely the observation that the ratio of the temperatures perpendicular and parallel to the field $T_\perp/T_\parallel$ is >1. Without the wave-particle interactions, the conservation of the magnetic moments of the ions would require $T_\perp/T_\parallel$ <1. These waves could be remnants of the wave field responsible for the acceleration of the wind in coronal holes. Axford also suggested that these waves may be the low-frequency tail of the coronal wave spectrum which survives in the wind because it is less readily damped than the higher frequency part of the spectrum. But not all the waves originate at the Sun; additional waves must also be created in situ by the cyclotron instability of a $T_\perp/T_\parallel > 1$ configuration.

It was demonstrated in Figure 3 that for the flow from coronal holes the proton temperature is positively correlated with the solar wind speed and in Figure 4 that the proton thermal speed is close to the Alfvén speed $= B/\sqrt{4\pi F/v}$, where F is the ion flux and v is speed. It follows that the Alfvén speed should be correlated with the flow speed and that the flow speed should correlate inversely with f_{max}, the areal expansion factor of the streamlines or flux tubes between the photosphere and interplanetary space. Both those correlations are observed; the dependence of speed on f_{max} is discussed further in the paper by Wang in these Proceedings.

5. Minor Ions

The helium abundance in the solar wind is notoriously variable with the alpha particle density n_α ranging from nearly zero to almost half the density of protons n_p. CH flow is an exception, with n_α/n_p remaining quite steady at $\approx$0.048 in pCH. In seCH, n_α/n_p is also quite steady on the scale of minutes to hours, but its average value correlates with the average speed of the flow (Neugebauer 1992); i.e., n_α/n_p is lower in seCH than in pCH flows.

The abundances of elements heavier than helium also differ from one type of flow to another, although their variations are smaller than those observed for helium. For example, the Mg/O ratio observed over a 6-month period by the Solar Wind Ion Composition Spectrometer on Ulysses varied by more than an order of magnitude with a standard deviation of $\approx$40% [R. von Steiger, personal communication]. The variations are not random. Those ions with first ionization potentials (FIP) less than the energy of Ly-α photons are over-abundant by factors up to $\approx$5 in the slow solar wind, in plasma from coronal mass ejections, and in gradual-event solar energetic particles. The abundances of these low-ionization-potential ions are, however, significantly less enhanced in the wind from coronal holes — both pCH and seCH [R. von Steiger, G. Gloeckler, personal communication]. In other words, the FIP effect is much weaker for the flow from coronal holes than it is for the slow wind or the plasma released from closed loops in the corona.

The charge-state of the ions heavier than helium is also different for coronal hole flows than for the slow or transient wind. The ionization temperature as determined, for example, by the ratio of O^{7+} to O^{6+} ion densities is lowest in coronal hole flows, higher in the slow solar wind, and highest of all in the plasma released in CMEs.

The distribution functions of the minor ions show the effect of the high wave flux in CH flows. In the wind from coronal holes, the different ion species have nearly equal thermal speeds rather than equal temperatures; i.e., their kinetic temperatures are proportional to their masses. In the pCH streams near 0.3 AU, Helios observed alpha- particle temperatures of $\approx 3\times10^6$ K, which is hotter than the corona (Marsch et al. 1982). And even at 1 AU the temperatures of some of the heavier ions, such as oxygen or

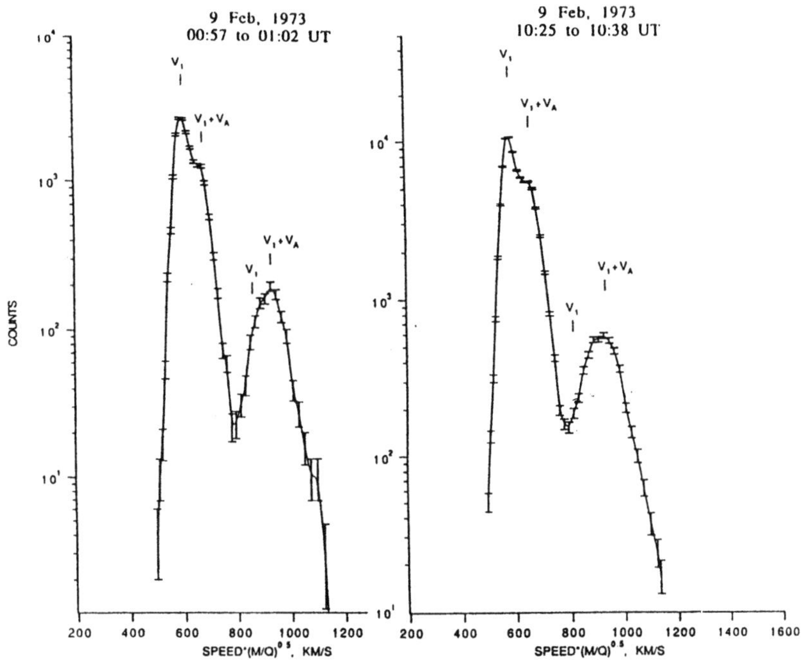

Fig. 6. Solar-wind ion spectra measured in plasma from a polar coronal hole observed by IMP (Feldman, et al. 1993).

iron, often exceed 10^7 K (Ogilvie et al. 1980).

Another probable effect of the intense wave field is the differential streaming between the heavy ions and the protons in CH flows. Despite their greater mass and the greater amount of energy required to remove them from the solar gravitational field, the alphas stream away from the Sun faster than the protons. Figure 6 (Feldman et al. 1993) shows two one-dimensional ion spectra of the high-speed wind from a coronal hole. The plots show counts versus energy per unit charge, converted to units of $v\sqrt{M/Q}$, with protons and alpha-particles constituting the peaks at the lower and higher values of $v\sqrt{M/Q}$, respectively. Above the left-most peak with $v = V_1$ is a secondary proton peak close to $V_1 + V_A$, where V_A is the Alfvén speed. The alpha-particle peak is at $V_1 + V_A$, rather than at V_1, which would be the case if the two ion species had the same flow speed. A simple view is that the alphas are surfing through the proton sea on the crests of the outward propagating Alfvén waves, but a detailed theory has not yet been developed to explain quantitatively why this happens (Isenberg 1984). Feldman et al. (1993) have interpreted this phenomenon as evidence that there are two sources of the solar wind in coronal holes: (1) a thermally driven, low-speed wind whose flux is so low that it cannot drag alpha-particles out into the wind and (2)

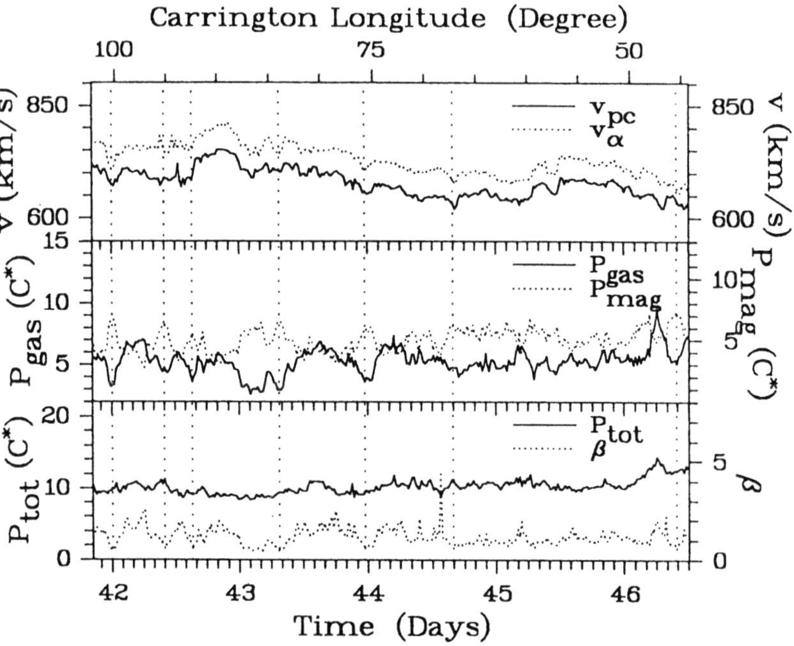

Fig. 7. Time profiles of various plasma parameters observed by Helios (Thieme et al. 1990). Further explanation is given in the text. Pressures are given in units of $C^* = 10^{-10}$ dyn cm^{-2}.

fast, helium-rich plasma jets, driven by reconnection, which interact with and accelerate the low-speed, helium-poor wind to high speed. In that scenario, the jets would also be the principal source of the waves observed in CH flows.

It is interesting to note that the Alfvén speed V_A decreases rapidly with solar distance. If the data from Helios and other spacecraft (e.g., Figure 6) are extrapolated inward, say to the solar-wind critical point, the energy lost by the alpha-particles $\Delta(m_\alpha V_A^2/2)$ is so large that it must be an important source of energy for accelerating the wind from coronal holes.

6. Small-Scale Structures

At distances from the Sun of 1 AU or greater, one of the distinguishing features of CH flow is the relative constancy of all parameters other than the directions of **v** and **B**, which are affected by the waves; i.e., the density, field magnitude, ion and electron temperatures, and n_α/n_p do not fluctuate very much. But it was possible to find some fine structure in the Helios data inside 1 AU. Figure 7, from a paper by Thieme et al. (1990), shows time profiles of v_{pc} = proton speed after subtraction of fluctuations due to Alfvén waves,

TABLE II

Summary of the properties of the solar wind flowing from large polar coronal holes (pCH) and from small, equatorial coronal holes (seCH), at 1 A.U..

Parameter	pCH	seCH	
V [km/s]	750 – 800	can be < 400	
nV $[10^8\,cm^{-2}\,s^{-1}]$	2.5	2.0	
Energy flux $[erg\,cm^{-2}\,s^{-1}]$	2.0	1.7	
Proton, electron temp.	$T_p > T_e$	T_p can be $< T_e$	*
$V_{thermal,p}/V_A$	≈ 1	can be < 1	*
Waves	Strong flux of outward propagating Alfvén waves	Strong flux of outward propagating Alfvén waves	*
n_α/n_p	≈ 0.048	≤ 0.048 (increase with V)	
FIP effect	small	small	
$V_\alpha - v_p$	$\approx V_A$	?	*
Heavy ion temperature	$T_i \sim m_i$	$T_i \sim m_i$	*
Fine structure	Supergranule scales	?	

v_α = alpha-particle speed, P_{gas} = gas pressure, P_{mag} = magnetic pressure, P_{tot} = total pressure = $P_{gas} + P_{mag}$, and the plasma β. These data were obtained by Helios 1 in 1975 at a solar distance of 0.66 to 0.60 AU. The vertical dotted lines mark occasions of pronounced anticorrelation between the gas and magnetic pressures, even though the total pressure remains very constant. At the closest distances to the Sun, 0.3 to 0.5 AU, the sizes of the blobs of extremes of gas or magnetic pressures correspond to angular dimensions on the Sun between 1 and 2°. Thieme et al. suggest that these structures are remnants of fine structures in coronal holes, such as plumes and macrospicules which are connected to the chromospheric network.

7. Summary

Table II summarizes the properties of pCH and seCH discussed above. An asterisk on the right denotes a property that indicates the presence or effect of a large flux of outward propagating Alfvén waves in the flow from coronal holes. An outstanding question for future research is the relationship between and the relative roles of waves versus jets in heating the corona and accelerating the wind in coronal holes.

Acknowledgements

This review was compiled at the Jet Propulsion Laboratory, California Institute of Technology, under a contract with the US National Aeronautics and Space Administration.

References

Bame, S.J., Goldstein, B.E., Gosling, J.T., et al.: 1993, *Geophys. Res. Lett.* **20**, 2323

Belcher, J.W., Davis, L., Jr.: 1971, *J. Geophys. Res.* **76**, 3534-3563

Denskat, K.U., Neubauer, F.M.: 1983, in *Solar Wind Five*, M. Neugebauer (Ed.), NASA, Washington, p. 81-91.

Feldman, W.C., Asbridge, J.R., Bame, S.J., et al.: 1977, in *The Solar Output and its Variation*, O.R. White (Ed.), Colorado Assoc. Univ. Press, Boulder, p. 351-382.

Feldman, W.C., Gosling, J.T., McComas, D.J., et al.: 1993, *J. Geophys. Res.* **98**, 5593

Isenberg, P.A.: 1984, *J. Geophys. Res.* **89**, 6613

Marsch, E., Mühlhäuser, K.-H., Rosenbauer, H., et al.: 1982, *J. Geophys. Res.* **87**, 35

Marsch, E., Richter, A.K.: 1984, *J. Geophys. Res.* **89**, 6599

Neugebauer, M.: 1992, in *Solar Wind Seven*, E. Marsch, R. Schwenn (Eds.), Pergamon Press, Oxford, p. 69-78.

Neugebauer, M., Alexander, C.J.: 1991, *J. Geophys. Res.* **96**, 9409

Ogilvie, K.W., Bochsler, P., Coplan, M.A., et al.: 1980, *J. Geophys. Res.* **85**, 6069

Schwenn, R., Montgomery, M.D., Rosenbauer, H., et al.: 1978, *J. Geophys. Res.* **83**, 1011

Thieme, K.M., Marsch, E., Schwenn, R.: 1990, *Ann. Geophys.* **8**, 713

UNSOLVED PROBLEMS OF SOLAR WIND EXPANSION: CAN WE LEARN ANYTHING FROM SOHO?

R. ESSER
Harvard-Smithsonian Center for Astrophysics, 60 Garden St, Cambridge, MA 02138, USA

Abstract. Some of the main problems of solar wind expansion are addressed. Emphasis is placed on solar wind acceleration and the mass flux problem. It is demonstrated how these two properties of the flow depend on other plasma parameters such as temperature, density and helium abundance. The importance of placing constraints on a given solar wind flow in the inner corona and at larger distances from the sun simultaneously, is also shown. Whether and how these constraints can be derived from observations carried out by SOHO instruments is then discussed.

Key words: Solar Corona, Solar Wind, Observational Constraints

1. Introduction

To solve the problems of solar wind expansion we have to determine the physical processes that are responsible for the behavior of the solar wind plasma, and we have to find the physical models that can be used to predict the properties of the plasma flow, or at least certain aspects of it. Before we can answer the question what the SOHO instruments can contribute to accomplish this task, we first have to discuss what kind of observations we need, and to what accuracy we need them.

In this brief review we address some aspects of placing observational constraints on solar wind plasma parameters, and the accuracy that theoretical models require from the observations and their interpretations. Several of the instruments on board SOHO will measure spectral line intensities to derive coronal temperatures and densities. Among the suggested spectral lines are Fe XIV 5303 Å and Fe X 6374 Å. We therefore choose these two lines as examples to demonstrate why it is so difficult to meet the requirements of the solar wind models. We conclude with recommendations for future observations that can be carried out with SOHO.

2. Boundary Conditions and Solar Wind Models

The parameters observed at 1 AU which are most commonly used to constrain solar wind models are the outflow velocity, proton flux, electron and proton temperatures, and in addition, if one chooses multi-fluid models, the helium abundance and outflow velocity have to be considered. As boundary conditions in the inner corona the electron and proton temperatures and the electron density are the most important parameters. From studies of

Space Science Reviews **70**, 331–340.

multi-fluid models we know that the helium abundance in the inner corona is not necessarily the same as the helium abundance at 1 AU (e.g. Joselyn and Holzer 1978). For multi-fluid models a value for the helium abundance at the coronal base is therefore also needed. To match velocities calculated from the models with velocities observed at 1 AU in high speed streams, an additional energy source and corresponding constraint is necessary (e.g. Parker 1965; Hollweg 1973).

While the outer boundary conditions are known to great detail and relatively high accuracy from in situ observations (e.g. Feldman et al 1976), the boundary conditions in the inner corona are all model dependent, and are therefore much less certain (e.g. Esser and Habbal 1994; Habbal et al 1993). Plasma parameters derived from observations in the inner corona depend for example on atomic data, such as the electron temperature derived from spectral line intensities, or they depend on solar wind expansion models such as the electron temperature derived from in situ charge state measurements (e.g. Owocki 1983; Owocki and Scudder 1983), or the helium abundance (e.g. Leer et al 1992). In addition, all these parameters are averaged over the line of sight. Comparing the results of solar wind models with in situ observations and observations in the inner corona one has always to keep in mind that one set of parameters is averaged over sometimes completely different solar wind structures, while the other set consists of locally measured quantities. The question is what accuracy can we expect from remote measurements, and is this accuracy high enough to constrain solar wind models. As a reminder for what accuracy is needed for the models we use the example given by Leer and Holzer (1991). In their two-fluid model the mass flux increases by about one order of magnitude when the temperature at the coronal base increases from 1.0 to 1.3 $\cdot 10^6$K. Although the temperature sensitivity of the mass flux is reduced when a helium abundance at the coronal base of 0.3 is included in their model, the mass flux still increases by a factor of 2 for the same increase of the coronal temperature. That means to compare mass fluxes measured in situ with the mass flux calculated from solar wind models requires an accuracy of the coronal temperature of the order of 10% or less. In the following we will examine, as an example, how the line of sight affects the accuracy of the coronal temperatures derived from observations.

3. The Effect of the Line of Sight on Coronal Temperature Inferences

Even though coronal holes persist over many solar rotations, their appearance usually changes from day to day due to the quiet sun emission in the foreground or background along the line of sight. This is demonstrated in Figure 1 which shows Yohkoh images from 1993 March 27 (upper panel) and

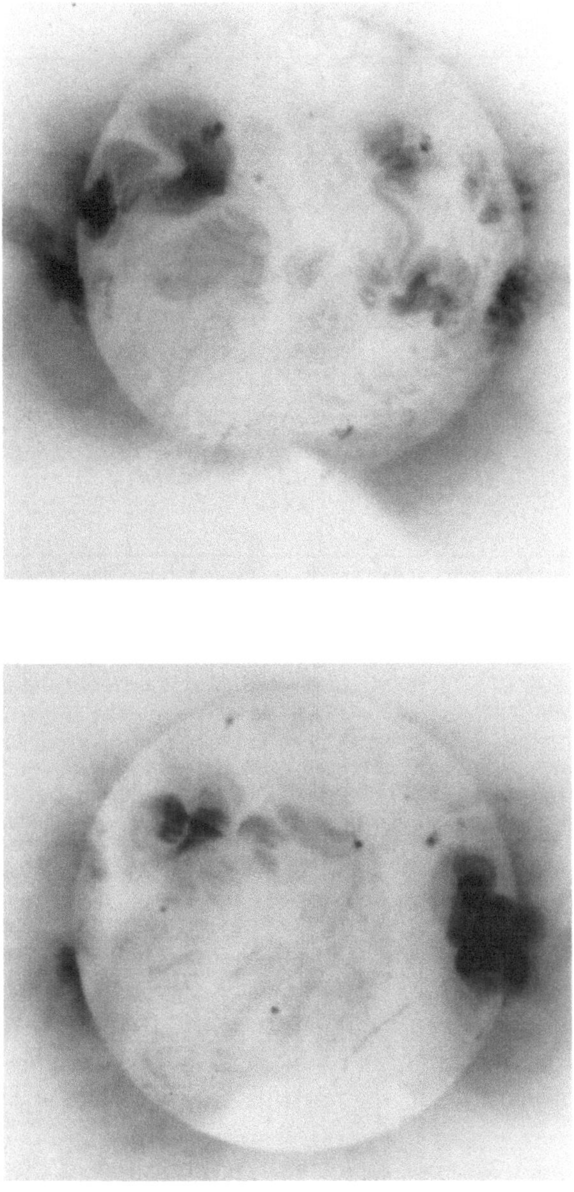

Fig. 1. Soft X-ray images of the sun from 1993 March 27 (upper panel) and 1993 April 10 (lower panel) provided by the Yohkoh team (Courtesy H. S. Hudson).

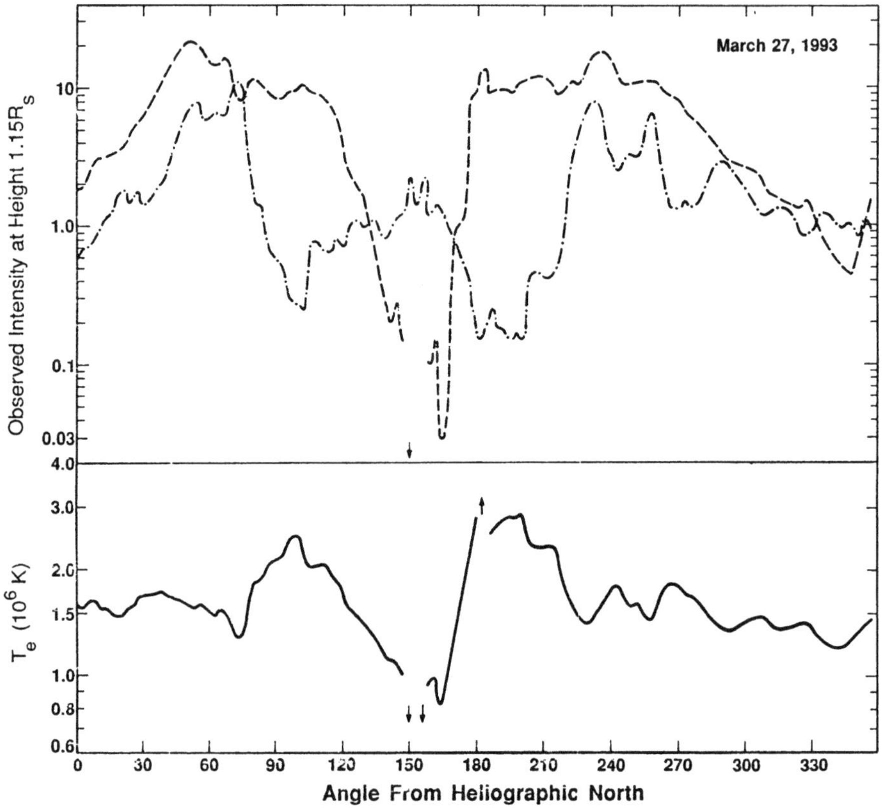

Fig. 2. Intensities of the red Fe X 6374 Å (dash-dotted line) and the green Fe XIV 5303 Å (dashed line) observed at 1.15 R_S at the National Solar Observatory at Sacramento Peak on 1993 March 27 (Courtesy R. C. Altrock). The intensities are measured every 3° around the sun from heliographic north. The minimum in the green line between about 120° and 180° corresponds to the southern coronal hole visible in the Yohkoh images in Figure 1. Also plotted is the temperature (solid line) derived from the intensity ratio using an atomic model provided by N. S. Brickhouse and and J. C. Raymond.

1993 April 10 (lower panel). While the southern coronal hole appears small on the disk on March 27, above the limb there is relatively little contribution from the surrounding regions in this case. About half a solar rotation later, on April 10, the southern coronal hole appears large on the disk while there is a significant contribution from the background emission on that day. Note also that the coronal hole in the north is hardly visible at all on these two days.

In Figures 2 and 3 we have plotted the intensities of the red Fe X 6374 Å and green Fe XIV 5303 Å spectral lines observed at the National Solar Observatory at Sacramento Peak on March 27 and April 10. The intensities

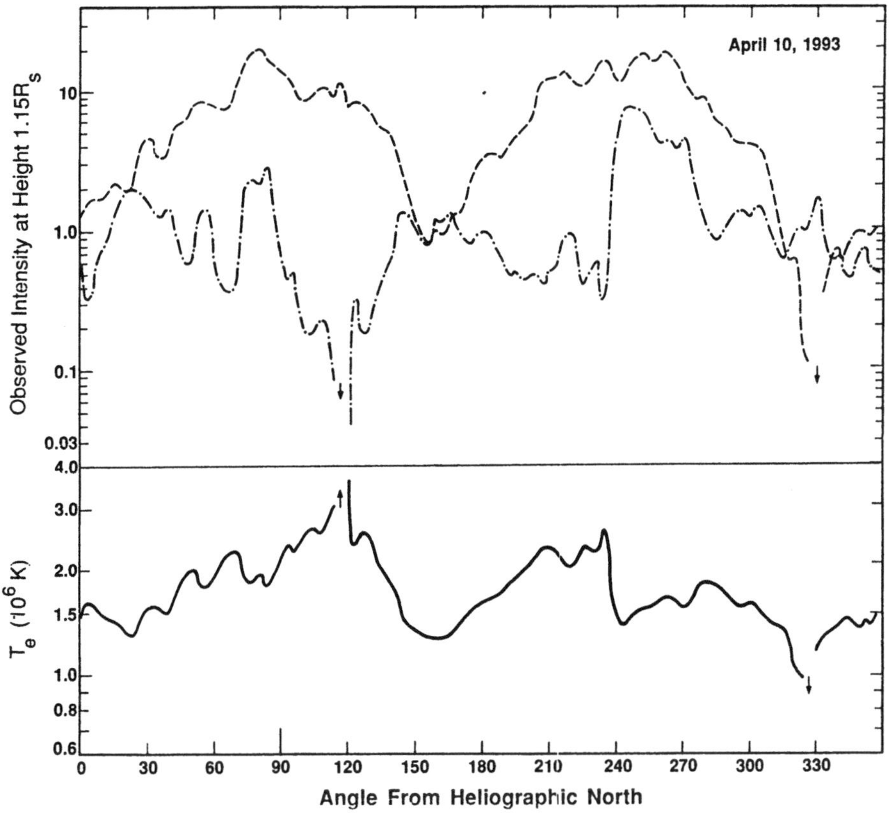

Fig. 3. Same as in Figure 2, except that the intensities were measured about half a solar rotation later on 1993 April 10.

are measured every 3° around the sun at 1.15 R_S. The minimum in the green line between about 120° and 180° corresponds to the southern coronal hole visible in the Yohkoh images. In Figure 3 the intensity of the green line is still significant at the poles. Although the green line should be absent in a coronal hole, a fact which has been known for a very long time (e.g. Koutchmy et al 1987). In Figure 2 the intensity of the green line falls almost to zero in the southern hole. This is consistent with the low emission visible in the Yohkoh image above the limb on that day. To demonstrate the errors that can be expected from the line of sight effect if no attention is paid to the possible foreground and/or background contribution, we have calculated the temperatures derived from the two lines using the line ratio technique (e.g. Habbal et al 1993, and references therein). The computer code for the calculations was provided by N. S. Brickhouse and J. C. Raymond. The minimum temperature inside the hole deviates by more than 0.5 $\cdot 10^6$ K in

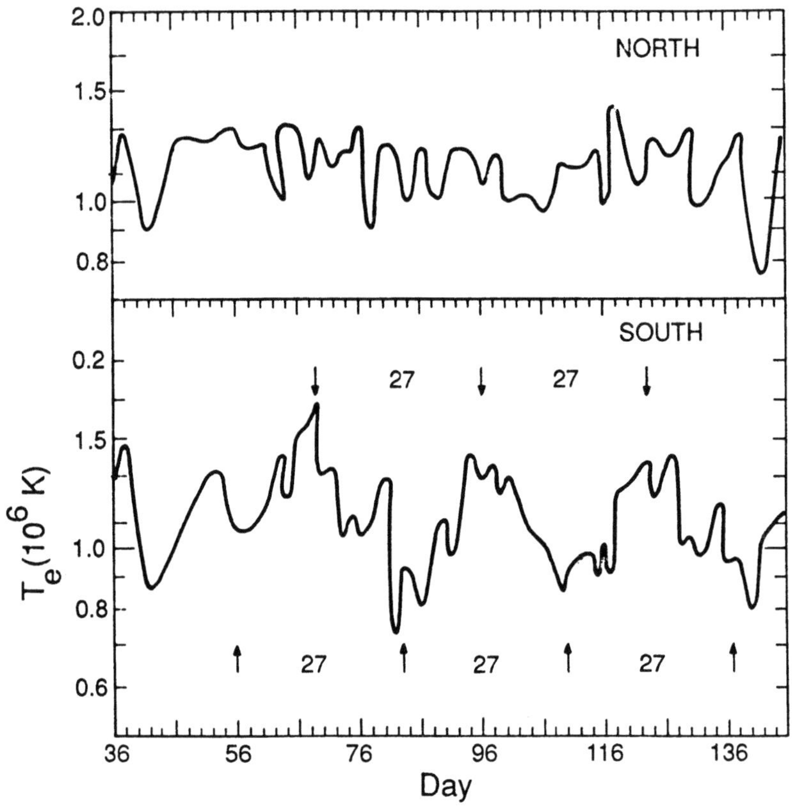

Fig. 4. Temperatures at 1.15 R_S derived from the red and green line intensities observed at the national Solar Observatory at Sacramento Peak. The temperatures across the northern and southern coronal hole were calculated every 3° for each day when data were available. The temperatures plotted in the figure are the minimum temperatures inside the polar coronal holes on each day.

these two cases. The variability of the minimum temperature in the polar holes as a function of solar rotation can also be seen in Figure 4. Here we have plotted the temperatures derived from the red and green line measurements for about four solar rotations, for each day when data were available from Sacramento Peak. In the southern coronal hole the temperature variations are of the order of $1.0 \cdot 10^6$ K and show a certain periodicity of roughly 27 days between maximum and minimum temperatures which shows even more clearly that these variations are due to line of sight effects. In the north pole where the regions surrounding the coronal hole are much more structured, no such periodicity is found.

The reason why the contribution from surrounding regions affects the

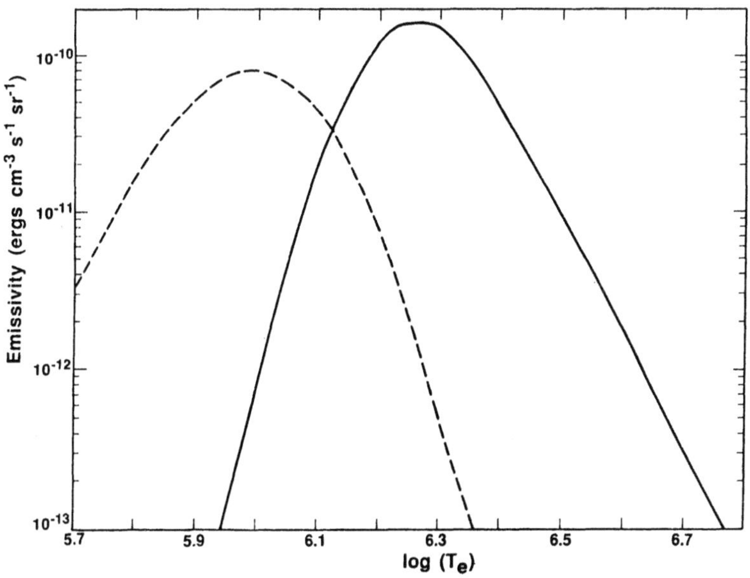

Fig. 5. Emissivity of the forbidden Fe X 6374 Å (dashed line) and Fe XIV 5303 Å (solid line) spectral lines versus temperature calculated from a computer code provided by N. S. Brickhouse and J. C. Raymond.

temperatures derived from the line ratio technique so strongly becomes immediately clear from the shape of the emissivity curves of the two Fe lines. Figure 5 shows the emissivity as a function of temperature calculated from the Brickhouse and Raymond computer code for the Fe X 6374 Å (dashed line) and Fe XIV 5303 Å (solid line) spectral lines. The emissivity of the green line has a maximum at about $2 \cdot 10^6$ K, and decreases by more than two orders of magnitude when the temperature decreases to 10^6 K. Previous measurements show that the temperature in coronal holes is 10^6 K or less (e.g. Doschek and Feldman 1977; Mariska 1978) close to the coronal base. Since the surrounding regions have both high temperatures and high densities relative to the coronal holes, the contributions from the foreground and background to the line intensity is particularly strong for lines that have their maximum at temperatures high compared to the coronal hole temperatures.

4. Summary and Discussion

The observational techniques that have been used so far to derive plasma parameters in the inner corona give usually the right order of magnitude, but are far from accurate enough to satisfy the needs of solar wind modeling. The temperature derived in the last section using the line ratio technique is just one example. Examples for other techniques can be found in

Habbal and Esser (1994), Esser and Habbal (1993), and Esser and Habbal (1994) and references in these papers. The line of sight effect is only one of the problems that one encounters in converting measurements into plasma parameters useful for solar wind modeling. Another major source of uncertainty is for example that some techniques have to rely on solar wind models to derive plasma parameters, such as the electron temperature derived from in situ charge state measurements or the outflow velocity derived from the Doppler dimming technique (Withbroe et al 1982). To infer the electron temperature in the inner corona using charge state measurements, assumptions about the relative flow speeds of the ions with different charge states, and about the electron velocity distribution have to be made. Each of these assumptions can lead to an over/under estimation of the temperature of $5 \cdot 10^6$ K (Owocki 1983; Owocki and Scudder 1983). When the intensities of resonantly scattered spectral lines are used to infer flow velocities, assumptions on the geometry of the flow tube expansion and on the mass flux have to be made (Esser and Habbal 1993). Some of these uncertainties can be reduced by long term studies over several solar rotations (tomographic techniques), such as the line of sight effect. The effect of the model dependence of many techniques can be minimized by coordinated observations, such as the uncertainties of the velocities derived from Doppler dimming due to the unknown mass flux. In this case coordinated observations with Ulysses will be necessary in order to guarantee that both the mass flux and the Ly-α line intensities are carried out in the same solar wind stream.

The parameter that defines the solar wind mass flux in the models is the average temperature below the sonic point which is expected to be located at a few solar radii from the sun (e.g. Leer and Holzer 1980). It is therefore important to measure the temperature not only close to the coronal base, but also as a function of heliocentric distance. Spectral lines that have been suggested for temperature and density diagnostics in polar coronal holes as a function of height and across the holes are for example Mg X 625 Å and Mg IX 368 Å, Si IX 342/349 Å (for density diagnostic) and Si XI 608 Å, He II 608 Å and He I 584 Å, and O IV 554 Å (e.g. Harrison 1993, page 100). The He lines are clearly lower transition region lines and therefore less relevant for the solar wind. The O IV line has its maximum temperature of formation at $2 \cdot 10^5$ K and its emission falls off very fast for increasing temperatures. The line is therefore presumably too cold. The two Mg lines have their maximum around 10^6 K, and have both been observed previously from Skylab. However, it has also been known since Skylab that the Mg IX line is almost totally absent in polar holes, except for the polar plumes (Bohlin 1977). In addition the intensity of these lines if observed decreases extremely fast with height. The Si IX line is sensitive to densities in the range 10^8 to 10^{10} cm^{-3}. The densities inferred from observations in coronal holes so far are of the order a few times 10^8 cm^{-3} and fall off almost exponentially with height. The Si IX lines are therefore at the limit for measuring in coronal holes.

We conclude that more emphasis has to be placed on selecting spectral lines suitable for temperature and density diagnostics in coronal holes, than has been done in the past. This is required by the special situation of coronal holes, i.e. extremely low densities, and temperatures that are significantly lower than the temperatures in the surrounding regions, a combination which greatly increases the misleading contribution of hot and dense plasmas along the line of sight. In addition long term observations of these regions have to be carried out since foreground and background emission can obscure a coronal hole either completely or partly, and the the contribution from these surrounding regions can not be estimated from a few isolated measurements. Whether the plasma parameters derived from such extensive studies are accurate enough to constrain solar wind models remains to be seen.

Acknowledgements

This work was supported by NASA grants NAGW-249 and NAGW-3513. The solar X-ray images are from the Yohkoh mission of ISAS, Japan. The X-ray telescope was prepared by the Lockheed Palo Alto Research Laboratory, the National Astronomical Observatory of Japan, and the University of Tokyo with the support of NASA and ISAS. We are thankful to R. C. Altrock for providing the coronal data from the Sacramento Peak Observatory. The NSO/Sacramento Peak data used here are produced cooperatively by USAF/AFMC/PL/GPSS and NSF/NOAO/NSO/SP. We are grateful to N. S. Brickhouse and J. C. Raymond for providing the computer code for the line ratio calculations, and for many helpful discussions.

5. References

Bohlin, J. D.: 1977, *In Coronal Holes and High Speed Wind Streams*, ed. J. B. Zirker, 27.

Doschek, G. A. and Feldman, U.: 1977, *Astrophys. J.*, **212**, L143.

Esser, R., and Habbal, S. R.: 1993, *Solar Phys.*, **147**, 241.

Esser, R. and Habbal, S. R.: 1994, *In Proceedings of Cosmic Winds and the Heliosphere, Space Science Series*, ed. J. R. Jokipii, submitted.

Feldman, W. C., Asbridge, J. R., Bame, S. J. and Gosling, J. T.: 1976, *J. Geophys. Res.*, **81**, 5054.

Habbal, S. R., Esser, R., and Arndt, M. B.: 1993, *Astrophys. J.*, **413**, 435.

Harrison, R. A. and Moore, V.: 1993, *The Coronal Diagnostic Spectrometer for SOHO:Scientific Report*, , ed. R. A. Harrison.
100

Hollweg, J. V.: 1973, *Astrophys. J.*, **181**, 547.

Joselyn, J.-A., and Holzer, T. E.: 1978, *J. Geophys. Res.*, **83**, 1019.
Koutchmy, O., Koutchmy, S., Nitschelm, C., Sykora, J. and Smartt, R. N.: 1988, *in Proceedings of the 9th Sacramento Peak Summer Symposium*, ed. R. C. Altrock, 256.
Leer, E., and Holzer, T. E.: 1991, *Ann. Geophys.*, **9**, 196.
Leer, E., Holzer, T. E., and Shoub, E. C.: 1992, *J. Geophys. Res.*, **97**, 8183.
Mariska, J. T.: 1978, *Solar Phys.*, **60**, 67.
Owocki, S. P.: 1983, *In Proceedings of the Solar Wind Five Conference*, ed. M. Neugebauer, NASA Conference Publication 2280. 623
Owocki, S. P., and Scudder, J. D.: 1983, *Astrophys. J.*, **270**, 758.
Parker, E. N.: 1965, *Space Sci. Rev.*, **4**, 666.
Saito, K.: 1965, *Publ. Astron. Soc. Japan*, **17**, 1.
Withbroe, G. L., Kohl J. L., Weiser, H. and Munro R. H.: 1982, *Space Sci. Rev.*, **33**, 17.

DIFFUSIVE FRACTIONATION IN THE CHROMOSPHERE

R. VON STEIGER
Physikalisches Institut, Universität Bern, Sidlerstr. 5, CH-3012 Bern, Switzerland

and

E. MARSCH
Max-Planck-Institut für Aeronomie, D-37191 Katlenburg-Lindau, Germany

Abstract. A new mechanism for the FIP fractionation in the solar wind in the form of a stationary diffusion model is proposed. It is based on a weakly stratified chromospheric layer of constant density and temperature, permeated everywhere by ionizing photons and a homogeneous magnetic field. Our model does not invoke any particular geometry or special set up of the system and is founded solely on robust and well understood atomic collisional physics. Technically, a boundary value problem of four coupled differential equations is solved for each chemical element, i. e. a continuity equation and a momentum equation for both atoms and singly ionized particles. For the main gas (hydrogen), an analytical solution can be found. This then serves as a background for the numerical integration of each trace gas system (several elements from He to Fe). We find that, after a few hydrogen diffusion lengths, each minor species asymptotically approaches a constant density. The ratios of these density values to some reference element reproduce the observed FIP fractionation pattern remarkably well.

Key words: solar wind abundances – chromosphere – solar wind

1. Introduction

It is observationally well established that element abundances of the solar wind obey a systematic fractionation pattern with respect to the photospheric abundances. Elements with low first ionization potential (FIP) are enriched by a factor of ~ 4, e.g. (Meyer, 1985; Geiss & Bochsler, 1985; Anders & Grevesse, 1989). Apparently, the elements are enriched or depleted by a process that depends on the FIP, or perhaps even more clearly, as emphasized by Geiss & Bochsler (1986), on the characteristic ionization time and related diffusion length for the neutrals of the minor species colliding with the major hydrogen atoms. The element separation process was first suggested by Geiss (1982) to operate at higher chromospheric levels in the solar atmosphere with a typical density N of 10^{16} m^{-3} and temperature T below 10^4 K. Now it is generally believed that the observed low-FIP element enrichment is caused by a chromospheric neutral-ion separation process, supplying a fractionated gas mixture from the upper chromosphere through the transition region, where it becomes multiply ionized, into the corona and the solar wind.

Space Science Reviews **70**, 341–346.

2. Element fractionation by diffusion

The present work advances a stationary diffusion model, which is based on a weakly stratified chromospheric layer, that is permeated from one side by ionizing photons, and contains a homogeneous magnetic field. Otherwise, the model does not invoke any particular geometry or special set up of the system and is founded on robust and well understood atomic physics.

The basic time scale in the ion-neutral separation process for the atomic species j is the ionization time τ_j, as derived and discussed by Geiss & Bochsler (1985). This typically ranges between 1 s and 200 s; it is to be compared with the different collision times for neutral-neutral, ion-neutral and ion-ion collisions (see below).

We briefly introduce the basic scales involved in the discussion. The barometric scale height of element j is $h_j = v_j^2/g_\odot$. The fundamental scale length in the problem is the neutral diffusion length, defined by means of the collision frequency of the neutrals with hydrogen atoms, $\nu_{j\mathrm{H}}$, and by the mean thermal speed $v_j = \sqrt{k_\mathrm{B}T/m_j}$ of species j. The hard-sphere collision frequency in units of s^{-1} is

$$\nu_{j\mathrm{H}} = 6.09\,\frac{r_{j\mathrm{H}}^2}{\sqrt{A_j(A_j+1)}}\,\sqrt{T_4}\,n_{\mathrm{H},16}, \tag{1}$$

which involves the atomic radius $r_{j\mathrm{H}}$ in Ångstrøm and atomic mass number A_j as characteristic parameters of atomic species j. The neutral-hydrogen diffusion constant is $D_{j\mathrm{H}} = v_j^2/\nu_{j\mathrm{H}}$, and the ionization-diffusion length is $\lambda_j = \sqrt{D_{j\mathrm{H}}\tau_j}$. Other diffusion lengths might be constructed by replacing the collision frequency in $D_{j\mathrm{H}}$. A basic natural diffusion speed of species j can be defined as $w_j = \sqrt{D_{j\mathrm{H}}/\tau_j}$, All the characteristic parameters are compiled in Table 1 for a number of prominent elements. The basic macroscopic parameters in our model are the density and temperature of the upper solar chromosphere.

3. Basic equations

Our starting point are the general fluid equations for each species. We assume T = const., so we do not require an energy equation and thus have a closed system for the variables density n_j and velocity $\mathbf{u}_j$ of each species j. We make the following assumptions: 1) Time stationarity, which eliminates the time derivatives. 2) The particle densities are functions of the coordinate s along the field only, which decouples the parallel and perpendicular kinetic degrees of freedom. 3) Subsonic flows, which entitles us to disregard the convective derivatives and to consider a diffusive equilibrium.

The relevant component of the momentum equation ($\parallel \mathbf{B}$) then reads

TABLE I
Atomic data (for $N_{16} = 1$ and $T_4 = 1$).

	FIP [eV]	τ_j [s]	$r_{j\mathrm{H}}$ [Å]	α_j [Å^3]	v_j [km/s]	λ_j [km]	w_j [km/s]
$^{1}_{1}$H	13.60	65	—	0.667	9.09	8.59	0.132
$^{4}_{2}$He	24.59	227	1.56	0.205	4.54	37.6	0.166
$^{12}_{6}$C	11.26	27	2.34	1.76	2.62	8.35	0.309
$^{14}_{7}$N	14.53	68	2.11	1.10	2.43	14.7	0.216
$^{16}_{8}$O	13.62	62	2.26	0.802	2.27	13.0	0.210
$^{20}_{10}$Ne	21.56	81	1.75	0.396	2.03	19.2	0.237
$^{24}_{12}$Mg	7.65	0.78	3.09	10.6	1.85	1.06	1.36
$^{28}_{14}$Si	8.15	1.1	2.91	5.38	1.72	1.34	1.22
$^{32}_{16}$S	10.36	11.6	2.58	2.90	1.61	4.90	0.423
$^{40}_{18}$Ar	15.76	50	2.07	1.64	1.44	12.7	0.253
$^{56}_{26}$Fe	7.87	0.91	3.09	8.4	1.21	1.14	1.26

$$0 = v_j^2 \frac{d}{ds} \ln n_j + Q_j c_j^2 \frac{d}{ds} \ln n_e + g(s) + \sum_k \nu_{jk}(u_j - u_k). \tag{2}$$

Q_j is the charge, and c_j is the ion acoustic speed of species j. We further assume: 4) Cold electrons, i.e. $T_e \ll T_j$. The simplified continuity equation with ionization rate γ_{jk} reads

$$\frac{d}{ds}(n_j u_j) = \sum_k \gamma_{jk} n_k. \tag{3}$$

For a more thorough discussion of these equations and the solutions, see Marsch *et al.* (1993).

Equation (1) gives the neutral-neutral collision frequency which is essential in the initial diffusion phase. After ionization has occurred, the freshly created ions will collide between themselves and with their parent atoms. The associated collisions frequencies differ by orders of magnitude, a fact which is essential for the ion-neutral separation process. We only need three frequency ratios to describe the basic four collision processes: Neutrals with H, neutrals with protons, ions with H, and ions with protons. These normalized ratios contain the atomic polarizabilities of hydrogen and the neutral minor atoms, α_H and α_j, the atomic radius in hard-sphere collisions between neutrals and H, $r_{j\mathrm{H}}$, and the Coulomb logarithm ($\ln\Lambda \simeq 12$ under the conditions considered here). The first ionization times are also compiled in Table 1. Our assumptions regarding the collision/reaction terms in Eqs. (2–3) are: 5) Simply photoionization is considered and recombination or other ionization processes are neglected. 6) For the main gas only collisions between hydrogen atoms and protons are included. 7) For the trace gases only collisions with the partly ionized hydrogen gas are considered.

4. Solutions

With these simplifying assumptions, we find, after some algebra, an analytical solution to Eqs. (2–3) for H and protons in the form

$$n_{\mathrm{p}}(s) = N\left(1 - e^{-s/\lambda_{\mathrm{H}}}\right) \quad ; \quad n_{\mathrm{H}}(s) = N e^{-s/\lambda_{\mathrm{H}}} \tag{4}$$

$$u_{\mathrm{p}}(s) = \left(U - \frac{\lambda_{\mathrm{H}}}{\tau_{\mathrm{H}}} e^{-s/\lambda_{\mathrm{H}}}\right)\left(1 - e^{-s/\lambda_{\mathrm{H}}}\right)^{-1} \quad ; \quad u_{\mathrm{H}}(s) = \frac{\lambda_{\mathrm{H}}}{\tau_{\mathrm{H}}}. \tag{5}$$

where $\lambda_{\mathrm{H}} = D/(\sqrt{D/\tau_{\mathrm{H}} + U^2/4} - U/2)$, and U is the mean speed of the bulk gas, which is an integration constant.

The diffusive fractionation of minor elements is modeled separately for a given hydrogen background. One obtains four coupled differential equations for the neutrals and ions, $n_j, \phi_j, n_j^+, \phi_j^+$. The basic equations (2–3), with axis $x = s/\lambda_j$, can be further reduced to a normalized form:

$$\frac{d}{dx}\hat{\phi}_j(x) = -\hat{n}_j(x) \quad ; \quad \frac{d}{dx}\hat{n}_j(x) = -\frac{\hat{n}_j(x)}{\hat{h}_j(x)} - \frac{\hat{\phi}_j(x)}{\hat{D}_j(x)} \tag{6}$$

$$\frac{d}{dx}\hat{\phi}_j^+(x) = +\hat{n}_j(x) \quad ; \quad \frac{d}{dx}\hat{n}_j^+(x) = -\frac{\hat{n}_j^+(x)}{\hat{h}_j^+(x)} - \frac{\hat{\phi}_j^+(x)}{\hat{D}_j^+(x)}. \tag{7}$$

$\hat{h}_j$ and $\hat{D}_j$ are the generalized normalized scale heights and diffusion constants, respectively.

At the lower boundary, $s = 0$, we specify a neutral gas at normalized density, $\hat{n}_j(0) = 1, \hat{n}_j^+(0) = 0$. At the upper boundary, $s = S$, we choose for the fluxes $\hat{\phi}_j(S) = 0, \hat{\phi}_j^+(S) = \hat{\phi}_{j,S}^+$. For $\hat{\phi}_{j,S}^+$ we choose the *same* constant value for all minor species, thus not favouring one species to the other. The boundary value problem is then solved numerically.

5. Results and Discussion

In Fig. 1, we give the result of a typical model run for Si. The stationary diffusion pattern can be divided into three regions: Near the lower boundary, the neutrals diffuse in hydrogen. Next, there is a mixed region, where either the minor ions diffuse in neutral H (as for Si) or the minor neutrals diffuse in protons. Finally, near the upper boundary, all species are ionized, and minor ions diffuse in protons.

In Fig. 2, we collect the fractionation across the region for the elements actually observed in the solar wind. Note that in the lower region, where

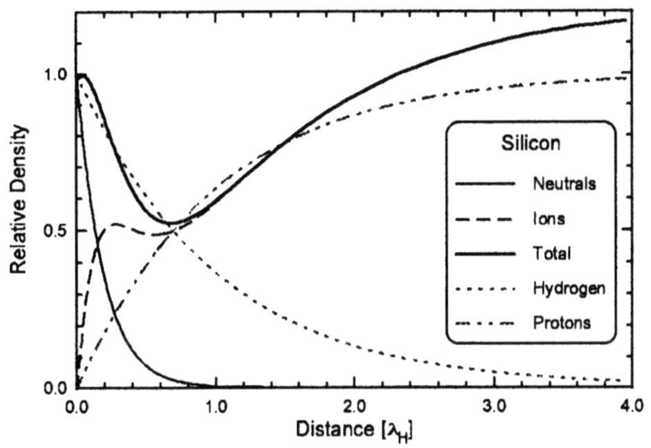

Fig. 1. Density as a function of height for Si in a typical model run. The analytical solution for the main gas is also indicated. Model parameters are $N_{16} = 1$, $T_4 = 1$, $U = -100$ m/s, and $\hat{\phi}^+_{j,S} = -0.1$.

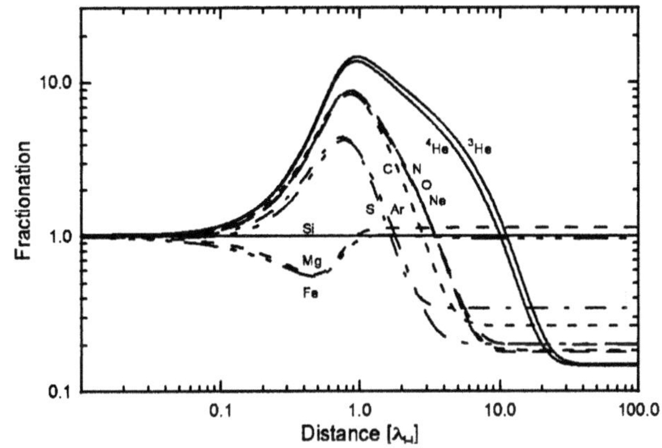

Fig. 2. Model results for the fractionation of solar wind elements relative to Si. The ratios reach constant values towards the upper boundary, which agree well with the observations. Near the lower boundary, the model predicts an inverse fractionation.

hydrogen is predominantly neutral, the fractionation between the low- and high-FIPs is reversed! This is caused by the ion diffusion in the predominantly neutral main gas, which produces an approximate fractionation

$$f_{jk} = \frac{\hat{n}_j}{\hat{n}_k} \simeq \frac{\nu^+_{j\mathrm{H}}}{\nu^+_{k\mathrm{H}}} \frac{\nu_{k\mathrm{H}}}{\nu_{j\mathrm{H}}} = \frac{r^2_{k\mathrm{H}}}{r^2_{j\mathrm{H}}}. \tag{8}$$

Since low-FIP atoms are larger, this ratio predicts an enrichment of *high*-FIP atoms. But as soon as the main gas is predominantly ionized, the situation reverses. For the fractionation near the upper boundary, we derive

$$f_{jk} = \frac{\hat{n}^+_j}{\hat{n}^+_k} = \frac{\hat{\phi}^+_{j,S}}{\hat{\phi}^+_{k,S}} \frac{u_k}{u_j} \frac{w_j}{w_k} \simeq \frac{w_j}{w_k} = \frac{r_{k\mathrm{H}}}{r_{j\mathrm{H}}} \sqrt{\frac{\tau_k}{\tau_j}} \left(\frac{A_j + 1}{A_j} \frac{A_k}{A_k + 1} \right)^{1/4} \tag{9}$$

For Si/Ne, this is $\simeq 5.2$, which agrees well with the observed values. The mass dependency in (9) nearly cancels, so the FIP fractionation is not disturbed by a mass fractionation, hence almost no isotopic fractionation is obtained.

In Fig. 3, we compare the model results at the upper boundary with the observations. The high-FIP elements ($12 \leq$ FIP[eV] ≤ 22) lie on a

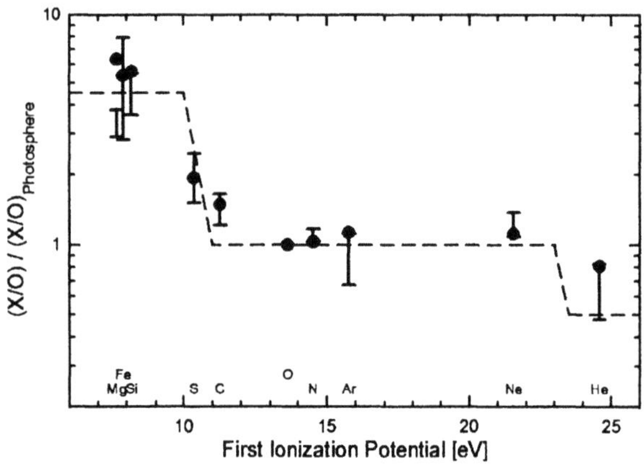

Fig. 3. Model results from Fig. 2 at the upper boundary, compared to the observed solar wind abundance ratios. The FIP fractionation pattern (dashed line) is well reproduced by the model: We obtain both a low-FIP and a high-FIP plateau, the intermediary elements (C, S) are slightly enriched, and He is slightly depleted relative to the high-FIPs.

plateau, a factor of $\sim$ 5 beneath the low-FIPs. This basic step function is well reproduced by the model. Furthermore, the observed values of C and S, which have a FIP near the critical value of $\sim$ 10 eV, are slightly enriched, and He is slightly depleted with respect to the high-FIPs. Both these observational features are also well reproduced by the model.

We have presented a new mechanism for the fractionation of elements in the solar chromosphere. One motivation for our approach was to free ourselves from the specific requirements on geometrical structures and chromospheric conditions, which are inherent in the model of von Steiger & Geiss (1989). We have shown that a purely diffusive, one-dimensional model with appropriately, but naturally, chosen boundary conditions, is able to reproduce the observed fractionation pattern between the elements in the solar wind. The mechanism operates under a wide range of model parameters, such as the chromospheric density and temperature. Such diffusive layers are likely to exist ubiquitously throughout the chromosphere. It is therefore conceivable that our mechanism is of primary importance in producing the observed fractionation, contributions from other mechanisms notwithstanding.

References

Anders, E., Grevesse, N.: 1989, *Geochim. Cosmochim. Acta*, **53**, 197.

Geiss, J.: 1982, *Sp. Sci. Rev.*, **33**, 201.

Geiss, J., Bochsler, P.: 1985, Ion composition in the solar wind in relation to solar abundances, *Rapports isotopiques dans le système solaire*, Toulouse: Cepadues-Editions, p. 213

Geiss, J., Bochsler, P.: 1986, Solar wind composition and what we expect to learn from out-of-ecliptic measurements, R. G. Marsden (ed.), *The Sun and heliosphere in three dimensions*, Dordrecht: Reidel, p. 173

Marsch, E., von Steiger, R., Bochsler, P.: 1993, *A&A*, submitted.

Meyer, J.-P.: 1985, *A&AS*, **57**, 151.

von Steiger, R., Geiss, J.: 1989, *A&A*, **225**, 222.

COUPLING OF THE CORONAL HE ABUNDANCE TO THE SOLAR WIND

V. H. HANSTEEN and E. LEER*
Institute of Theoretical Astrophysics, University of Oslo, Norway

and

T. E. HOLZER
*High Altitude Observatory, National Center for Atmospheric Research**, Boulder CO, USA*

Abstract. Models of the transition region – corona – solar wind system are investigated in order to find the coronal helium abundance and to study the role played by coronal helium in controlling the the solar wind proton flux. The thermal force on α-particles in the transition region sets the flow of helium into the corona. The frictional coupling between α-particles and protons and/or the electric polarization field determines the proton flux in the solar wind as well as the fate of the coronal helium content.

Key words: Transition Region – Solar Wind – Helium Abundance

1. Introduction

Parker (1958) treated the solar wind as a fully ionized hydrogen plasma. His assumption of an isothermal, spherically symmetric, radial flow was appropriate for illustrating the basic dynamics of an electron-proton solar wind. Much of the following theoretical work on the solar wind has been focused on the acceleration of high speed streams. Less attention has been offered the observations of a relatively constant solar wind proton flux.

In thermally driven electron-proton models the proton flux at is essentially a measure of the coronal scale height and thus varies exponentially with coronal temperature. The nature of the "solar wind mass flux problem" is clear when we recall that the proton flux at 1AU is observed to lie in the range $(nu)_E = (2-5) \times 10^8 \mathrm{cm}^{-2}\mathrm{s}^{-1}$ (e.g. Neugebauer 1992) while the theoretically derived mass flux increases by two orders of magnitude if the coronal temperature increases from 10^6K to 2×10^6K. There is no independent observational evidence for a constant temperature in the source regions of the solar wind. This so called "solar wind mass flux problem" led Leer & Holzer (1991) to suggest that coronal helium acts as a regulator for the proton flux. Studies of the out-flow from a hydrogen – helium corona were also carried out by Bürgi (1992) and Leer, Holzer & Shoub (1992).

The solar wind α-particle flux is about 5% of the proton flux, but it is much more variable than the proton flux (e.g. Neugebauer 1981). These *in-*

* Associate scientist High Altitude Observatory, NCAR
** NCAR is sponsored by the National Science Foundation

Space Science Reviews **70**, 347–352.

situ observations of the helium abundance in the solar wind cannot exclude the possibility that helium is very abundant in the solar corona (Joselyn & Holzer, 1978). Observations of the HeII λ304Å and HI λ1215Å spectral lines from the solar disk and from the corona can be used to deduce the coronal helium abundance (e.g. Ahmad 1977). Such observations have been carried out by the Coronal Helium Abundance Experiment (CHASE) on Spacelab 2 (cf. Patchett et al. 1981; Parkinson & Gabriel 1986), but it has not been possible to derive values of the helium abundance in different coronal regions. Theoretical studies of the chromosphere – corona transition region have shown that the upwards directed so-called "thermal force" on helium ions is large compared to gravity. In diffusive equilibrium models the thermal force is balanced by a pressure gradient force, resulting in a helium filled corona (e.g. Delache 1967, Nakada 1969). In a recent study Hansteen, Holzer & Leer (1993) allowed helium to flow upwards relative to hydrogen and found that the coronal helium abundance could be significantly enhanced relative to the photospheric value on time-scales short in comparison to the time it takes the solar wind to carry away one coronal mass.

In this paper, combining the study by Hansteen et al. (1993) with the study of Leer et al. (1992), we will consider a hydrogen – helium plasma in the upper chromosphere/lower transition region — corona — solar wind system consistently.

2. He in the Transition Region – Corona – Solar Wind System

Within the framework of steady-state solar wind models there are several mechanisms that may be introduced in order to reduce the variability of the in the solar wind proton flux. The first and most obvious is to introduce some coronal thermostat so that the coronal temperature remains fairly invariant. Another possibility lies in allowing the cross-sectional area of various outflow regions beyond the critical point to vary (Withbroe 1989, Wang 1993). A third possibility is to introduce additional forces in the corona so that the proton scale height is not solely a function of temperature. The presence of helium in the corona gives rise to a frictional force on protons due Coulomb collisions as well as altering the electric polarization field. It is this third possibility that we will consider further in this paper.

2.1 He and the Solar Wind Proton Flux

The force balance on protons in a corona containing helium has been discussed by Leer et al. (1992). The frictional force on protons from helium (α-particles) scales as $F_{\alpha\mathrm{p}} \propto n_\alpha(u_\alpha - u_\mathrm{p})\nu_{\alpha\mathrm{p}}$ where the collisional frequency is $\nu_{\alpha\mathrm{p}} \propto n_\mathrm{p}T^{-\frac{3}{2}}$. This force acts as a drag force decreasing the coronal proton scale height. The polarization electric field scales with the electron scale height as $eE = -(1/n_\mathrm{e})(\partial p_\mathrm{e}/\partial r)$. The electron scale height decreases with increasing coronal helium abundance; the effect of adding α-particles is therefore to increase the outward electric force on protons. The inward drag force dominates the electric field for large values of the proton flux, for small values of the proton flux it is the outward electric force that dominates.

The presence of helium in the corona can therefore act as a regulator for the solar wind proton flux by tending to increase the proton scale height at low values of the proton flux and decreasing the proton scale height when the proton flux is large. We can see this clearly by considering the extreme case of a isothermal corona where helium is the major constituent, in this case the scale heights of protons and α-particles are nearly the same. Electrons and α-particles form a quasi-static background where the force balance on protons is determined by the inward drag force, an outward pressure gradient force, and an outward electric force. Gravity plays a minor role and can be considered as a correction to the electric force. The proton flux at 1AU is then

$$(n_\mathrm{p}u_\mathrm{p})_\mathrm{E} \approx \frac{GM_\odot}{r_\mathrm{E}^2}\left(\frac{n_\mathrm{p}}{\nu_{\alpha\mathrm{p}}}\right)\frac{5n_\mathrm{pc}}{4(3n_{\alpha\mathrm{c}}+2n_\mathrm{pc})}. \tag{1}$$

Notice that the temperature dependence of the proton flux is reduced to $T^{\frac{3}{2}}$ as opposed to the exponential dependence that held in the electron – proton solar wind solution.

2.2 Transport of He into the Corona

The steep transition region temperature gradient gives rise to the thermal force which plays a dominant role in the α-particle force balance allowing helium to flow up from the chromosphere into the corona. This force is proportional to the temperature gradient and is a consequence of Coulomb collisions between ions of differing masses.

In considering the importance of the thermal diffusion of helium up the transitions region temperature gradient into the corona the relevant parameter to consider is the time scale with which helium can be brought up into the corona. Hansteen et al. (1993) considered the scenario where the corona is replenished at irregular intervals with material of photospheric abundance, i.e. by a spicule or similar event. The evolution of the helium abundance following such a mixing event was modeled by assuming that the upper chromosphere – transition region – coronal system originally had a uniform helium abundance and was in overall hydrodynamic equilibrium. A result of the models was that the force balance of α-particles is initially dominated by the upward directed thermal force and the downward directed Coulomb drag force. This force balance is valid until a significant amount of helium is transported into the corona and the downward directed pressure gradient force can replace the drag force in balancing the thermal force. Furthermore it was found that the time-scale for increasing the helium abundance in the corona to 30% was on the order of $24h$ for transition region temperature gradients consistent with the observed emission measure (Woods et al. 1989).

2.3 Solar Wind Models with He

In order to model the coronal helium abundance we must consider the balance of processes that enhance and deplete helium in the corona. We have constructed consistent solar wind models reaching from the upper chromosphere/lower transition region through the corona and into interplanetary space. The inner boundary is set at the level where the temperature is $T_0 = 1.5 \times 10^4$K. The outer boundary, at a heliocentric distance $r_1 \approx 25R_\odot$, is well into the supersonic region of the solar wind flow but close enough to the sun to allow for a parameter study. The energy equations have not been solved, we have instead used representative temperature profiles. The winds generated have not been additionally accelerated by Alfvén waves or other mechanisms.

In most of the model studies we consider a corona and solar wind with a constant temperature, T_c. We specify the hydrogen density and the helium abundance at the inner boundary, n_{H0} and $(n_{He}/n_H)_0$, respectively. We also specify the heat conductive flux entering the upper transition region, q_0, which sets the transition region temperature gradient. In this study we keep the helium abundance $(n_{He}/n_H)_0 = 10\%$ fixed and vary n_{H0}, T_c and q_0. The source and sink terms in the ionization rate equations are assumed to be determined by electron collisional ionization and radiative recombination. The details of the numerical method are outlined in Hansteen et al. (1993a) and Hansteen et al. (1993b).

With high enough coronal temperatures the proton flux and thus the frictional coupling between proton and α-particles is sufficient to ensure a Parker type solution to the problem. We therefore proceed as follows in the parameter study described below: For each pair of chromospheric density and transition region temperature gradients studied, we initially create a high temperature model that is in steady state. From this model the temperature is reduced to a new level and the model is allowed to relax for a time that is long compared to the time it takes for the solar wind to carry away one coronal mass. We find that two classes of solutions arise depending on the magnitude of T_c, i.e. on the amplitude of the proton flux.

For values of the proton flux above $(n_p u_p)_E \approx 10^9 \text{cm}^{-2}\text{s}^{-1}$ a steady state is achieved with properties similar to the Parker solar wind. The amount of helium in the corona increases with increasing proton flux, varying roughly from 4% to 8%. The helium abundance in the corona is set by the amount of frictional coupling between helium and hydrogen in the lowest parts of the atmosphere, before helium is ionized, and by the amount of helium specified at the lower boundary. Since the hydrogen flow speed in this region is larger than the flow speed of helium the neutral helium scale height increases with increasing hydrogen drag, i.e. increasing hydrogen flux. This result implies that the position of helium ionization may be important in determining the scale height of helium, a more sophisticated treatment of helium ionization than that carried out in the models described here is therefore certainly warranted.

When the proton flux falls below $(n_p u_p) \approx 10^9 \text{cm}^{-2}\text{s}^{-1}$ the helium content of the corona continually increases and no time independent solution is found on the timescales of this study. We may understand this by considering the force balance on α-particles in the corona. In the steady state models the frictional force on α-particles from protons is sufficient to pull helium out of the corona. When the coronal temperature is reduced so is the solar wind proton flux. This reduced proton flux entails a corresponding reduced drag on α-particles and the α-particle velocity decreases. As the coronal temperature falls below a certain limit the α-particle velocity becomes negative and since helium no longer is pulled out in the wind helium starts accumulating in the corona. We can provide a rough estimate of when this occurs by equating the frictional and gravitational forces on α-particles, ignoring the electric polarization field, which gives (in cgs units)

$$n_\alpha u_\alpha \approx \frac{n_\alpha}{n_p}[n_p u_p - \frac{m_{\text{He}} G M_\odot T^{3/2}}{1.2 \times 10^{-23} r^2}]. \tag{2}$$

For an isothermal corona with $T_c = 2 \times 10^6$K the limiting proton flux at 1AU is $(n_p u_p)_E = 9.3 \times 10^8 \text{cm}^{-2}\text{s}^{-1}$ which is in close agreement with the values found in our model calculations. For lower values of the proton flux helium builds up in the corona. The time-scale of coronal helium buildup ranges from several hours to several hundred hours depending on the parameters chosen; varying roughly proportionally with the hydrogen density at the base of the atmosphere and inversely proportionally to the transition region temperature gradient. This time scale is too large, compared to a typical coronal mixing time, to allow for the formation of a helium filled corona.

Since the transport of helium into the corona by thermal diffusion is only marginally altered by the decreasing proton flux, helium will continue to build up in the corona until the pressure gradient force is sufficient to balance the thermal force. Time dependent solutions with a continual decrease in the proton flux result. This can be understood by examining the arguments of Leer et al. (1992). When the helium abundance exceeds approximately 20%, the coronal scale heights for hydrogen and helium are nearly the same. The balance of forces on protons is then determined by gravity, the inward frictional force, the outward pressure gradient force and the outward electric force. The acceleration mechanism for the solar wind then resembles the acceleration mechanism for the "polar wind". This leads to the relation given in Eq.(1) between the proton flux at 1AU and the coronal helium abundance ($n_{\alpha c}/n_{Hc}$). The time scale of the decreasing proton flux in the solar wind is thereby coupled to the time-scale of the increasing coronal helium abundance.

Though this paper is primarily concerned with elucidating the physical processes in a hydrogen-helium wind rather than comparison with actual observations it is interesting to note that, in this spherically symmetric solar wind model, the value we find for a limiting proton flux between the two classes of solution is fairly large in comparison with proton fluxes observed in the quasi-steady solar wind. It should also be noted that the mechanism discussed here should also operate in magnetically closed regions where the solar wind loss is negligible. In loops characterized by a large transition region temperature gradient we expect that helium buildup may proceed on fairly short time-scales.

3. Conclusions

Though the calculations presented here are schematic they show that the presence of helium in the corona plays an important role in the dynamics of the solar wind, specifically on the determination of the proton flux. On the basis of the models presented here we expect that, in the absence of an efficient mixing mechanism, the coronal helium abundance will vary from region to region and measurements of the coronal helium abundance with spatial resolution would certainly be welcome. An extended version of this work will appear in The Astrophysical Journal (Hansteen et al. 1993b).

References

Ahmad, I.A. 1977, Sol. Phys. 53, 409.
Bürgi, A. 1992, J. Geophys. Res. 97, 3137.
Delache, P. 1967, Ann. Astrophys. 30, 827.
Hansteen, V. H., Holzer, T. E., & Leer, E. 1993a, ApJ 402, 334.
Hansteen, V. H., Leer, E., & Holzer, T. E. 1993b, ApJ (accepted).
Joselyn, J.A. & Holzer, T.E. 1978, J. Geophys. Res. 83, 1019.
Leer, E. & Holzer, T.E. 1991, Ann. Geophysicae 9, 196.
Leer, E., Holzer, T.E., & Shoub, E. C. 1992, J. Geophys. Res. 97, 8183.
Nakada, M.P. 1969, Sol. Phys. 7, 302.
Neugebauer, M. 1981, in Solar Wind Four, ed. H. Rosenbauer, p. 425, Max Planck Institut für Aeronomie.
Neugebauer, M. 1992, in Solar Wind Seven, ed. E. Marsch & R. Schwenn, p. 69, Max Planck Institut für Aeronomi.
Parker, E.N. 1958, ApJ 128, 664.
Parkinson, J.H. & Gabriel A.H. 1986, Adv. Space Res. 6, 243.
Patchett, B.E., Norman, K., Gabriel, A.H., & Culhane, J.L. 1981, Space Science Reviews 29, 431.
Wang Y.-M., 1993, ApJL 410, L123.
Woods, D.T., Holzer, T.E., & MacGregor, K.B. 1989, ApJ 355, 295.
Withbroe, G.L. 1989, ApJL 337, L49.

ELECTRON CORONAL DENSITY IRREGULARITY $\overline{n^2}/(\overline{n})^2$ FROM MEASUREMENTS OF K-CORONAL AND LYMAN LINES BRIGHTNESSES

SILVANO FINESCHI *

Dipartimento di Astronomia e Scienza dello Spazio
Università degli Studi di Firenze, 50125 Firenze (ITALY)

and

MARCO ROMOLI **

Harvard-Smithsonian Center for Astrophysics,
60 Garden St, MS 50. Cambridge, MA 02138 (U.S.A.)

Abstract. We propose a technique to derive the coronal density irregularity factor $\overline{n^2}/(\overline{n})^2$, where n is the electron density. The absolute photometric comparison between the intensity of UV lines and the white-light K-coronal polarized brightness (pB) provides an unique constraint on the inhomogeneity of the corona. The ratio of the measured H I Lyman β (Ly-β) line intensity to the resonant-scattering dominated H I Lyman α (Ly-α) intensity can be used to extract the collisional component of the Ly-β. This component yields an estimate of $\overline{n^2}$. The quantity $(\overline{n})^2$ is then obtained from white-light K-coronal measurements. The use of lines of the same atomic species minimizes the effects due to outflow velocities (i.e., Doppler dimming), and reduces the errors introduced by the uncertainties in the ionization balance, the atomic parameters, and the solar abundances. The UVCS/SOHO unique capability of performing cotemporal and cospatial measurements of the Ly-α and Ly-β lines, and of pB makes this instrument ideal for implementing this technique.

Key words: White Light Corona – Electron Density – UV Spectroscopy

1. Introduction

A proper interpretation of remote observations of the solar corona must take into account the effects due to the integration along the line-of-sight. The assumption of an unstructured spherically symmetric corona can only give a first order approximation of the distribution along the line-of-sight of the plasma contributing to the measured signal. A better approximation would be the use of a more structured model of the corona selected with the help of some observational constraint on the coronal inhomogeneity. An observational parameter that can provide such a constraint is the ratio of the intensity of a collisionally excited emission line to the square of the white-light K-coronal polarized brightness pB. The former quantity is proportional to $\overline{n^2}$, the latter to $(\overline{n})^2$, where n is the coronal electron density.

Allen (1963) introduced the term "coronal irregularity factor" designated by X to describe the coronal density inhomogeneity, where X is defined

* Also: Harvard-Smithsonian Center for Astrophysics, Cambridge, MA 02138 (U.S.A.)
** ESA non-resident fellow

Space Science Reviews **70**, 353–358.

to be the ratio $\overline{n^2}/(\overline{n})^2$. Withbroe (1972), Allen (1975), and Orrall *et al.* (1990) have given estimates of X in the inner corona [i.e., at heliocentric heights ≤ 1.5 solar radii ($R_\odot$)] based on XUV spectroheliograms and white-light K-corona measurements. The results of these derivations of X are not convergent. Besides the difficulty of an accurate photometric comparision among data from different instruments [e.g., Sime and Fisher, (1993)], there are also the uncertainties on coronal temperatures and chemical abundancies that make the interpretation of the measurements not straightforward. Moreover, the estimates of X have been limited to the inner corona because at higher altitudes, either the intensity of purely collisional lines is too weak, or, in lines where the emission is still measurable, the resonantly scattered component becomes comparable to the collisional one.

Here we describe a method for deriving the collisional component of the H I Lyman β (Ly-β) coronal line (λ 1025 Å) from observations of the Ly-β and H I Lyman α (Ly-α) line (λ 1216 Å) in corona and chromosphere. This method is weakly sensitive to uncertainties on coronal temperatures and chemical abundancies, and can be used for estimates of X at heliocentric heights higher than 1.5 $R_\odot$. We propose to implement this technique with the UVCS instrument, which can simultaneously measure pB and the two UV lines, reducing in this way the photometric intercalibration problems.

2. The coronal Ly-α and Ly-β lines

Under coronal conditions, the total observed intensity (erg cm^{-2} s^{-1} sr^{-1}) of an optically thin line from a two-level atom is the sum of a collisional and a resonantly scattered component

$$I_{obs} = I_{res} + I_{coll} = 0.8 \cdot \left[\langle q(T_e) R(T_e) n^2 \rangle + B_{12} I_\odot \langle D(T_H, W) R(T_e) n \rangle \right], \quad (1)$$

where q is the collisional excitation rate coefficient, B_{12} is the Einstein absorption coefficient, $I_\odot$ is the solar disk's line brightness, $D(T_H, W)$ accounts for Doppler dimming and geometrical dilution factors and is function of the hydrogen temperature, T_H, and wind velocity, W. $R(T_e)$ is the hydrogen ionization balance, and $0.8 \cdot R(T_e)$ gives the hydrogen to electron density ratio in a fully ionized plasma with 10% helium, and electron temperature T_e. The quantities in brackets $\langle ... \rangle$ are integrated along the line of sight.

The line intensities, $I_{obs}^{(\alpha)}$, and $I_{obs}^{(\beta)}$, of both collisional and resonant components for the Ly-α and Ly-β coronal lines have been numerically derived by solving Eq. 1 with simple coronal density models: Withbroe et al. (1986) electron density profile for a coronal hole and Saito's (1970) for a quiet region. The wind velocity as a function of the heliocentric height is derived from the conservation of flux and its profile is normalized to a given velocity at 4 $R_\odot$. The plasma temperature is assigned as a parameter, $T \equiv T_e \simeq T_H$,

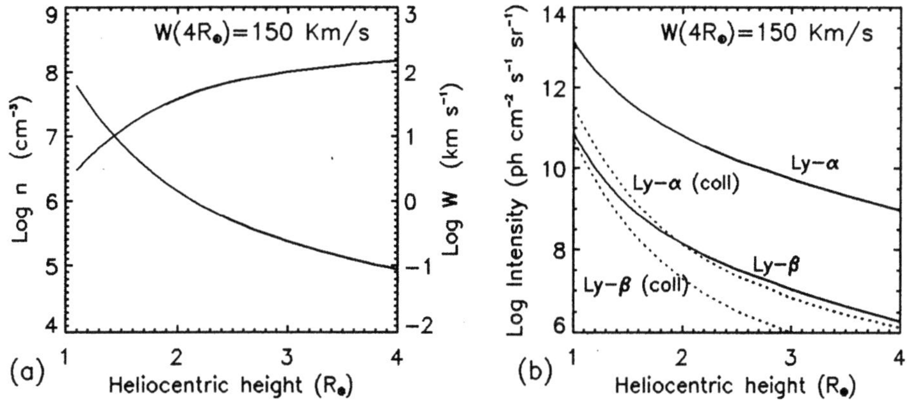

Fig. 1. (a) Wind velocity (conservation of flux) and electron density profiles [Withbroe et al. (1986)] used in the simulation of a isotropic coronal hole. (b) Coronal hole intensity profiles of total Ly-α and Ly-β and their collisional components versus heliocentric height.

and maintained constant throughout the coronal structure. Fig. 1b gives an example of the line intensities calculated for an isotropic coronal hole with the wind velocity and the electron density profiles plotted in Fig. 1a. The results in Fig. 1b show that, unlike the Ly-β, the Ly-α line has a small collisional component, compared to the total line intensity. This is shown in a more quantitative way in Fig. 2, where the ratios between the collisional component and the total intensity of Ly-α and Ly-β are plotted versus the heliocentric height, for four different coronal morphologies.

From the analysis of these ratios, it is clear that the observed Ly-α intensity can be pratically all attributed to resonance scattering (with an error less than 10% above 1.5 $R_\odot$ from sun center)

$$I_{obs}^{(\alpha)} \simeq I_{res}^{(\alpha)} . \tag{2}$$

As a consequence of this approximation, the resonant component of the Ly-β can be written in this form

$$I_{res}^{(\beta)} \simeq k \cdot \frac{B_{12}^{(\beta)}\, I_{\odot}^{(\beta)}}{B_{12}^{(\alpha)}\, I_{\odot}^{(\alpha)}}\, I_{obs}^{(\alpha)} , \tag{3}$$

where

$$k \equiv \left[\langle D^{(\beta)}(T,W)\, R(T)\, n\rangle\right] / \left[\langle D^{(\alpha)}(T,W)\, R(T)\, n\rangle\right] . \tag{4}$$

The result in Eq. 3 will be used to derive the Ly-β collisional component.

3. Derivation of the Ly-β collisional component

The value of the ratio k in Eq. 4 has been calculated with a model of an isotropic corona and plotted as a function of the heliocentric height for

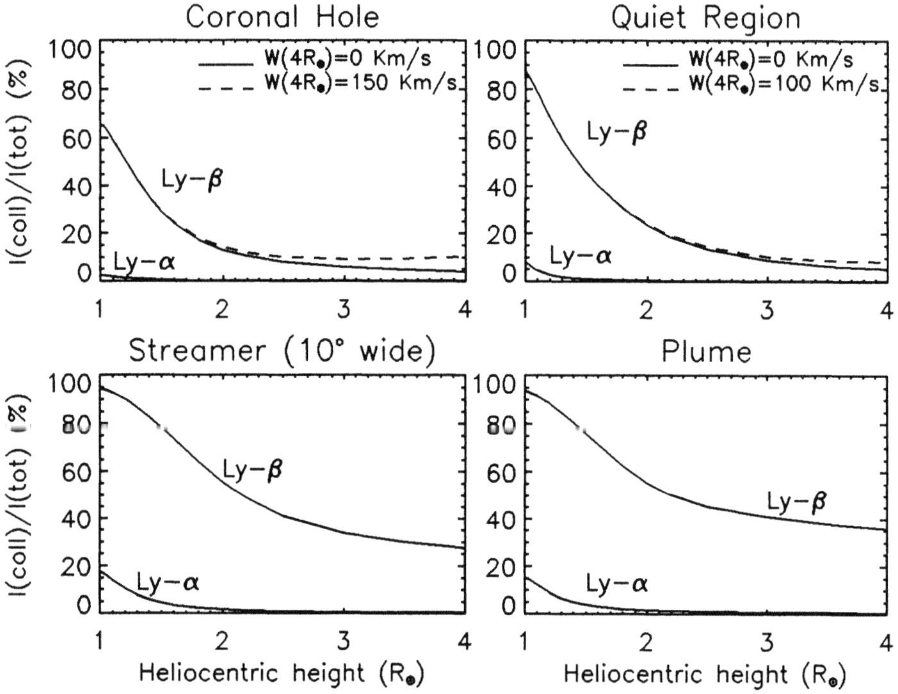

Fig. 2. Ratio between the intensity of the collisional component and the total intensity in the coronal Ly-α and Ly-β lines in four different morphologies. Coronal hole [Withbroe et al. (1986)] with and without the presence of solar wind; quiet region [Saito (1970)], with and without solar wind; radial streamer with 10° angular width and ten times the electron density of the embedding coronal hole; and radial and periodic plumes with a 0.2 filling factor and ten times the electron density of the embedding coronal hole.

different wind velocities (Fig. 3a) and plasma temperatures (Fig. 3b). The result is $k \simeq 1.2$, with variation from this value never larger than 5% in the 1 – 4 $R_\odot$ heliocentric height range, and for all the temperatures and velocities considered.

In other words, the ratio k can be considered, for all purposes, as a constant in the physical condition typical of the coronal plasma. This result can be used to derive the Ly-β collisional component from observable quantities

$$I_{coll}^{(\beta)} = I_{obs}^{(\beta)} - k \cdot \frac{B_{12}^{(\beta)} I_\odot^{(\beta)}}{B_{12}^{(\alpha)} I_\odot^{(\alpha)}} I_{obs}^{(\alpha)} \propto \overline{n^2} \, . \tag{5}$$

The Ly-β collisional component is proportional to $\overline{n^2}$ via the ionization balance and the collisional excitation rate, which are dependent on the electron temperature. This physical parameter can be measured or derived with well known spectroscopic techniques [Withbroe *et al.* (1982)].

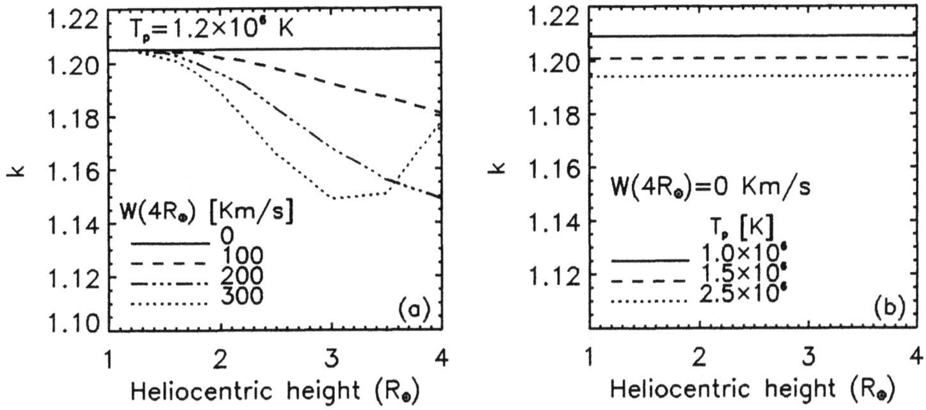

Fig. 3. Values of the constant, k, as a function of heliocentric height (a) for different values of the wind velocity at 4 $R_\odot$ and fixed plasma temperature, and (b) for different values of the plasma temperature with fixed zero wind velocity.

4. Derivation of the irregularity factor

The density irregularity factor is derived from the ratio between the intensity of the extracted Ly-β collisional component and the square of the observed K-corona polarized brightness*

$$\frac{I_{coll}^{(\beta)}}{(pB_{obs})^2} \propto \frac{\overline{n^2}}{(\overline{n})^2} \equiv Coronal\ Irregularity\ Factor\ \ (X)\,. \tag{7}$$

The actual use of this technique is constrained by the total intensity of the Ly-β and by the fraction of the line intensity which is collisionally excited. In fact, in order to evaluate the Ly-β collisional component with an accuracy better of, say, 50%, it is important for this component to be a significant fraction of the total intensity. Assuming a 3% statistics on the measured quantities, and propagating this error in Eq. 5, it comes out that the collisional component must be at least 25% of the total. With this requirement, the simulations in Fig. 2 show that the irregularity factor, X, can be derived up to 1.7 $R_\odot$ in a coronal hole, up to 2.2 $R_\odot$ in a quiet corona, and up to the Ly-β sensitivity limit of the instrument for the two cases of structured corona.

Expected values of X, derived with the technique proposed here, have been computed for typical coronal structures by Romoli and Fineschi (1993).

* The observed polarized brightness is given by:

$$pB_{obs} = I_{\odot}^{(WL)} \cdot \sigma_T \cdot \langle G n\rangle \propto \overline{n}\,, \tag{6}$$

where $I_{\odot}^{(WL)}$ is the average brightness of the solar disk in the white-light wavelength band of observation; σ_T is the Thomson scattering cross-section; and G is a geometrical factor [see van de Hulst (1950) for details].

5. Conclusions

We presented a method that from cospatial and cotemporal observations of the coronal and chromospheric Ly-α and Ly-β intensities derives the collisionally excited component of Ly-β. The comparision of this quantity with the K-corona polarized brightness gives strong constraints on the inhomogeneity of the corona. The UVCS can perform these observations, by simultaneously measuring pB and the two UV lines [Kohl, and Noci (1992)].

The Ly-β collisional component can be extracted if $I_{coll}^{(\beta)} \geq 0.25 \cdot I_{obs}^{(\beta)}$, that is, up to 1.7 $R_{\odot}$ in a coronal hole, up to 2.2 $R_{\odot}$ in a quiet region, and up to the instrumental sensitivity limit in a structured corona. The UVCS sensitivity is limited by the instrument stray light. Laboratory measurements indicate a UV stray light level of 5×10^{-8} [Weiser, and Romoli (1990)]. The ratio Ly-β corona to disk intensity is about ten times larger than this stray light level up to 2 $R_{\odot}$ in a coronal hole, and 3 $R_{\odot}$ in an active region.

Unlike previous cases where only one UV line was used [e.g., Orral *et al.* (1990)], now the use of two lines of the same atomic species minimizes the effects due to Doppler dimming, and reduces the errors introduced by the uncertainties in the ionization balance, the atomic parameters, and the solar abundances. The ratio of the collisionally excited Ly-β, derived with this method, to the cospatial and cotemporal measurement of pB gives, therefore, strong constraints on $\overline{n^2}/(\overline{n})^2$ in a less model-dependent way.

Acknowledgements

S. F. acknowledges the support of the University of Firenze and the Italian Space Agency (ASI) under a *Dottorato di Ricerca in Astronomia* fellowship.

M. R. acknowledges the support of the European Space Agency under a non-resident post-doctoral fellowship.

References

Allen, C. W.: 1963, 'Local Physics of the Corona' in J. W. Evans, ed(s)., *IAU Symposium 16, the Solar Corona*, Academic: New York, 1

Allen, C. W.: 1975, *Monthly Notices of the RAS* **172**, 159

Kohl, J. L., and Noci, G.: 1992, 'UVCS – The Ultraviolet Coronagraph Spectrometer for SOHO' in A.I.Poland,. and V.Domingo, ed(s)., *1st SOHO Workshop*, ESA SP-348, 23

Orrall, F. Q., Rottman, G. J., Fisher, R. R., and Munro, R. H.: 1990, *Ap. J.* **349**, 656

Romoli, M., Fineschi, S.: 1993, *Space Sci. Rev.* **this issue**,

Saito, K.: 1970, *Ann. Tokyo Astr. Obs.* **12**, 53

Sime, D. G., and Fisher, R. R.: 1993, *Solar Phys.* **144**, 75

Weiser, H., Romoli, M.: 1990, 'UVCS Stray Light Suppression', *UVCS Technical Report* ,

Withbroe, G. L.: 1972, *Solar Phys.* **25**, 116

Withbroe, G. L., Kohl, J. L., Weiser, H.: 1982, *Space Sci. Rev.* **33**, 17

Withbroe, G. L., Kohl, J. L., Weiser, H.: 1986, *Ap. J.* **307**, 381

van de Hulst H. C.: 1950, *Bull. Astron. Inst. Neth.* **11**, 135

SIMULATED OBSERVATIONS OF THE ELECTRON CORONAL DENSITY IRREGULARITY $\overline{n^2}/(\overline{n})^2$

MARCO ROMOLI *
Harvard-Smithsonian Center for Astrophysics,
60 Garden St, MS 50. Cambridge, MA 02138 (U.S.A.)

and

SILVANO FINESCHI **
Dipartimento di Astronomia e Scienza dello Spazio
Università degli Studi di Firenze, 50125 Firenze (ITALY)

Abstract. A technique to derive the coronal density irregularity factor $\overline{n^2}/(\overline{n})^2$, where n is the electron density, has been proposed by Fineschi and Romoli (1993). This technique will exploit the unique UVCS capability of cotemporal and cospatial measurements of both UV line radiation and K-coronal polarized brightness, pB.

The ratio of the measured H I Lyman β (Ly-β) line intensity to the resonant-scattering dominated H I Lyman α (Ly-α) intensity can be used to extract the collisional component of the Ly-β. This component yields an estimate of $\overline{n^2}$. The quantity $(\overline{n})^2$ is then obtained from the UVCS white-light K-coronal measurements.

We present simulated observations of the UVCS for coronal atmosphere models with different filling factors and electron density profiles, and for different coronal structures (e.g., coronal holes, streamers). These simulations will show how the proposed technique may be used to probe inhomogeneities of the solar corona.

Key words: White Light Corona – Electron Density – UV Spectroscopy

1. Introduction

The SOHO/UVCS cospatial and cotemporal observation of broadband visible polarized brightness, Ly-α, and Ly-β line intensities in the solar corona [Kohl and Noci (1992)] provides a unique constraint on the inhomogeneities of the electron density along the line of sight. The definition of irregularity factor, and a technique, weakly dependent on atomic parameters and Doppler dimming effects, to derive this parameter from the polarized brightness and the collisional component of the Ly-β intensity is described by Fineschi and Romoli (1993).

Here, we present the results of computer simulations based on the electron density profile given by Withbroe et al. (1986) for a polar coronal hole. Three different morphologies are analyzed: isotropic corona, streamers and polar plumes. The parameters of each coronal structure are changed to monitor their effects on the irregularity factor. The present analysis is performed from 1.5 up to 4 $R_\odot$ from Sun center, where the collisional component of

* ESA non-resident fellow
** Also: Harvard-Smithsonian Center for Astrophysics, Cambridge, MA 02138 (U.S.A.)

Space Science Reviews **70**, 359–363.

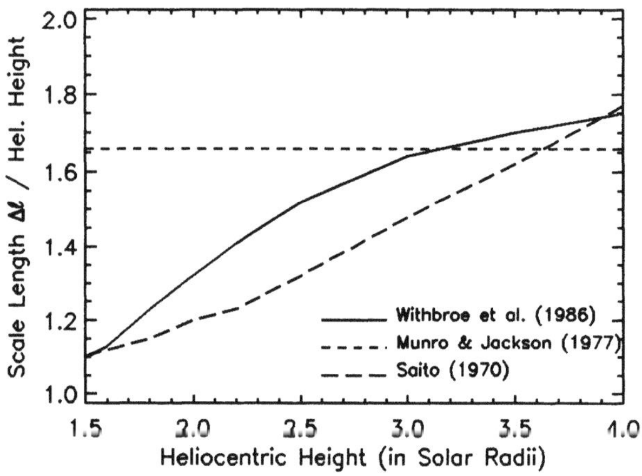

Fig. 1. Characteristic scale length/heliocentric height vs. heliocentric height for Ly-β with different electron density profiles.

Ly-β is still 25% of the total line intensity [see Fineschi and Romoli (1993)] and can be separated from the resonant component with a 50% accuracy.

Comparison between coronal data and irregularity factor models have been performed by Withbroe (1972), Orrall et al. (1990), and Sime and Fisher (1993), but they are all limited to the lower corona (< 1.3 R$_\odot$).

2. Derivation of the irregularity factor

The observed irregularity factor, given by the ratio of the collisional Ly-β intensity to the squared polarized brightness, can be written, in a first order approximation, as:

$$\frac{I_{coll}^{(\beta)}}{pB^2} \simeq \frac{H(\overline{T})\,\overline{n^2}\,\Delta\ell}{\left[I_{\odot}^{(WL)}\sigma_T\,\langle G\,n\rangle\right]^2}\,, \tag{1}$$

where $\Delta\ell$ is a characteristic scale length along the line of sight, from which most of the signal is originated.* This scale length is weakly dependent on electron density models that are homogeneous, and is about equal to the heliocentric height, r, of observation: $\Delta\ell \simeq r$ (see Fig. 1); $H(\overline{T})$ is the product

* The characteristic scale length is defined numerically from:

$$\frac{\int_0^{\Delta\ell/2} H(T)n^2dy}{\int_0^{\infty} H(T)n^2dy} = 0.9\,. \tag{2}$$

of the collisional excitation coefficient, q, and the hydrogen ionization balance, R, which are calculated for a plasma temperature, $\overline{T} \equiv \overline{T}_e \simeq \overline{T}_H$, averaged over $\Delta\ell$; $I_{\odot}^{(WL)}$ and σ_T are the average white-light brightness of the solar disk and the Thomson scattering cross-section, respectively; G is a geometrical factor [see coefficients A and B in van de Hulst (1950)]; and $\overline{n^2}$ is defined by: $\overline{n^2} \equiv \langle n^2 \, H\rangle/\langle H\rangle$ [in first approximation $\langle H\rangle \simeq H(\overline{T}) \cdot \Delta\ell$].

Thus, the theoretical irregularity factor is given by:

$$X \equiv \frac{\overline{n^2}}{(\overline{n})^2} \simeq \frac{I_{coll}^{(\beta)}/[H(\overline{T})\, r]}{\left[pB/\left(I_{\odot}^{(WL)}\sigma_T \, \langle G\rangle\right)\right]^2} \simeq \frac{(\sigma_T \, \langle G\rangle)^2}{H(\overline{T})\, r} \cdot I_{coll}^{(\beta)} \cdot \left(\frac{I_{\odot}^{(WL)}}{pB}\right)^2 , \quad (3)$$

where $\overline{n}$ is defined by $\overline{n} \equiv \langle n \, G\rangle/\langle G\rangle$.

In a Log – Log plot, this equation represents a straight line with slope 1 and y-axis intercept that gives the irregularity factor (e.g., see Fig. 2b).

3. Numerical simulation

We have used the Withbroe et al. (1986) model for the electron density distribution in an isotropic corona, and in the 1.5 – 4 $R_{\odot}$ range.

In order to simulate the irregularities, we have assumed several radial morphologies: streamer and plumes. The streamer is embedded in a lower density corona, and is characterized by an angular width, γ, and by a displacement off the plane of the sky, ϕ. The plumes are modeled as a periodic structure, with period p, in which regions of coronal hole density alternate plumes with higher density, characterized by a filling factor, α. The electron density distribution inside the structures is assumed to have the same profile but scaled by a factor 10. The temperatures are set at 1.2×10^6 K in the ambient corona, and at 2.0×10^6 K inside the structures.

Fig. 2a shows four different morphologies corresponding to the values plotted in Fig. 2b. The lines with slope 1 are the best fit of data calculated for heliocentric heights ranging between 1.5 and 4 $R_{\odot}$.

4. Conclusions

Tab. 1 gives the irregularity factor for all the modeled morphologies. Fig. 3 plots the irregularity factor as a function of the filling factor (x-axis) and of the density scale factor of the structures (parameter on the curves), as determined with the two-component model described by Orrall et al. (1990). It is worth noting that: $X \propto \alpha^{-1}$. The figure also displays the theoretical values of the irregularity factor for all the morphologies used in the simulation, and shows that there is a similar trend between the theoretical model and the simulated irregularity factors.

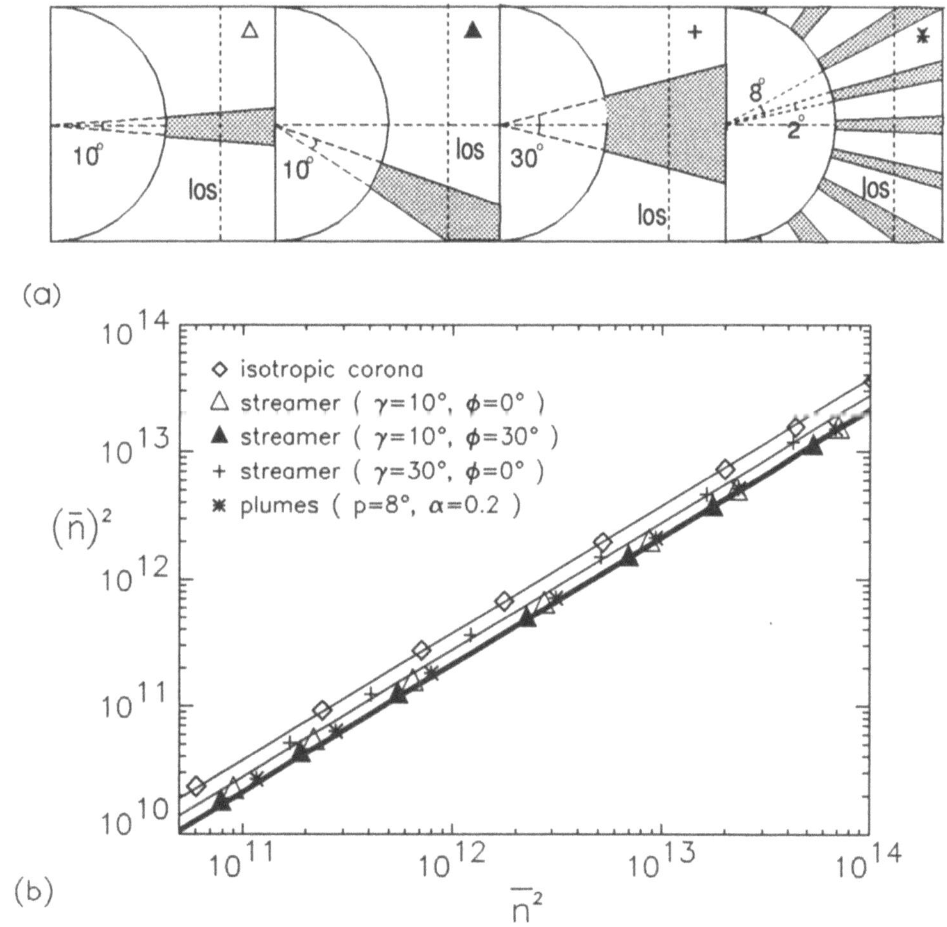

Fig. 2. (a) Coronal morphologies used in the simulation (streamer or plumes are the shaded areas; *los*: line of sight); (b) $(\overline{n})^2$ vs. $\overline{n^2}$.

TABLE I

Simulated irregularity factors

X	$X/X_{\text{isotropic}}$	Morphology
2.66	1	isotropic corona
4.69	1.76	streamer with $\gamma = 10°$ and $\phi = 0°$
4.89	1.84	streamer with $\gamma = 10°$ and $\phi = 30°$
3.61	1.36	streamer with $\gamma = 30°$ and $\phi = 0°$
4.53	1.70	plumes with $p = 8°$ and $\alpha = 0.2$

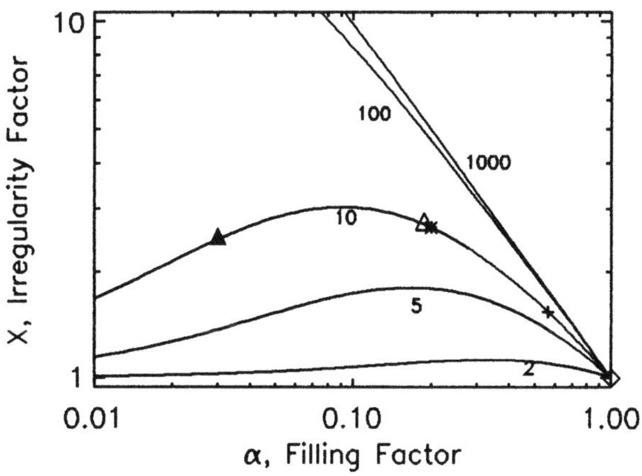

Fig. 3. Irregularity factor vs. filling factor, as a function of the ratio between densities inside and outside the structure. The data points corresponds to the morphological parameters used in the simulation (see Fig.2).

The irregularity factor of the isotropic corona is not unity because the electron density function is not constant along the line of sight.

The irregularity factor determined with the observations cannot give a complete description of the coronal morphology integrated along the line of sight, but it can be a very useful indicator of inhomogeneities (setting, for exemple, a lower limit to the density scaling factor inside the structures), and give constraints to the applicability of an isotropic corona model.

Acknowledgements

Marco Romoli acknowledges the support of the European Space Agency under a non-resident post-doctoral fellowship. Silvano Fineschi acknowledges the support of the University of Firenze and the Italian Space Agency (ASI) under a *Dottorato di Ricerca in Astronomia* fellowship.

References

Fineschi, S., and Romoli, M.: 1993, *Space Sci. Rev.* **this issue**,

Kohl, J. L., and Noci, G.: 1992, 'UVCS – The Ultraviolet Coronagraph Spectrometer for SOHO' in A.I.Poland,. and V.Domingo, ed(s)., *1st SOHO Workshop*, ESA SP-348, 23

Munro, R. H., and Jackson, B. V.: 1977, *Ap. J.* **213**, 874

Orrall, F. Q., Rottman, G. J., Fisher, R. R., and Munro, R. H.: 1990, *Ap. J.* **349**, 656

Saito, K.: 1970, *Ann. Tokyo Astr. Obs.* **12**, 53

Sime, D. G., and Fisher, R. R.: 1993, *Solar Phys.* **144**, 75

van de Hulst, H. C.: 1950, *Bull. Astr. Inst. Netherlands.* **11**, 135

Withbroe, G. L.: 1972, *Solar Phys.* **25**, 116

Withbroe, G. L., Kohl, J. L., Weiser, H.: 1986, *Ap. J.* **307**, 381

UV SPECTRAL LINES FROM CORONAL TRANSIENTS

D. SPADARO, R. VENTURA
Osservatorio Astrofisico di Catania, Italy

and

R. MARTIN
Istituto di Fisica Generale, Università di Torino, Italy

Abstract. We investigated the UV emission expected from solar coronal transients, selecting some spectral lines which will be observed with the UVCS spectrocoronagraph onboard the SOHO spacecraft. The line intensities were calculated starting from a representative, simplified model of coronal transient. We discuss how the considered intensities depend on the physical parameters of the examined structures. This work is aimed to give a contribution in defining and preparing the future observations of coronal transients and coronal mass ejections by the UVCS/SOHO.

1. Introduction

UV spectroscopic diagnostics with UVCS/SOHO (Kohl et al. 1991) are expected to provide useful information on solar coronal transients, in particular on their thermal state and on the interplay between the motion of the material ejected during the transient and the shock compression and heating of the ambient coronal gas. The knowledge of the amount of heating that accompanies the transients may provide clues to the physical processes driving them. Other information on densities and expansion velocities should also become available (Kohl et al. 1991).

In view of the preparation and interpretation of the UVCS/SOHO future observations, we investigated the UV emission expected from coronal transients, selecting some spectral lines which will be observed with this instrument. We calculated the H I Ly–α and the O VI λ1032 and λ1038 line intensities starting from a representative, simplified model of coronal transient, based both on the theoretical and empirical information presently available.

We discuss the effect of coronal transients on the UV emission lines originating in the extended solar corona and the sensitiveness of these lines to the physical parameters of the coronal regions where the transient propagates.

2. Adopted model

The geometry of the perturbed region in the corona has been modelled by a cone, with the vertex in the centre of the Sun, the axis of symmetry located in the plane of the sky and an angular extent of 40 degrees, according to the typical values measured in white light images detected by coronagraphs (Hundhausen 1993). Within this region radial symmetry is assumed.

Space Science Reviews **70**, 365–368.

The values of plasma temperature, density and velocity used in the calculation of spectral line intensities are those derived by Steinolfson and Nakagawa (1977) in a study of the dynamical response of the solar corona induced by a transient energy input, with a spherically symmetric, single fluid, adiabatic model, described in Nakagawa and Steinolfson (1976). Starting from a steady–state quiet corona with a solar wind included, a perturbation is introduced near the coronal base.

In our study, we have considered for the perturbed solar corona the radial profiles of the plasma parameters corresponding to the "base simulation" of Steinolfson and Nakagawa. These profiles, reported in Fig. 4 of their paper, are calculated for unperturbed coronal conditions ($t = 0$ s), and at a time $t = 90$ min after the perturbation has been introduced.
The perturbed coronal region is surrounded by a steady–state quiet corona, whose plasma parameters have radial profiles corresponding to the initial, i.e. calculated at $t = 0$ s, profiles.

While this model of coronal transient is rather schematic, neglecting all the effects due to the presence of magnetic fields and dissipation processes, it represents adequately some relevant properties of the observed coronal transient, e.g., the dynamics of the ejected plasma and the presence of a compression region followed by a region where rarefaction occurs (Hildner et al. 1975).

3. Mechanisms of line formation

We considered the two following main mechanisms of line formation:
- resonant scattering of chromospheric photons by coronal ions;
- collisional excitation by electron impact (only for the oxygen lines).

The effect of the outflow velocity on the resonantly scattered component of each line, known as Doppler dimming (Hyder & Lites 1970; Beckers & Chipman 1974), has been taken into account. We refer the reader to Kohl & Withbroe (1982), Withbroe et al. (1982) and Noci et al. (1987) for an extensive description of these processes.
The intensity and profile of the chromospheric O VI exciting lines considered in this work are reported in Noci et al. (1987), while those of the Ly–α line are reported in Gouttebroze et al. (1978). We assume the lower atmosphere to be uniformly bright in the exciting radiation. We use the ionization balance of hydrogen calculated by Gabriel (1971), while that of oxygen is taken from Noci et al. (1988). We assume the O VI kinetic temperature and outflow speed to be equal to the plasma temperature and velocity, respectively, calculated by Steinolfson and Nakagawa (1977).

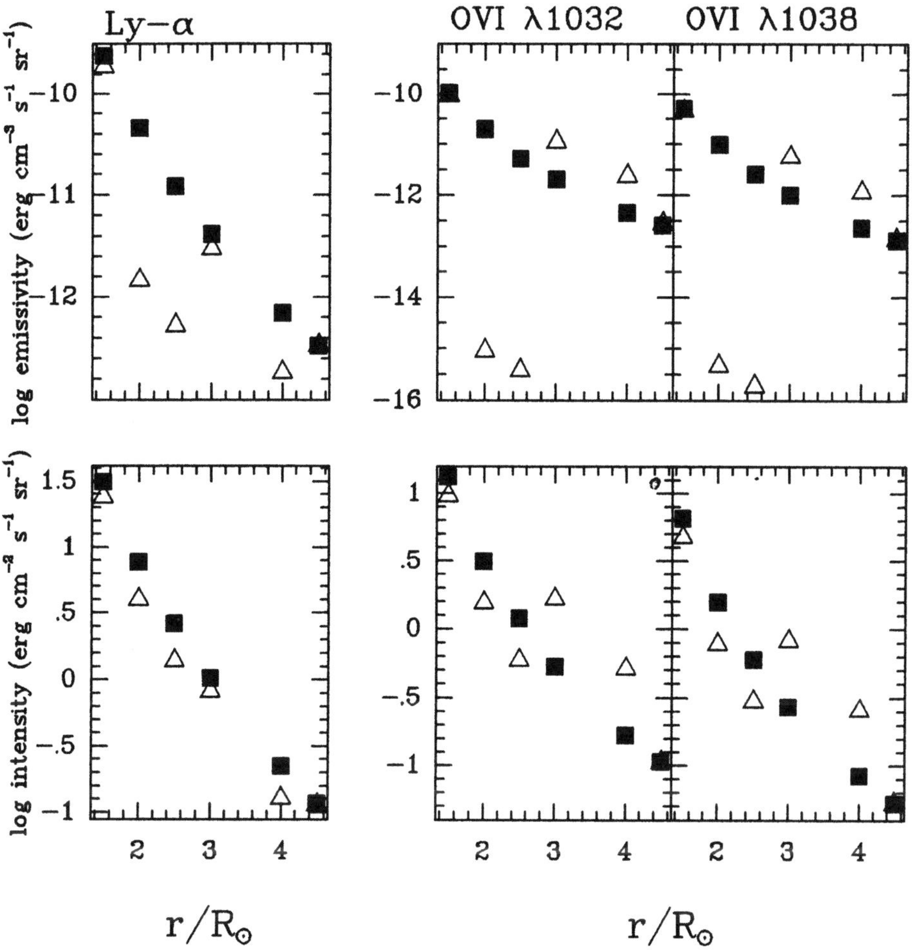

Fig. 1. Total line emissivities calculated at the point where the line of sight intersects the axis of symmetry of the coronal transient (top panels) and line intensities calculated considering the contributions of the transient and the surrounding quiet corona (bottom panels). These spectral quantities are obtained at different heliocentric distances for the unperturbed corona (filled squares) and at $t = 90$ *min* after the perturbation has been introduced (triangles)

4. Results and discussion

Fig. 1 reports the total emissivities and intensities of the Ly−α and O VI lines calculated at different heliocentric distances for the unperturbed corona and at $t = 90$ *min* after the perturbation has been introduced. The emissivity is calculated at the point where the line of sight intersects the axis of symmetry of the coronal transient, while the intensity includes the contributions along the line of sight of the transient and the surrounding

quiet steady–state corona.
Both the emissivities and the intensities significantly change as a consequence of the variation of the plasma parameters, in particular the density. In fact within 2.5 $R_\odot$, where the density is considerably reduced, these spectral quantities decrease at $t = 90\ min$, also due to the strong Doppler dimming caused by the increase in the expansion velocity.
Beyond 3 $R_\odot$ the O VI emissivity and intensity increase, due to the increase in the plasma density which enhances particularly the collisional component of the lines. The Ly–α line, which has a collisional component negligible at coronal conditions (Withbroe et al. 1982), is depleted all the same, owing to the Doppler dimming caused by the high expansion velocity.

From these preliminar results we note that the effect of the transient on the considered line intensities appears detectable, even if we consider the quiet unperturbed corona surrounding the region where the transient occurs.

We believe that the simulations of the UV emission from coronal transients can be furtherly developed, adopting improved models and also considering a better temporal resolution and more extended temporal intervals. Another important point is the ionization balance of the considered elements: possible deviations from the ionization equilibrium in the rapidly outflowing plasma should be taken into account in the calculation of the spectral intensities. However, the results of this work may give useful information for planning the mission operations and the observation sequences of the UVCS/SOHO space instrument.

References

Beckers, J.M., & Chipman, E. 1974, *Solar Phys.*, 34, 151.
Gabriel, A.H. 1971, *Solar Phys.*, 21, 392.
Gouttebroze, P., Lemaire, P., Vial, J.C., & Artzner, G. 1978, *ApJ*, 225, 655.
Hildner, E., Gosling, J.T., MacQueen, R.M., Munro, R.H., Poland, A.I., & Ross, C.L. 1975, *Solar Phys.*, 42, 163.
Hundhausen, A.J. 1993, *J. Geophys. Res.*, 98, 13177.
Hyder, C.L., & Lites, B.W. 1970, *Solar Phys.*, 14, 147.
Kohl, J.L., & Withbroe, G.L. 1982,*ApJ*, 256, 263.
Kohl, J.L. et al. 1991, *Adv. Space Res.*, 11, No. 1, 359.
Nakagawa, Y., & Steinolfson, R.S. 1976, *ApJ*, 207, 296.
Noci, G., Kohl, J.L., & Withbroe, G.L. 1987, *ApJ*, 315, 706.
Noci, G., Spadaro, D., Zappalà, R.A., & Zuccarello, F. 1988, *A&A*, 198, 311.
Steinolfson, R.S., & Nakagawa, Y. 1977, *ApJ*, 215, 345.
Withbroe, G.L., Kohl, J.L., Weiser, H., & Munro, R.H. 1982, *Space Sci. Rev.*, 33, 17.

MODELING THE CORONAL MAGNETIC FIELD IN A POLAR HOLE

XUEPU ZHAO and J. TODD HOEKSEMA
CSSA, ERL 328, Stanford University, Stanford, CA 94305, USA

Abstract. The coronal magnetic field in the northern polar coronal hole in 1986 is predicted on the basis of the photospheric magnetic field observations and the horizontal current–current sheet coronal model (Zhao and Hoeksema, 1993). The predicted magnetic field intensity is stronger near the center of the hole than near the edge. The calculated expansion factor for the entire hole does not match the expansion factor of any flux tube in the hole, suggesting that it would not be appropriate to use the expansion factor for entire hole to represent the divergence of the flux tube in analyzing the acceleration and heating of the plasma in coronal holes.

Key words: Sun – Corona – Coronal Holes – Magnetic Field Structure

1. Introduction

Coronal holes are the major source of the solar wind. The knowledge of magnetic fields in coronal holes is vital to the development of an understanding of the mechanisms that heat the coronal plasma and accelerate it outward to form the solar wind (e.g., Wang, 1993). There are no direct measurements of the coronal magnetic field in holes. The evidence which suggests that the field has an open and diverging geometry comes from the observed diverging, approximately radial orientation of emission features in holes (Bohlin and Sheeley, 1978) and from the potential field-source surface (PFSS) modeling (Wang and Sheeley, 1992 and the references therein).

Observations suggest that large-scale current systems do exist in the corona. We have developed a coronal magnetic field model with horizontal currents and current sheets (HCCS). This model allows us to map the observed photospheric field into the corona above $2.5R_\odot$, the heliocentric distance of the source surface. Preliminary modeling of the coronal magnetic field for Carrington Rotation 1647, when a total solar eclipse in the minimum phase of the sunspot cycle was observed, showed that with the optimum choice of free parameters this model reproduces polar plumes and the axes of coronal streamers better than the PFSS model (Zhao and Hoeksema, 1993).

The objective of the present study is to predict the coronal magnetic field and electron density in the 1986 northern polar hole from the HCCS model using the photospheric magnetic field observed at the Wilcox Solar Observatory.

Space Science Reviews **70**, 369–372.

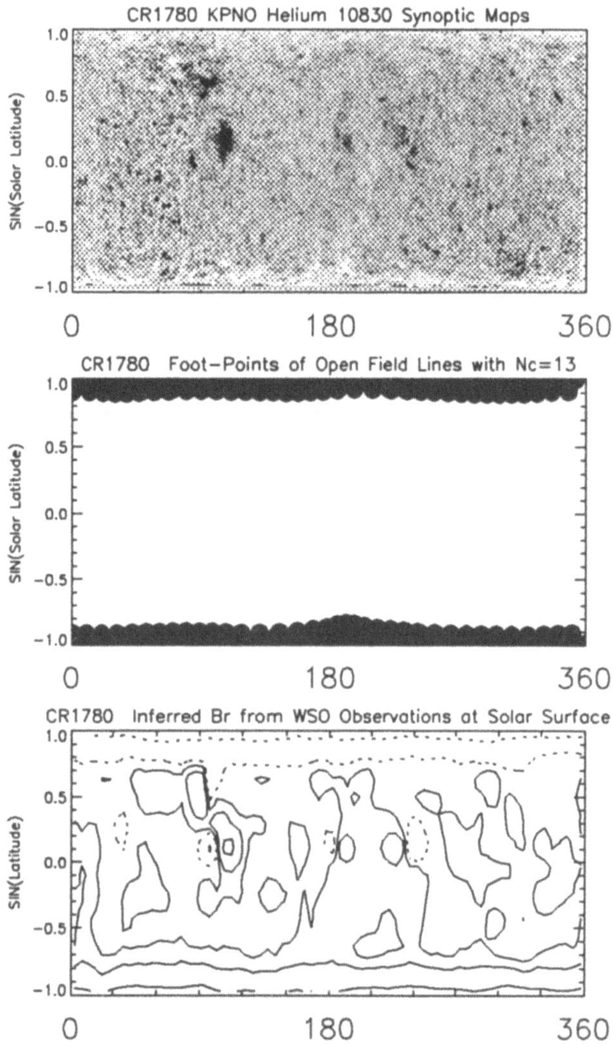

Fig. 1. Comparison of the location of coronal holes in 10830 synoptic chart (1a) with the location of foot-points of open field lines (1b), which is calculated using radially pointed photospheric fields (1c). The contours are scaled with 0, ±1, ±5 and ±10 Gauss. Solid and dotted line styles denote $B_r \geq 0$ and $B_r < 0$, respectively.

2. Results

Figure 1 shows the location of the 1986 polar coronal holes at the coronal base. The calculated location of foot-points of open field lines (Figure 1b) matches very well with the location of 10830 hole observations (the lightest areas around the poles in Figure 1a). The holes are located in strongest large-scale field regions (Figure 1c).

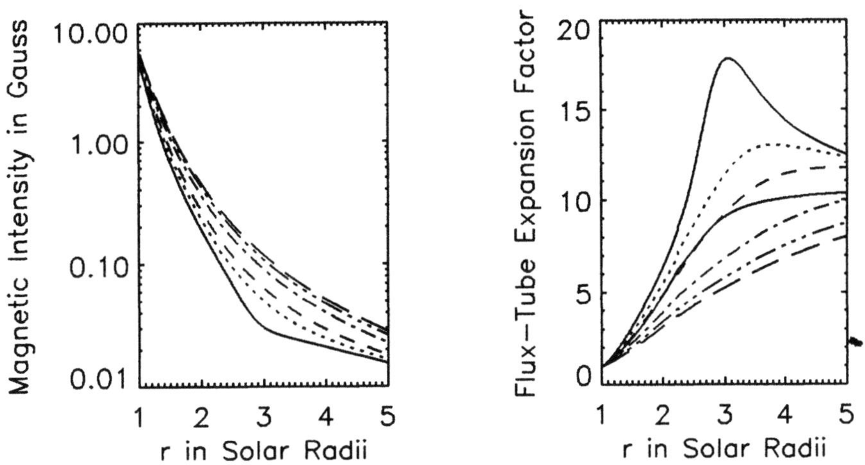

Fig. 2. The radial variation of the total magnetic intensity (2a, the left panel) and the expansion factor (2b, the right panel) along flux tubes.

The polar coronal hole studied here is nearly symmetric around the Sun's rotation axis.. We select six foot-points with polar angles of $4°, 9°, 14°, 19°, 21°$ and $23°$ in the meridian plane of the Carrington longitude of 180°. Starting from these six foot-points, we follow each field line to calculate the magnetic field intensity and electron density. Figure 2 displays the radial variation of the total magnetic intensity (Figure 2a) and the flux tube expansion factor (Figure 2b). The line styles from long dashed to solid correspond to the foot-points from near the center of the hole (4°) to near the edge of the hole (23°). Figure 2 shows that the total magnetic intensity decreases as the angular distance increases from the center of the hole. The expansion factor changes nonmonotonically along flux tubes near the edge of the hole. The decline in the expansion factor implies that the radial component of magnetic fields falls off less rapidly than r^{-2}. The solid line between the dashed and dash-dotted lines in Figure 2b denotes the expansion factor for entire hole, showing that the expansion factor for entire hole does not match the expansion factor of any flux tube in the hole.

Figure 3 displays the radial variation of the calculated electron density in the 1986 northern polar hole (3a). The electron density increases as the angular distance increases from the center of the hole, which is qualitatively consistent with the observed density distribution for a 1973 polar coronal hole (Munro and Jackson, 1977) as shown in Figure 3b.

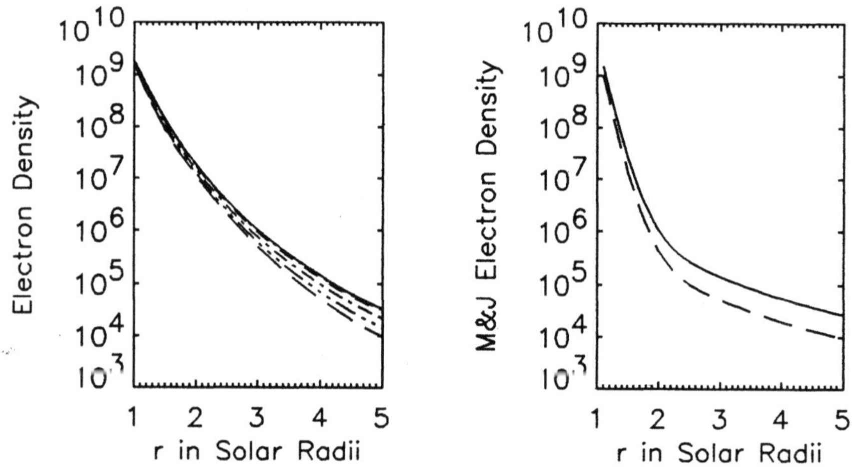

Fig. 3. The radial variation of electron densities in the 1986 (3a, the left panel) and the 1973 (3b, the right panel) polar coronal holes.

3. Conclusion

The horizontal current–current sheet coronal model allows us to obtain the geometry of entire holes and the divergence of magnetic flux tubes within holes. The expansion factor of entire hole may not be a good prototype for analyzing the physical processes taking place in a flux tube. The predicted magnetic intensity near the center is stronger than near the edge of the hole. Using the simultaneous SOHO observations of the magnetic field from SOI/MDI and the coronal density obtained from UVCS and LASCO we should be able to obtain the really optimum free parameters in the HCCS model and calculate more reliable magnetic field in coronal holes.

References

Bohlin, J.D. and Sheeley, N.R.,Jr.: 1978, *Solar Phys.* **56**, 125
Munro, R.H. and Jackson, B.V.: 1977, *Astrophys. J.* **213**, 874
Wang, Y.-M.: 1993, *this issue* ,
Wang, Y.-M. and Sheeley, N.R.,Jr.: 1992, *Astrophys. J.* **392**, 310
Zhao, X.P. and Hoeksema, J.T.: 1993, *Solar Phys.*, submitted

BROADENING OF FE X (6374 Å) PROFILES ABOVE THE LIMB IN A CORONAL HOLE

D.M. HASSLER *
Harvard-Smithsonian Center for Astrophysics Cambridge, MA, 02138, U.S.A.

and

T.G. MORAN
NASA Goddard Space Flight Center Greenbelt, MD, 20771, U.S.A.

Abstract. Profiles of the visible Fe X (6374 Å) coronal emission line as a function of height above the limb were obtained out to 1.16 solar radii in a coronal hole using the NSO/Sacramento Peak Observatory Coronagraph, Universal Spectrograph and a CCD camera. These are the first coronal line profiles obtained as a function of height in a coronal hole from the ground. Analysis of the line widths suggests a large component of nonthermal broadening which increases with height ranging from 40 to 60 km/s, depending upon the assumed temperature or thermal component of the profile.

Key words: Solar corona – Coronal holes – Solar wind – UV spectroscopy

1. Introduction

The analysis of coronal emission line profiles is an important technique for studying the physical conditions in the solar corona. The majority of these coronal emission lines appear in the ultraviolet (UV) and extreme ultraviolet (EUV) wavelength regimes requiring access to space in order to observe them. However, visible coronal emission lines from forbidden transitions of Fe X (6374 Å), Fe XIV (5303 Å), and Ca XV (5694 Å) have been studied by many researchers from the ground with coronagraphs (Billings 1963, Billings et al. 1963, Tsubaki 1975, 1977) and during eclipses (Kim et al. 1975, Kim 1992, Delone and Makarova 1988, Singh et al. 1982). The primary limitations with the previous coronagraph observations have been associated with signal-to-noise, dynamic range (non-linear response of the film), and background sky subtraction resulting from the use of photographic film. As a result, these observations were limited primarily to very bright regions of the corona and heights very close to the limb. Although line profile measurements during eclipses extend out to 1.6 to 1.7 solar radii, most of these measurements are in bright structures above active regions. There have been no systematic studies of line widths as a function of height in coronal holes. We discuss here observations made at the National Solar Observatory/Sacramento Peak with the Evans Facility 40 cm coronagraph, the universal spectrograph and a cooled RCA-504 CCD detector. With a

* Now at the High Altitude Observatory/NCAR, Boulder, CO, 80307, U.S.A.

Space Science Reviews **70**, 373–377.

linear response over a wide dynamic range, CCD detectors have been the single most important factor in improving the ability to extract low signal-to-noise profiles from observations dominated by sky background scattered light. It is this improvement which has permitted these first observations of Fe X (6374 Å) profiles in a coronal hole out to 1.16 solar radii.

2. Observations

The spatial resolution of these observations is 1 arcsecond in height and 40 arcseconds tangent to the limb and the spectral dispersion is 5.7 Å/mm or 0.114 Å/pixel. The entrance slit of the spectrograph was placed tangent to the limb and scanned radially outward in approximately 0.02 solar radii steps, with slit jaw images taken at each height to record the precise position of the slit with respect to the limb. Images from the Soft X-ray Telescope (SXT) (Tsuneta et al. 1991) on the Yohkoh satellite indicate that there was a large coronal hole at the south pole during these observations. This coronal hole was large enough to be uncontaminated by foreground or background emission from coronal structures along the line-of-sight, as can be seen by the complete disappearance of the limb in the SXT images. The observations on 23 September 1992 were made along a radial scan line at 190° heliographic latitude (190° counter-clockwise from heliocentric north), near the edge of the coronal hole. The observations on 24 September 1992 were made along a radial scan line at 200° heliographic latitude, at the center of the coronal hole. Exposures were taken every 16 seconds, integrating for up to 15 minutes at 1.15 solar radii. Short (5 millisecond) exposures were taken just inside the limb with the partially occulted 20 cm aperture to approximate the sky background or disk spectrum scattered by the earth's atmosphere. The measured disk spectrum is then scaled to the Fraunhofer lines in the observed spectrum and subtracted to yield the coronal spectrum. The profiles are also corrected for pixel-to-pixel sensitivity variations (with a "flat field") and for dark counts.

Sample profiles of the Fe X 6374 Å line at 1.04 and 1.09 solar radii in the south polar coronal hole on 24 September 1992 are shown in Figure 1. The solid line in Figure 1 shows the observed spectrum and the dashed line represents the sky background. The resulting Fe X profiles after the sky background has been subtracted are shown in Figure 2. For this analysis, each of the observed profiles has been fit to a Gaussian profile (shown by the dashed lines in Figure 2) and the resulting line width has been plotted as a function of height, as shown in Figure 3. The dashed line in Figure 3 represents data taken on 23 September 1992 at the edge of the coronal hole and the solid line shows the line widths obtained on 24 September 1992 in the center of the coronal hole. The error bars represent a one sigma uncertainty in the Gaussian line width. The total observed Gaussian widths

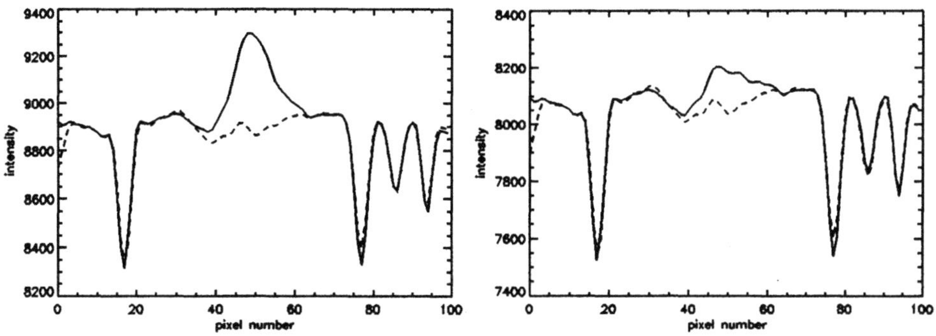

Fig. 1. a) Sample spectrum of the observed Fe X 6374 Å line at 1.04 solar radii in the south polar coronal hole on 24 September 1992. The solid line shows the observed spectrum and the dashed line shows the sky background. b) Same as Figure 1a but at 1.09 solar radii.

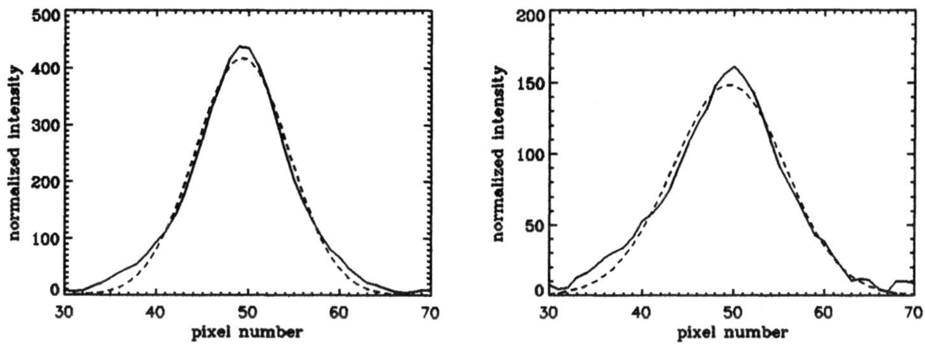

Fig. 2. The resulting Fe X profiles shown in Figure 1a at 1.04 solar radii and Figure 1b at 1.09 solar radii after the sky background has been subtracted. The dashed lines represent Gaussian fits to the profiles.

(both thermal and nonthermal components) range from 34 to 62 km/s.

3. Discussion

The line widths in Figure 3 suggest a large component of nonthermal broadening which increases with height ranging from 30 to 60 km/s depending upon the assumed temperature or thermal component of the line width. These results are in agreement with the UV line profile measurements of Hassler et al. (1990) and may suggest the presence of hydromagnetic waves, which are thought to play an important role in heating the corona and accelerating the solar wind. Hollweg (1973) and Esser et al. (1987) showed that hydromagnetic waves will produce significant broadening of high temperature lines observed above the limb if the waves have velocity amplitudes of more than 10 km/s at the coronal base. Since the density falls off exponen-

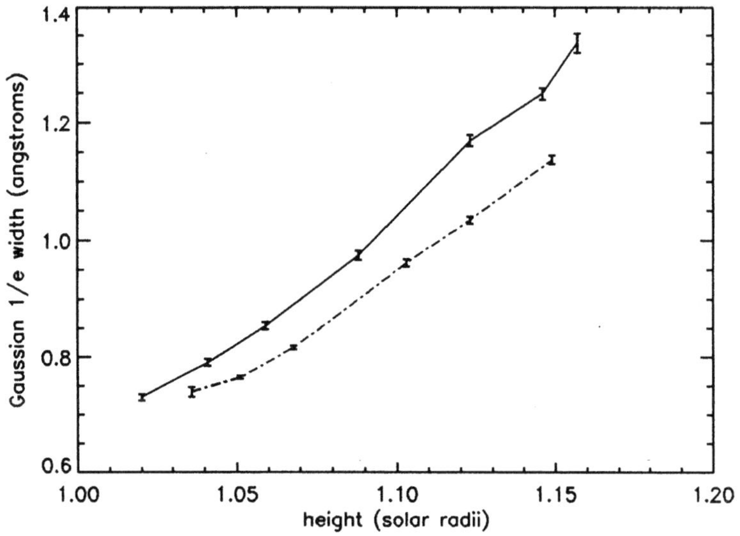

Fig. 3. Observed line width of Fe X (6374 Å) as a function of height in the south polar coronal hole, assuming a Gaussian fit for each of the profiles. The dashed line represents data taken on 23 September 1992 at the edge of the coronal hole and the solid line shows the line widths obtained on 24 September 1992 in the center of the coronal hole. The error bars represent a one sigma uncertainty in the Gaussian line width.

tially with increasing radial distance, the Alfvén wave velocity amplitude (if undamped) increases exponentially with radius, which causes the observed line width to increase rapidly, similar to the increase in line width shown in Figure 3.

It is interesting to note that the profiles in Figure 2 appear to deviate slightly with height from a simple Gaussian profile. It is quite likely, for example, that the observed profiles could be better fit to a double Gaussian, as done by Dere and Mason (1993) with profiles from HRTS. However, justification for doing so would require a priori knowledge of the coronal structure along the line of sight at the temperature of formation of the line. Yohkoh images provide some insight into this structure, revealing for example in this case, a large unobstructed coronal hole with very little foreground or background emission to contaminate the line-of-sight profile. However, it must be remembered that the sensitivity of the Yohkoh SXT instrument is primarily between 2-4 million degrees and may or may not correspond to the coronal structure at one million degrees, where the Fe X line is formed. Therefore, we also measured the Fe X emission in a nearby streamer and found the intensity to decrease rapidly with height, making the likely contribution to the coronal hole profile small.

One must also consider the possibility of a non-Maxwellian velocity distribution as described by Scudder (1993) and others (i.e. Shoub 1988) which might account for a slight broadening in the wings of an otherwise Gaussian distribution. However, it is just as likely that the wave spectrum, which may be contributing to the extreme nonthermal widths of these lines, may not be Gaussian, leading to the observed profiles. Further consideration of these and other possibilities is ongoing and will be discussed in a future publication (Hassler and Moran, 1994).

One of the exciting prospects with SOHO will be the ability of the EIT instrument, for example, to provide high resolution images in Fe IX/X (171/173 Å) to complement, not only this type of ground-based observation, but also the spectroscopic capabilities of SUMER, CDS, LASCO and UVCS. Such coordinated studies will improve the interpretation of coronal line profiles based on observed coronal structures along the line-of-sight at the temperature of formation of the lines.

Acknowledgements

A portion of the work of DMH has been supported by the Research Opportunities Fund of the Smithsonian Astrophysical Observatory. TGM is being supported by a NAS/NRC Postdoctoral Fellowship. The National Solar Observatory/Sacramento Peak is funded by the National Science Foundation. We would like to thank Lou Gilliam, Brian Armstrong, and Joe Elrod for their support at the Evans Facility/Sac Peak which made these observations possible. We would also like to thank the members of the Yohkoh team for providing us with daily SXT images during these observations.

References

Billings, D.E.: 1963, *Ap. J.* **137**, 592

Billings, D.E., Lilliequist, C.G.: 1963, *Ap. J.* **137**, 16

Delone, A.B., Makarova, E.A.: 1969, *Solar Phys.* **9**, 116

Dere, K.P., Mason, H.E.: 1993, *Solar Phys.* **144**, 217

Esser, R., Holzer, T.E., Leer, E.: 1987, *J. Geophys. Res.* **92**, 13377

Hassler, D.M., Rottman, G.J., Shoub, E.C., Holzer, T.E.: 1990, *Ap. J. Lett.* **348**, L77

Hassler, D.M., Moran, T.G. : 1994, *Ap. J.* , in preparation

Hollweg, J.V.: 1973, *Ap.J.* **181**, 547

Kim, I.S., Nikolsky, G.M.: 1975, *Solar Phys.* **43**, 351

Kim, I.S.: 1992, *Proceedings of the First SOHO Workshop*, ESA SP-348, p. 149

Scudder, J.D.: 1992, *Ap. J.* **398**, 319

Shoub, E.C.: 1988, *Proc. 6th Int. Solar Wind Conf.*, NCAR TN-306, p. 59

Singh, J., Bappu, M.K.V., Saxena, A.K.: 1982, *J. Astrophys. Astron.* **3**, 249

Tsubaki, T.: 1975, *Solar Phys.* **43**, 147

Tsubaki, T.: 1977, *Solar Phys.* **51**, 121

Tsuneta, S., L. Acton, M. Bruner, J. Lemen, W. Brown, R. Caravalho, R. Catura, S. Freeland, B. Jurcevich, M. Morrison, Y. Ogawara, T. Hirayama, J. Owens: 1991, *Solar Phys.* **136**, 37

SIGNATURE OF CORONAL HOLES AND STREAMERS IN THE INTERPLANETARY SPACE

YURI I. YERMOLAEV
Space Research Institute, Russian Academy of Sciences, Profsoyuznaya 84/32, 117810 Moscow, Russia

Abstract. The properties of different solar wind streams depend on the large scale structure of the coronal magnetic field. We present average values and distributions of bulk parameters (density, velocity, temperature, mass flux, momentum, and kinetic and thermal energy, ratio of thermal and magnetic pressure, as well as the helium abundance) as observed on board the Prognoz 7 satellite in different types of the solar wind streams. Maximum mass flux is recorded in the streams emanating from the coronal streamers while maximum thermal and kinetic energy fluxes are observed in the streams from the coronal holes. The momentum fluxes are equal in both types of streams. The maximum ratio of thermal and magnetic pressure is observed in heliospheric current sheet. The helium abundance in streams from coronal holes is higher than in streams from streamers, and its dependences on density and mass flux are different in different types of the streams. Also, the dynamics of α-particle velocity and temperature relative to protons in streams from coronal holes and streamers is discussed.

Key words: Sun – Solar Wind – Coronal Holes - Streamers

1. Introduction

Direct measurements in the solar wind allow us to obtain valuable information about the solar corona regions where the solar wind streams are formed. Parameters such as the relative abundance and ionization state of different ion components do not change in the interplanetary space and may be used for diagnostics of those solar corona regions from which solar wind streams originate (Hundhausen, 1972; Veselovsky, 1984; Geiss, 1985; Yermolaev and Zastenker, 1990; Yermolaev, 1992b). The hydrodynamical parameters (density, bulk velocity and kinetic temperature) of different ion components change in the interplanetary space and may be used for studies of solar wind dynamics (Parker, 1963; Hundhausen, 1972; Veselovsky, 1984). Nevertheless, the large-scale variations in the hydrodynamical parameters arise from solar coronal structures and may be used for the study of the formation and dynamics of the solar wind as well as for investigations of the structure and dynamics of the solar corona (e.g. see reviews by Neugebauer, 1982; Veselovsky, 1984; Withbroe, 1986; Schwenn, 1986; Yermolaev, 1991).

On the basis of the Prognoz 7 (November 1978 – July 1979) measurements, we studied the distributions of several solar wind parameters for five types of solar wind streams for which their connection to solar coronal structures and non-stationary events was known (see Fig. 1): (1) the heliospheric current sheet, (2) streams from streamers, (3) streams from coronal holes,

Space Science Reviews **70**, 379–386.

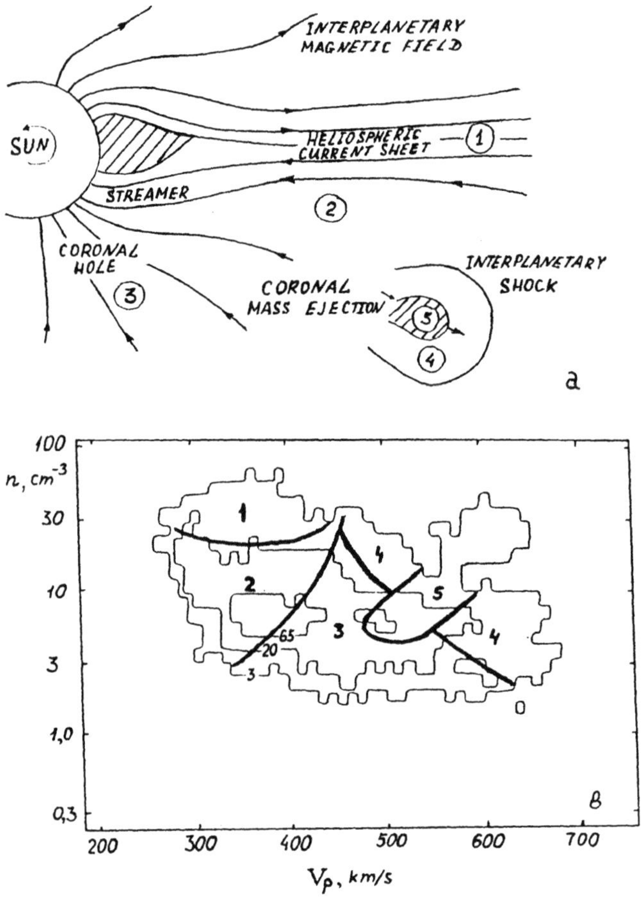

Fig. 1. a) A schematic view of different types of solar wind streams and their connection with solar corona structure and phenomena, and b) the distribution of different types of streams on the "density-velocity" plane.

(4) solar wind streams disturbed by non-stationary events in the corona, and (5) disturbed streams including the matter of coronal mass ejections (Yermolaev, 1991).

2. Experimental Results

We assume that the relationships between the proton density n_p, the proton bulk velocity V_p, the proton temperature T_p, the absolute value of the interplanetary magnetic field B and the abundance of helium relative to protons n_α/n_p cannot be arbitrary and we have investigated the two-dimensional distributions of these parameters on the density-velocity plane on the basis of the Prognoz 7 data. The behaviour of the parameters B, T_p, β, and n_α/n_p

TABLE I
Parameters in 5 types of solar wind

SW parameters (70% intervals or average)	Types of solar wind HCS	CS	CH	Shocked SW plasma	CME
V_p [km/s]	< 450	270-450	380-650	450-700	450-700
n_p [cm^{-3}]	15-50	5-30	2-10	5-30	5-30
T_p [10^4 K]	1-6	2-10	5-20	10-50	5-20
β	0.5-3.0	0.1-1.0	0.2-1.0	0.2-1.0	0.2-2.0
n_α/n_p [%]	< 2	2-5	3-7	3-7	> 6
B [nT]	5-8	6-10	5-10	8-14	7-15
Time of obs. [%]	8	33	49	6	4
nV_p [10^8 cm^{-2} s^{-1}]	10.5±3.9	3.4±1.5	2.7±1.5	4.5±3.9	4.5±3.4
$m_p n V_p^2$ [10^{-8} dyn cm^{-2} s^{-1}]	6.3±2.9	2.1±1.0	2.1±1.2	4.0±3.4	4.1±3.6
$1/2 m_p n V_p^3$ [erg cm^{-2} s^{-1}]	1.1±0.6	0.4±0.2	0.5±0.3	1.1±1.0	1.1±1.1
nkT_pV_p [10^{-2} erg cm^{-2} s^{-1}]	0.9±0.8	0.3±0.5	0.4±0.5	1.1±1.8	1.3±1.4

on the n_p–V_p plane (see 70% intervals of these parameters in Table 1) allows the selection of five characteristic regions indicated in Fig. 1b by the numbers 1 to 5 which, as mentioned above, may be connected with different solar coronal structures and phenomena (Yermolaev, 1991).

Histograms of proton fluxes of particles nV_p (or mass mnV_p), momentum nV_p^2 or mnV_p^2, kinetic energy ($\frac{1}{2}mnV_p^3$), and thermal energy ($nkT_pV_p^3$) as well as the ratio of thermal to magnetic pressure $nkT_p/(B^2/8\pi)$ in 5 types of solar wind streams were presented in Figs. 2-6 in Yermolaev (1992a). The average magnitudes and standard deviations (dispersions) for each histogram are presented in Table 1.

The analysis of these data shows that the average mass flux nV_p decreases with increasing bulk velocity in quasi-stationary solar wind streams: from 10.5×10^8 in heliospheric current sheet to 3.4×10^8 and $2.7\times10^8 cm^{-2} s^{-1}$ in the streams from streamers and coronal holes, respectively. The observed mass flux in disturbed streams is about $4.5\times10^8 cm^{-2} s^{-1}$. Nevertheless, taking into account the duration and appearance rate of different solar wind streams during the Prognoz 7 observations, the Sun's loss of mass is almost equal in different quasi-stationary solar wind streams and is higher by approximately a factor 5 compared to the disturbed streams.

The momentum fluxes mnV_p^2 are equal (2.1×10^{-8} dyn cm^{-2} s^{-1}) in quasi-stationary solar wind streams (from streamers and coronal holes), and the momemtum flux increases in the disturbed solar wind streams up to ≈

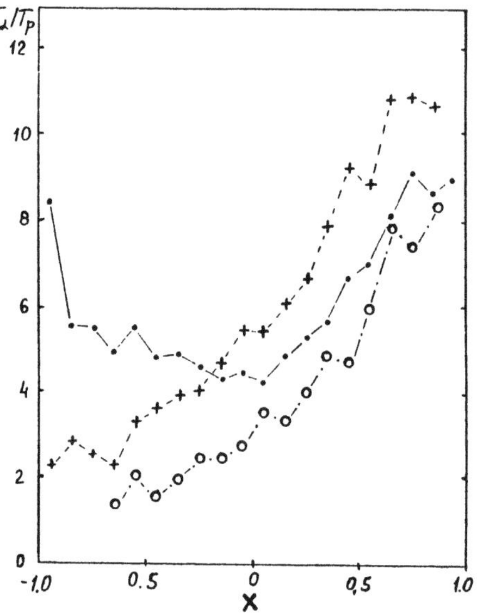

Fig. 2. Dependence of α-particle and proton temperature ratio T_α/T_p on the relative velocity difference X (see text) for heliospheric current sheet (open circles), streams from coronal streamers (crosses) and holes (closed circles).

4.0×10^{-8} and in heliospheric current sheet up to $6.3\times10^{-8}\,\mathrm{dyn\,cm^{-2}\,s^{-1}}$.

The thermal and kinetic ion energy fluxes ($nkT_pV_p^3$ and $nmV_p^3/2$) in disturbed streams and in heliospheric current sheet are higher by a factor 2-3 compared to quasi-stationary streams. Nevertheless, taking into account the duration and appearance rate of the different streams, the Sun's loss of thermal and kinetic energy in streams from coronal holes is higher by a factor 2-5 in the heliospheric current sheet and disturbed streams and by a factor 1.5-2.0 higher than in streams from streamers.

A high ratio of thermal to magnetic pressures $\beta = nkT/(B^2/8\pi)$ is often observed in heliospheric current sheet ($\approx$2) and in the streams from coronal mass ejections ($\approx$1). In other types of the solar wind streams the beta-parameter is $\approx$0.5.

The selection of solar wind streams made it possible to investigate the relationship between different solar wind parameters in different types of streams. In the streams from coronal holes the abundance of helium relative to protons slightly increases (or is approximately constant) with increasing solar wind flux and proton density while in the streams from coronal streamers it decreases (Yermolaev, 1992b). These results indicate that the mechanisms of solar wind formation in different regions of the solar corona differ. In particular, the Prognoz 7 data confirm the hypothesis of Geiss et al. (1970) that by Coulomb friction the heavy ions may be dragged into interplanetary space only in streams from coronal holes (Yermolaev, 1992b).

The Prognoz 7 measurements allow us to study processes of preferential heating and acceleration of the α-particles in different types of solar wind streams. The dependence of the ratio of α-particle to proton temperatures T_α/T_p on their bulk velocity difference relative to the average thermal velocity $X=(V_\alpha-V_p)/[2k(T_\alpha/m_\alpha+T_p/m_p)]^{1/2}$ in different types of streams is presented in Fig.2. In the heliospheric current sheet (light circles) T_α/T_p monotonously increases from $\approx$ 1.5 to 7 with increasing X from −0.6 to +0.7. In streams from coronal streamers (crosses) the dependence is similar, but the temperature ratio is higher than one in heliospheric current sheet and increases from 2 to 11 with increasing X from −1.0 to +1.0. The dependence for streams from coronal holes (dark circles) is not monotonous and temperature ratio decreases from 8 to 4 with increasing X from −1 to 0 and increases from 4 to 9 with increasing X from 0 to +1. So, we can conclude that in streams from coronal holes preferential heating of α-particles does not depend on the sign of the velocity difference and increases with increasing absolute value of the velocity difference. Thus, the mechanisms of α-particle heating in streams from coronal holes and streamers differ from each other.

Because Coulomb collisions are suggested to result in an equalization of velocities and temperatures of α-particles and protons (Neugebauer, 1981; Yermolaev and Stupin, 1990), it is interesting to consider the relationship between the velocity differences $(V_\alpha-V_p)$ and $(V_\alpha-V_p)/V_A$ (V_A is the Alfvén velocity) and the ratio of the time of solar wind expansion and the time of momentum exchange due to Coulomb collisions τ_e/τ_s (Fig. 3), and the temperature ratio T_α/T_p with the ratio of solar wind expansion time to the time of energy exchange due to Coulomb collisions τ_e/τ_c (Fig.4) in the different types of solar wind streams. The parameters τ_e/τ_s and τ_e/τ_c were derived on the basis of Prognoz 7 measurements using the formulae presented in Neugebauer (1981) and Yermolaev and Stupin (1990). It is necessary to take into account that errors of the velocity difference averaged over abscissa bins are quite large, but not higher than ± 7 km/s for $(V_\alpha-V_p)$ and ± 0.15 for $(V_\alpha-V_p)/V_A$. The error of the temperature ratio is not more than $\pm 30\%$.

In Fig. 4a and 4b the dependences of the velocity differences $(V_\alpha-V_p)$ and $(V_\alpha-V_p)/V_A$ on τ_e/τ_s are presented for the heliospheric current sheet, streams from coronal holes and streamers. These figures show that in collisionless plasma ($\tau_e/\tau_s<0.3$) the discrepancies between the curves for streams from coronal holes and streamers are small, but in collisional plasma ($\tau_e/\tau_s >$ 0.3) it reaches $\approx$ 10 km/s, and for both types of streams the velocity differences are negative. An unexpected result is the fact that the velocity differences in the heliospheric current sheet are higher than in streams from coronal holes and streamers.

In contrast to the velocity differences, the temperature ratios in streams from coronal holes and streamers differ, and the curve for streams from

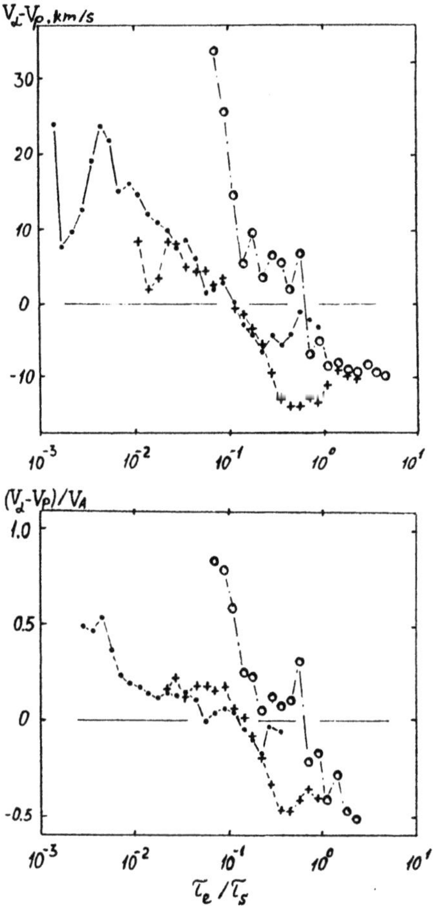

Fig. 3. Dependence of α-particles and protons: a) velocity difference (V_α–V_p) and b) velocity difference relative to the Alfvén velocity (V_α–V_p)/V_A on the ratio of the solar wind expansion time to the time of momentum exchange due to Coulomb collisions τ_e/τ_s. Symbols are the same as in Fig. 2.

coronal holes is higher than the one for streams from coronal streamers and is close to the one for the heliospheric current sheet. All curves reach ≈ 1 in collisional plasma (τ_e/τ_s>0.3).

3. Conclusions

Direct measurements of the solar wind on board the Prognoz 7 satellite (November, 1978 - July, 1979) allow us to reach several conclusions:

1. Different structures of coronal magnetic field and non-stationary coronal phenomena result in different types of solar wind streams which have particular distributions of basic plasma parameters and relations

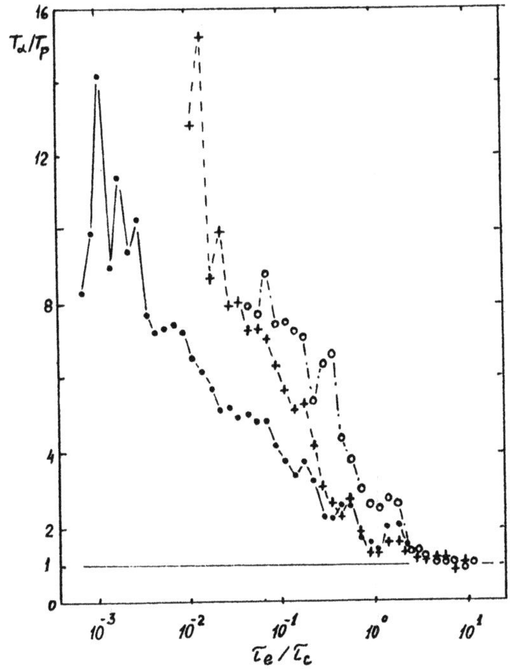

Fig. 4. Dependence of α-particle and proton temperature ratio T_α/T_p on the ratio of the solar wind expansion time to the time of energy exchange due to Coulomb collisions τ_e/τ_c. Symbols are the same as in Fig 2.

between them;

2. Maximum mass flux is observed in the heliospheric current sheet, mass flux in streams from coronal streamers is higher than the one in streams from coronal holes. Nevertheless, the Sun's loss of mass is equal in these types of streams and is higher than the one in disturbed streams;
3. Momentum fluxes in streams from coronal holes and streamers are equal, but they are lower than the one in the heliospheric current sheet and in disturbed streams;
4. Maximum kinetic and thermal energy flux is observed in disturbed streams and the heliospheric current sheet. Nevertheless, streams from coronal holes provide the maximum energy transport from the Sun to interplanetary space;
5. The maximum ratio of ion thermal pressure to magnetic pressure is observed in the heliospheric current sheet;
6. The discrepancies in the dependences of the helium abundance on the solar wind flux and the density of streams from coronal holes and streamers indicate that conditions and/or mechanisms of solar wind formation in these solar corona regions differ from each other;

7. No evidence has been obtained yet that the mechanisms of α-particle acceleration differ from each other in streams from coronal holes and streamers, but they differ from the one in the heliospheric current sheet;
8. The mechanisms of α-particle heating in streams from coronal holes and streamers are suggested to differ from each other, but they may be the same in streams from coronal streamers and the heliospheric current sheet.

Acknowledgements

The auther thanks V.V. Stupin for assisting the data processing and analysis and G.N. Zastenker for interesting and helpful discussions of the results. The work was partly supported by the International Science Foundation.

References

Geiss J.: 1985, in *Future Mission in Solar, Heliospheric and Space Plasma Physics*, Garmisch-Partenkirshen, Germany, ESA SP-235, p. 37.
Geiss J., Hirt, P., Leutwyler, H.: 1970, *Solar Phys.* **12**, 458
Hundhausen, A.J.: 1972, Coronal expansion and solar wind. Springer, Heidelberg, New York.
Neugebauer, M.: 1981, *Fund. Cosm. Phys.* **7**, 131
Neugebauer, M.: 1982, *Space Sci.Rev.* **33**, 127
Parker, E.N.: 1963, Interplanetary Dynamical Processes. Interscience, New York.
Schwenn, R.: 1986, in *Space Astronomy & Solar System Exploration*, ESA SP-268, p. 131.
Veselovsky, I.S.: 1984, in *Science and Technology Results*, Space Research 22, VINITI, Moscow, (in Russian).
Withbroe, G.L.: 1986, in *The Sun and the Heliosphere in Three Dimensions*, ed. R.G. Marsden, D.Reidel Publishing Company, p. 19.
Yermolaev, Yu.I., Stupin, V.V.: 1990, *Planet. Space Sci.* **38**, 1305
Yermolaev, Yu.I., Zastenker, G.N.: 1990, *Kosmich. Issled.* **28**, 103 (in Russian).
Yermolaev, Yu.I.: 1991, *Planet. Space Sci.* **39**, 1351
Yermolaev, Yu.I.: 1992a, in *Study of the Solar-Terrestrial System*, Proc. 26th ESLAB Symp., Killarney, 1992, ESA SP-346, p. 217.
Yermolaev, Yu.I.: 1992b, in *Coronal Streamers, Coronal Loops, and Coronal and Solar Wind Composition*, Proc. 1st SOHO Workshop, Annapolis, 1992, ESA SP-348, p. 339.

EFFECT OF AREAL EXPANSION AND CORONAL HEATING ON THE SOLAR WIND

Y.-M. WANG
Code 7672W, E.O. Hulburt Center for Space Research, Naval Research Laboratory, Washington, DC 20375-5352

Abstract. Empirical studies have shown that the solar wind speed at Earth is inversely correlated with the areal expansion rate of magnetic flux tubes near the Sun. Recent model calculations that include a self-consistent determination of the coronal temperature allow one to understand the physical basis of this relationship; they also suggest why the solar wind mass flux is relatively constant.

Key words: Solar Physics – Solar Corona – Solar Wind

1. Introduction

Empirical studies, in which current-free extrapolations of solar magnetograph data are compared with solar wind observations at Earth, indicate that solar wind speed is inversely correlated with magnetic flux-tube divergence near the Sun: the more slowly a coronal flux tube expands, the faster the terminal wind speed along that flux tube (Levine et al. 1977; Wang & Sheeley 1990).

This result may surprise some of those familiar with the study of Munro & Jackson (1977), who found that the areal cross-section of a polar coronal hole (known to be a source of fast solar wind) diverged as much as 7 times faster than a purely radial ($1/r^2$) expansion. However, the Munro-Jackson result referred to the *total* cross-section of the hole: in fact, the flux tubes near the center of an axisymmetric polar hole expand by less than a factor of 7, whereas the flux tubes at the boundary of the hole — where slow wind is observed — expand by *much more* than a factor of 7.

The empirical relationship accounts nicely for the prevalence of slow wind around sunspot maximum, when the large polar holes are absent and are replaced by small holes scattered over a wide range of latitudes. The magnetic flux tubes from these small open-field regions must fan out rapidly to fill the interplanetary volume, implying large expansion factors and hence slow wind. When the polar holes are present, slow wind tends to be concentrated around the polar hole boundaries and their interplanetary extension, the heliospheric current sheet.

The physical basis for the correlation has been explored by Wang & Sheeley (1991) and Wang (1993). The latter paper presents detailed model computations that show how the properties of the solar wind flow depend on the areal expansion factor; the calculations include a self-consistent determi-

Space Science Reviews **70**, 387–390.

nation of the coronal temperature, and are based on energy balance models developed by Hammer (1982), Hollweg (1986), Withbroe (1988), and others. I will now describe the results of these calculations from a physical perspective.

2. Physical Basis of the Correlation

In Wang (1993), a set of one-fluid equations is solved for the flow along a radially oriented flux tube (where all quantities depend only on the radial coordinate). The magnetic field along the flux tube is assumed to fall off as $r^{-\gamma}$ near the Sun and as r^{-2} far from the Sun (the larger the value of γ, the more rapidly the flux tube diverges near the Sun). The flow is driven by thermal and Alfvén-wave pressure gradients. The energy equation includes the effect of heat conduction, mechanical energy dissipation, Alfvén wave propagation, and radiative losses. The mechanical energy flux is assumed to be damped exponentially over a lengthscale H_m, thereby heating the corona.

The boundary conditions include the value of the temperature at Earth, the temperature gradient at Earth, and the temperature at the base of the corona, defined to be 5×10^5 K. In addition, following Hammer (1982) and Withbroe (1988), the downward heat flux at the coronal base is required to balance the total radiative and enthalpy losses from the transition zone (which is otherwise not explicitly included in the model). This condition yields a relationship between the thermal pressure, heat flux, and mass flux at the coronal base. It should be noted that this physical constraint on the mass flux is omitted in many wind models (such as those described by Leer et al. 1982), which generally do not take the energy balance requirements of the corona into account; the mass fluxes and other physical quantities derived in such models are thus essentially arbitrary.

The wind solutions are specified by choosing the areal expansion factor (or value of γ), the mechanical and Alfvén-wave energy fluxes at the coronal base, and the temperature at Earth. The mass flux at the coronal base and the dissipation length H_m are then no longer arbitrary, but have to be adjusted to satisfy the boundary conditions at the Sun (this usually requires a large number of iterations).

Before describing the solutions obtained in Wang (1993), let us first consider what happens if the coronal temperature is arbitrarily fixed (see Wang & Sheeley 1991). In that case, the mass flux at the coronal base increases as the areal expansion factor increases (see, e.g., Holzer & Leer 1980). (Because the corona is essentially in hydrostatic equilibrium below the sonic point, the density and mass flux at the sonic point are roughly independent of the expansion factor, being functions mainly of the coronal temperature; mass conservation along a flux tube then implies that the mass flux at the coronal base must increase with the expansion factor.) If the mechanical or wave energy flux that accelerates the wind is assumed to be constant at the coronal base, the energy per particle (energy flux divided by mass flux) must then decrease with increasing expansion factor, in agreement with the empirical result.

The situation is somewhat more complex when the coronal temperature is determined self-consistently rather than being arbitrarily fixed. Let us again assume that the mechanical energy flux (which heats the corona) and the Alfvén wave flux (which accelerates the wind to the observed speeds) are both constant at the coronal base. As we increase the flux-tube expansion factor, the same mechanical energy flux must heat a greater volume of plasma, so that the coronal temperature decreases. Despite the lower temperature, however, the model calculations (Wang 1993) show that the mass flux at the coronal base still *increases* in response to the increased expansion factor (although the increase is smaller than it would have been had the temperature remained fixed). Since the lower coronal temperature implies less thermal acceleration while the higher mass flux implies less Alfvén-wave energy per particle, the net effect of increasing the expansion factor is to decrease the asymptotic wind speed, as observed.

An important characteristic of the energy balance models is that the mass flux at the Sun is relatively insensitive to parameter variations (see also Withbroe 1989). This self-regulatory property is not present in models where the coronal temperature is arbitrarily specified: the mass flux is then extremely sensitive to the assumed temperature (see, e.g., Leer et al. 1982).

There is a remaining problem common to all non-MHD solar wind models. If we calculate the mass flux and the magnetic field strength at Earth, we find that the slow divergence geometries yield values that are higher than observed, whereas the rapid divergence geometries yield values that are too low. The source of this problem is easily understood. From empirical studies, we know that the interplanetary magnetic flux is distributed far more uniformly at 1 AU than it is near the Sun (see Wang & Sheeley 1988). This means that flux tubes that diverge rapidly near the Sun must later "reconverge" (in the sense that the field falls off less rapidly than r^{-2} at some intermediate distance), whereas flux tubes that diverge slowly near the Sun must later expand more rapidly. As pointed out by Withbroe (1989), this restructuring acts to maintain pressure balance between flux tubes by reducing the transverse gradients in the thermal pressure (at intermediate radii where the plasma β increases to order unity). Models that assume more realistic "non-monotonic" flux-tube geometries have been computed in Wang (1993); it was found that the wind speed at Earth is determined by the areal expansion rate near the Sun, but that the mass flux at Earth depends on the net expansion undergone by the flux tube between the Sun and Earth.

3. Conclusions

The implications of the coronal energy-balance models may be summarized as follows:

1. For a constant flux of mechanical and Alfvén-wave energy at the coronal base, greater flux-tube divergence produces lower coronal temperatures but higher mass fluxes at the Sun.
2. As a result, because of the decreased thermal acceleration and the decreased wave energy per particle, greater flux-tube divergence also produces lower asymptotic wind speeds, as observed. This agreement suggests that the mechanical and wave energy fluxes are indeed roughly constant within coronal holes.
3. The calculations imply that, near the Sun, the temperatures and thermal pressures are higher near the center of an axisymmetric polar hole (where the flux-tube divergence rates are relatively small) than at its boundary (where the divergence rates are very large).
4. The calculations indicate that the mass flux is remarkably constant at the Sun. The fact that the mass flux observed at Earth is also remarkably constant implies that the net expansion undergone by flux tubes between the Sun and Earth is roughly invariant, despite the large differences in their divergence rates near the Sun. MHD modeling combined with self-consistent treatment of the coronal energy balance may thus eventually resolve the well-known "solar wind mass flux problem."

Finally, we note that our empirical and theoretical results are based on the large-scale properties of the solar magnetic field, and do not take account of finer-scale structure such as that represented by polar plumes. On these finer scales, the assumption of constant mechanical and wave energy fluxes at the coronal base may break down.

References

Hammer, R.: 1982, *Astrophys. J.* **259**, 767

Hollweg, J.V.: 1986, *J. Geophys. Res.* **91**, 4111

Holzer, T.E., & Leer, E.: 1980, *J. Geophys. Res.* **85**, 4665

Leer, E., Holzer, T.E., & Flå, T.: 1982, *Space Sci. Rev.* **33**, 161

Levine, R.H., Altschuler, M.D., & Harvey, J.W.: 1977, *J. Geophys. Res.* **82**, 1061

Munro, R.H., & Jackson, B.V.: 1977, *Astrophys. J.* **213**, 874

Wang, Y.-M.: 1993, *Astrophys. J.* **410**, L123

Wang, Y.-M., & Sheeley, N.R., Jr.: 1988, *J. Geophys. Res.* **93**, 11227

Wang, Y.-M., & Sheeley, N.R., Jr.: 1990, *Astrophys. J.* **355**, 726

Wang, Y.-M., & Sheeley, N.R., Jr.: 1991, *Astrophys. J.* **372**, L45

Withbroe, G.L.: 1988, *Astrophys. J.* **325**, 442

Withbroe, G.L.: 1989, *Astrophys. J.* **337**, L49

CORONAL PLUMES AND FINE SCALE STRUCTURE IN HIGH SPEED SOLAR WIND STREAMS

M. VELLI
Dip. di Astronomia e Scienze dello Spazio, Università di Firenze, Largo E. Fermi 5, 50125 Firenze, Italy

and

S. R. HABBAL, R. ESSER
Harvard-Smithsonian CFA, 60 Garden St., Cambridge, MA 02138, USA

Abstract. We present a solar wind model which takes into account the possible origin of fast solar wind streams in coronal plumes. We treat coronal holes as being made up of essentially 2 plasma species, denser, warmer coronal plumes embedded in a surrounding less dense and cooler medium. Pressure balance at the coronal base implies a smaller magnetic field within coronal plumes than without. Considering the total coronal hole areal expansion as given, we calculate the relative expansion of plumes and the ambient medium subject to transverse pressure balance as the wind accelerates. The magnetic flux is assumed to be conserved independently both within plumes and the surrounding coronal hole. Magnetic field curvature terms are neglected so the model is essentially one dimensional along the coronal plumes, which are treated as thin flux-tubes. We compare the results from this model with white-light photographs of the solar corona and in-situ measurements of the spaghetti-like fine-structure of high-speed winds.

Key words: Solar Wind, Coronal Plumes

1. Introduction

Polar plumes are ray-like structures that were first observed in white light eclipse photographs of the solar corona (Van de Hulst 1950, Koutchmy 1977). They have also been observed at EUV, XUV and X-ray wavelengths and appear to have an average lifetime extending from hours to days. As described by Habbal (1992) coronal plumes appear in Mg IX 368 å as "searchlights overlying a thicker compact emission on the disk, identified as coronal bright points". Coronal bright points lie in the boundaries of supergranular cells and seem to consist of arcades of miniature loop-like structures which resemble small active regions. They also exhibit occasional flaring activity though their lifetime of a few hours is much smaller than that of active regions. Along with macrospicules, shorter life-time (around 45 minutes) jets of cooler (invisible in lines formed above 3.0 10^5 °K) chromospheric material, coronal plumes define the fine structure of the open field lines in coronal holes.

The idea that the solar wind flow could bear the signature of the fine-scale structure of coronal flux tubes was first put forward by Parker(1963), who argued that adjacent flow tubes with slight differences in their plasma

Space Science Reviews **70**, 391–396.

properties and separated by tangential magnetic field discontinuities should be swept by the solar wind along a spacecraft trajectory. This observational scenario was described by the expression 'spaghetti model'.

Thieme, Marsch, and Schwenn (TMS) (1992) analyzed data from the Helios spacecraft in the search of the remnants of coronal fine structures; after removing the contribution of Alfvén waves from the proton and α particle velocities they showed how an increase (decrease) of the proton temperature was often associated with an increase (decrease) of the densities, and frequently also of the velocities of protons and α-particles. These modulations are anticorrelated with magnetic field magnitude. Plots of the total kinetic (protons, electrons and α-particles) and magnetic pressures show that there is a strong anti-correlation between the gas and magnetic pressures, while the total pressure is almost constant over a period of four and a half days. A common feature in this data is the presence of a local maximum in the magnetic pressure and a minimum in the kinetic pressure separating regions of higher proton and α particle velocities. If the local magnetic maximum represented the boundary of the flow tube, such a feature would imply a stronger velocity within flow-tubes than outside. TMS also examine the angular dimensions of the hypothetical flow tubes and find that within 0.5 AU they are consistent with the angular dimensions of polar coronal plumes (2°) at R=1.1 $R_\odot$, though their subsequent expansion appears to be greater than radial.

The observational evidence described above is consistent with the idea that polar plumes expand into interplanetary space while retaining an overall pressure balance with the surrounding coronal hole. Because the plume lie above coronal brightpoints and hence the supergranule boundaries, we assume they occupy only a small area of the coronal surface, and that they are locally hotter and denser. Pressure balance then requires the magnetic field within the plumes to be less intense (which is consistent with their position above bright points). In this paper we explore a model for the solar wind which takes these properties into account. Although some general asymptotic results for great distances may be obtained analytically, the preliminary numerical results presented here concern only winds which are isothermal and one-fluid. Note that Parker (1965) discussed very similar ideas to describe the expansion of coronal streamers.

2. Model Equations

We shall divide the coronal hole in two different type of regions: the background coronal hole, with density ρ_1, temperature T_1, magnetic field B_1, covering an area A_1; and coronal plumes, which for simplicity we consider to have the same physical properties everywhere, and therefore model by a single density ρ_2, temperature T_2, magnetic field B_2 and overall area A_2. In addition we will consider the magnetic flux to be conserved separately in the coronal plumes and the background coronal hole $B_1A_1 = \Phi_1$, $B_2A_2 = \Phi_2$, while the total coronal hole expansion will be a given function of heliocentric distance $f(R)$. If the filling factor of coronal plumes at the coronal base is given as α_2, we may write

$$\alpha_1 A_1 + \alpha_2 A_2 = f(R), \qquad \alpha_1 + \alpha_2 = 1, \tag{1}$$

where we have normalized all areas to the total coronal area so that at the coronal base $A_1 = A_2 = f(R_\odot) = 1$.

Neglecting curvature terms, transverse pressure balance implies that

$$\rho_1 T_1 + \frac{B_1^2}{8\pi} = \rho_2 T_2 + \frac{B_2^2}{8\pi} \tag{2}$$

while conservation of magnetic flux gives $B_1A_1 = B_{01}$, $B_2A_2 = B_{02}$. It is convenient to introduce

$$\beta_{1,2} = 8\pi T_{1,2}/B_{01,2}^2, \qquad \delta = T_2/T_1, \qquad a_{1,2} = A_{1,2}/f(r). \tag{3}$$

Denoting the adimensional gravitational scale height by $g = GM_\odot/\mu R T_1 R_\odot$ which for $T_1 = 10^6$ °K is about $g = 20$, and normalizing the velocities by the sound speed at temperature T_1, the stationary isothermal wind equations become

$$u_1 - \frac{1}{u_1} = \frac{a_1'}{a_1} + \frac{f'}{f} - \frac{g}{r^2} \qquad \frac{u_2}{\delta} - \frac{1}{u_2} = \frac{a_2'}{a_2} + \frac{f'}{f} - \frac{g}{\delta r^2}. \tag{4}$$

In the following we will take the overall coronal expansion factor to be $f(r) = r^2$. When $\alpha_2 = 0$, a_2 is determined completely by the pressure balance equ.(2), the equations for u_1, u_2 are uncoupled and may be solved in the standard way. When $\alpha_2 \neq 0$ the equations are solved iteratively, starting from the $\alpha_2 = 0$ solution for u_2 to get u_1, imposing the total area constraint, and then updating the solution for u_2 with this guess for u_1 until convergence is achieved.

3. Results

3.1 Asymptotic results for large distances

Consider first the isothermal winds of the previous section. For large values of r, the kinetic and magnetic pressures evolve as

$$\rho_1 \sim 1/r^2 a_1 \log r \qquad P_m \sim 1/\beta_1 a_1^2 r^4, \tag{5}$$

so that pressure equilibrium at great heliocentric distances implies $\rho_{1\infty} = \delta\rho_{2\infty}$.

From mass flux conservation and the asymptotic behaviour of the wind speeds we can then obtain the relation $a_2/a_1 \sim \delta^{1/2}\rho_{20}u_{20}/u_{10}$ where the 0 subscript denotes quantities at the solar surface. Now at the solar surface the pressure balance constraint does not play a great role because the magnetic pressure is largely dominant, so that the ratio u_{20}/u_{10} is determined as in the standard uncoupled case, yielding $u_{20}/u_{10} \simeq \delta^{-3/2}\ e^{g(1-1/\delta)}$. Hence at great distances we find

$$a_2/a_1 \simeq \rho_{20}/\delta e^{g(1-1/\delta)}. \tag{6}$$

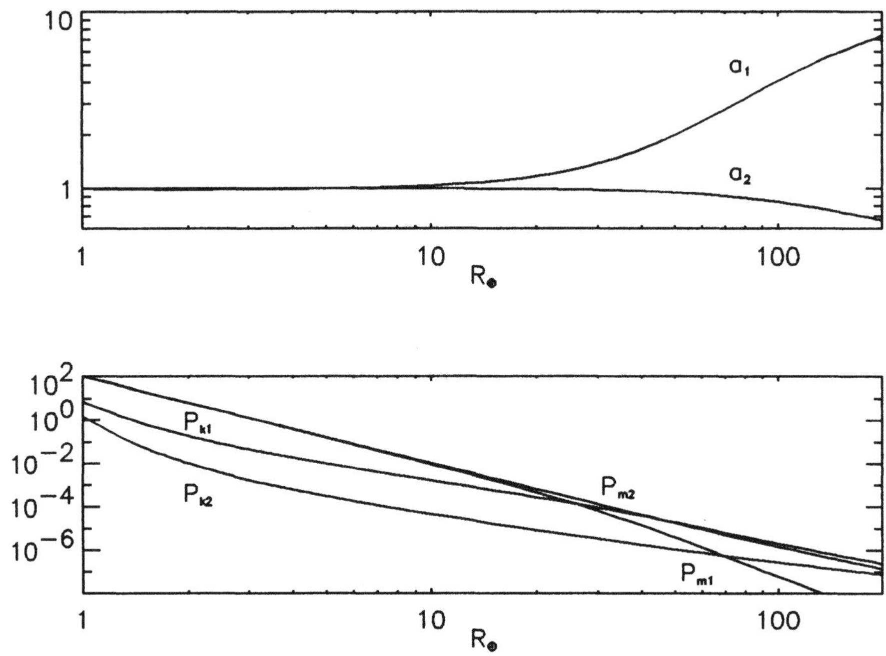

Fig. 1. Top panel: specific areas a_1, a_2, of plume and ambient hole respectively Bottom panel: kinetic pressures,P_k and magnetic pressures P_m as a function of heliocentric distance. Initial conditions are described in text.

On the other hand if we consider a more realistic wind in which (a) the speed tends to a constant at great distances (the ratio of speeds outside and inside the plumes also tending to a constant M_p of order 1) and (b) the temperature beyond a certain distance decreases adiabatically, then the density ratio goes as

$$\rho_{1\infty}/\rho_{2\infty} \simeq \delta(a_1/a_2)^{2/3} \tag{7}$$

where now δ is the (constant) temperature ratio in the inner heliospheric regions where the wind may be considered isothermal. Again from mass flux conservation we now obtain $(a_2/a_1)^{5/3} \sim \delta\rho_{20}/M_p\ u_{20}/u_{10}$. Since it is not unrealistic to consider the isothermal region extending to the sonic point we may use the previous values for u_{10}, u_{20}, to obtain finally

$$a_2/a_1 \simeq \rho_{20}^{3/5}\delta^{-3/10}M_p^{-3/5}\ e^{3/5g(1-1/\delta)}. \tag{8}$$

With the same arguments we can estimate the overall mass flux

$$F_m = \alpha_1 u_{10} + \alpha_2\rho_{20}u_{20} \simeq g^2/4e^{3/2-g}(\alpha_1 + \alpha_2\rho_{20}e^{g(1-1/\delta)}/\delta^{3/2}). \tag{9}$$

Since the term containing the exponential dependence of the mass flux on the plume temperature is multiplied by the filling factor α_2 which we assume to be small, we remark that changes in plume temperature do not necessarily lead to a large change in the overall mass flux, although for this statement to be true the overall mass input from coronal plumes should be small compared to that coming from the ambient coronal hole.

3.2 Numerical Results

The behaviour of the specific areas a_1, a_2 and the kinetic and magnetic pressures for one isothermal wind model as a function of distance is shown in fig. 1. The background coronal hole is taken to have a temperature $T_1 = 10^6$ °K, a density 10^8 cm^{-3} and a plasma beta of 1%. The filling factor of coronal plumes is $\alpha_2 = 5\%$, while their base parameters are given by $T_1 = 1.4\ 10^6$ °K, a density $4\ 10^8$ cm^{-3} and a plasma beta of 1.47%. Roughly speaking, a_1, a_2 remain constant out to $r \simeq 10R_\odot$ before contracting and expanding respectively. The plasma beta crosses 1 at about 30 $R_\odot$ for the plume while it remains smaller than 1 out to beyond 100 $R_\odot$ for the background coronal hole.

The specific areas also vary within the first $10R_\odot$, but in the opposite way, plumes contracting slightly relative to the overall coronal hole expansion. This behaviour is due to the fact that initially the kinetic pressure decays exponentially while the magnetic pressure only decays algebraically: a pressure equilibrium at the coronal base implies that regions with a stronger magnetic field will have a higher total pressure somewhat above and will therefore expand. This collimation effect is relatively weak however, the minimum value of a_2 being e.g. 0.93 for a base plasma beta of 10%, density $4\ 10^8$ cm^{-3} and temperature of $1.4\ 10^6$ °K.

When comparing with the results of TMS one major difference is that in the isothermal case considered here the ratio of plume to background speeds continues to increase with heliocentric distance and is much greater, $\delta u/u \sim$ 25%, than observed at 0.3 AU $\delta u/u \sim 5 - 10\%$. The run of the pressures and specific areas is encouraging however, so we are considering now a more realistic treatment of the two wind species. Observations in the ecliptic plane show that the coronal plume signatures disappear beyond about 0.5 AU, due to the increasing role of the magnetic field rotation and the influence of the radial compression in the velocity shears. This tends to destroy the boundaries of the structures. The Ulysses spacecraft might however be able to demonstrate the existence of such structures out to greater distances above the solar poles. One marker of the coronal plume plasma which should be looked for is a difference in ion abundance.

Acknowledgements

M. Velli would like to thank the Center for Astrophysics for the kind hospitality when this work was begun and financially supporting his visit.

References

Van de Hulst, H.C. : 1950, *Bull. Astron. Soc. Neth.* **11**, 150
Koutchmy, S.: 1977, *Solar Phys.* **51**, 399
Habbal, S.R.: 1992, *Ann. Geophys.* **10**, 34
Parker, E.N.: 1963, *Interplanetary dynamical Processes*, Interscience, New York
Parker, E.N.: 1965, *Space Science Reviews* **4**, 666
Thieme, K.M., Marsch, E. and Schwenn, R.: 1990, *Ann. Geophys.* **8**, 713

RADIAL PROFILE MEASUREMENTS OF THE SOLAR WIND SPEED USING RADIO SOUNDING TECHNIQUES

A.I. EFIMOV
Institute of Radio Engineering and Electronics of the Russian Academy of Sciences, Moscow, 103907, Russia

Abstract. Velocity measurements of the solar wind in the region of its acceleration were carried out in 1984 with Venera Orbiters using the following specific radio sounding techniques: (i) phase and frequency correlations from two or three widely-spaced ground stations; (ii) phase and frequency correlations from two spacecraft and two widely-spaced ground stations; (iii) two station two-way coherent Doppler measurements; (iv) determination of the break frequency in power spectra of intensity fluctuations. Our results are substantially lower than those derived from direct Doppler shift measurements of transition region and lower coronal resonance lines and those from measurements applying the Doppler dimming technique.

Key words: Sun – Solar Wind

1. Introduction

One of the major unsolved problems in solar physics is the solar wind acceleration and coronal heating mechanism. An important parameter in the various models is the distance of the critical point where the transition from subsonic to supersonic flows takes place. Scintillation measurements of radio emission from natural or artificial sources provide information on the outflow velocity both for small heliocentric distances (Coles et al., 1986) and for large distances (Watanabe and Schwenn, 1989). However, it is difficult to derive the outflow velocity from temporal spectra of intensity fluctuations as the velocity depends strongly on irregularities of the anisotropy which can reach high values in the acceleration region. Besides, in this region the plasma turbulence is enhanced (Lotova, 1988). More adequate is to observe the scintillations simultaneously from several ground stations (Armstrong and Woo, 1981; Coles et al., 1991).

2. Methods of Deriving Plasma Velocities

To measure the solar wind speed we applied various techniques (cf. Fig. 1):

1. Cross-correlation analysis of phase and/or frequency fluctuations at 32 cm at the ground stations Ussuriisk and Evpatoria. The time lag between similar radio wave extrema at the two ground stations can be explained as a result of an arrival of the disturbance at the two coronal ray paths at different times.

Space Science Reviews **70**, 397–402.

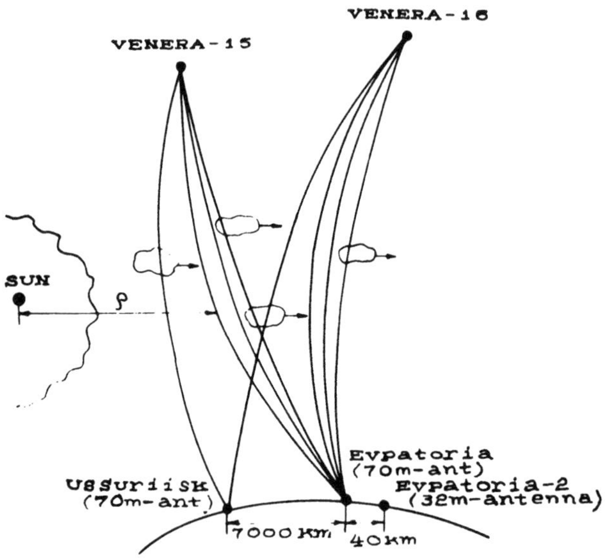

Fig. 1. Investigation method: multipath multifrequency radio sounding of the solar corona.

2. Cross-correlation analysis of phase and frequency fluctuations of decimetre signals radiated simultaneously by Venera-15, -16 spacecraft and received in each ground station (Ussuriisk and Evpatoria, Ussuriisk and Evpatoria-2). The radial distances between the two coronal ray paths are of the order of 5000-6000 km in this case.
3. Cross-correlation analysis of phase and frequency fluctuations for a case when the ground station Ussuriisk receives the Venera-15 signals and Evpatoria receives the Venera-16 signals. In this case, the coronal path distance can reach 10000-20000 km and thus it is possible to obtain the third independent value of the velocity v.
4. Cross-correlation analysis of frequency and intensity fluctuations for three wavelengths (32cm, 8cm and 5cm). Between similar extrema of the radio wave parameters a shift in time will appear which can be explained as the result of an arrival of the disturbances at two coronal ray paths at different times corresponding to the different refraction for different wavelengths in the solar corona.
5. The method of two-way coherent Doppler measurements which exploits the spatial separation of the up-link and down-link between ground station and spacecraft. In this case the coronal separation between the up-link and down-link ray path can reach 20000-30000 km.
6. Spectral analysis of intensity fluctuations. The break frequency (or the knee frequency) in temporal spectra is proportional to the velocity of the irregularities. However, an anisotropy in the irregularity spectrum

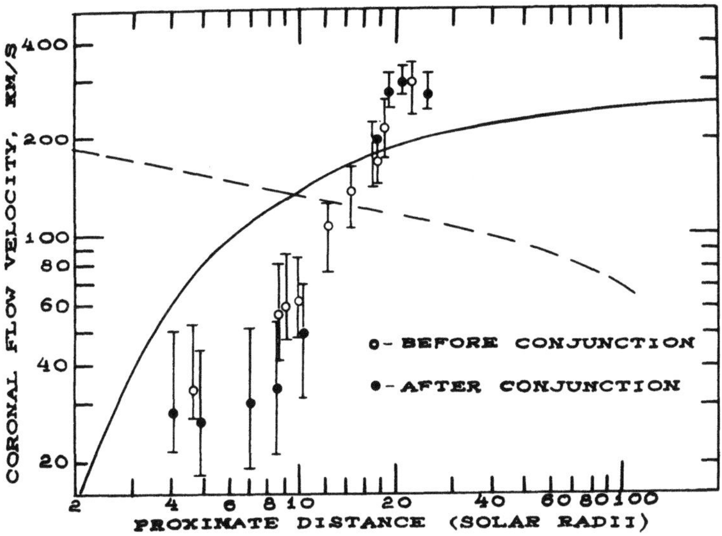

Fig. 2. An estimate of the coronal flow velocity as a function of the distance according to frequency measurements at two widely-spaced stations in Evpatoria and Ussuriisk. The solid line shows the radial velocity profile for a quiet, unstructured coronal region according to a model of Withbroe (1988). The dashed line shows the radial dependence of the sound speed (after Ogilvie, 1979).

can shift the knee to lower frequencies and thus decrease the measured velocity.

3. Experimental Data

From March 12 to September 21, 1984 radio occultation experiments were conducted in the solar corona using Venera Orbiters. The spacecraft were equipped with L-, S- and C-band (0.93, 3.72 and 5.93 GHz) coherent radio systems. The sampling rate for the signal amplitude data was usually 20 samples per 1 sec and for phase difference and frequency 1 sample per 1 sec. The ray path from Venera Orbiters to Earth moved from the solar west limb to the east one at an average speed of 1.05 solar radius per day. The solar latitudes were less than 15° except for a few days around solar conjunction when the spacecraft appeared to pass the north limb of the Sun. Figure 2 shows the results derived from observations of frequency fluctuations at 32 cm at two widely-spaced ground stations in Ussuriisk and Evpatoria.

The deduced velocities are displayed in Figure 2 as a function of radial distance. The observed velocities are compared with the velocities of a model of Withbroe (1988) for a quiet unstructured coronal region (solid line) and the radial dependence of the sound speed of Ogilvie (1978) which is shown as a dashed line. The critical point for the model of Withbroe is located at a distance of about 10 solar radii from the Sun or closer if the final solar wind velocity will be 400 km/s. In spite of the large scatter in this figure one may

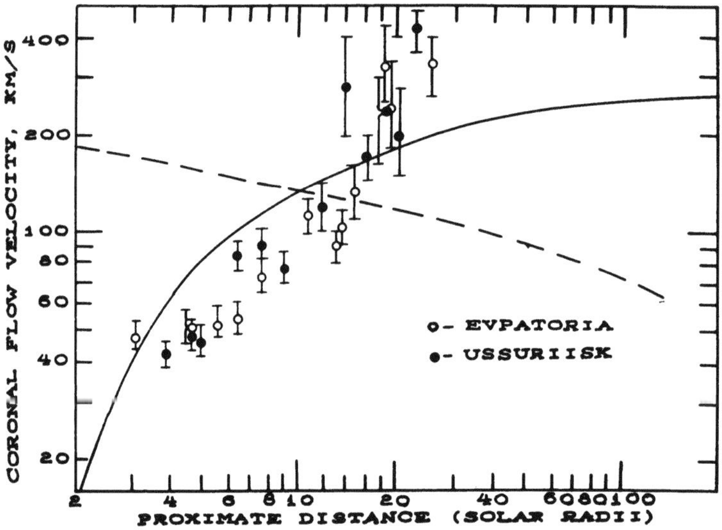

Fig. 3. An estimate of coronal flow velocity as a function of distance according to two-way coherent Doppler data.

state that the critical point is farther away from the Sun. Our estimated distance is near 15 solar radii.

In Figure 3 we present coronal plasma velocity estimates from two way, coherent Doppler measurements at 32 cm. The initial radio data were obtained at each ground station independently. These independent estimates of the solar wind velocity corroborate our conclusions from Fig. 2.

Multipath soundings of the solar corona were carried out with the radio signals radiated simultaneously by Venera-15 and Venera-16. The signals propagated along different radio lines and were received at three ground stations: Ussuriisk, Evpatoria and Evpatoria-2. The ground stations in Ussiriisk and Evpatoria are situated at a distance of about 7000 km. The distance between Evpatoria and Evpatoria-2 is about 60 km. The coronal ray path separation between ray paths changed from 2400 km to 10500 km. The cross-correlation coefficients of frequency fluctuations which have been determined for such coronal ray path separations are presented in Figure 4.

One can see that for all the cases when the coronal path separation exceeds 5000 km the time shifts of the cross-correlation maxima have large values. This means that the apparent velocity of irregularities derived from spaced frequency measurements has a very small value. We explained this unexpected fact by a peculiarity of the IPS velocity measurement technique which measures the velocity of the density fluctuations and not the bulk flow velocity. The apparent velocity could differ from the flow velocity even by the sound speed. This conclusion is confirmed by an analysis of the results of intensity measurements.

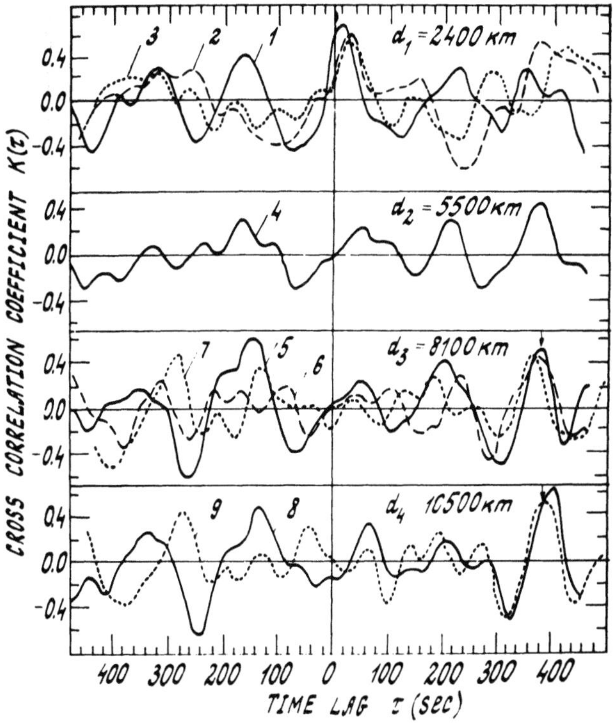

Fig. 4. Cross-correlation coefficients of the frequency fluctuations observed at three widely-spaced ground stations.

In Figure 5 we present irregularity velocity estimates from solar occultations of Venera-15 and 16. The main features of the dependence of the measured velocity on the heliocentric distance are: a small increase of the velocity at distances below 12 solar radii, a noticeable decrease between 12 and 14 radii and a steep increase above about 16 solar radii. For heliocentric distances larger than 25 radii the velocity becomes almost constant at a level of about 300 km/s. It is important that the velocity is smaller in the region where the frequency measurements (Fig.2 and Fig.3) indicate the transition from a subsonic to supersonic flow. It should be noted that a peculiar behaviour of the solar wind between 12 an 15 solar radii was first discovered by Dryer and Cuperman (1972) based on optical measurements.

4. Conclusions

The use of multifrequency and multipath coronal sounding permits us to determine one of the main properties of the solar wind, the radial profile of its velocity. The velocities measured with coronal sounding experiments are substantially smaller than the values derived from Doppler dimming mea-

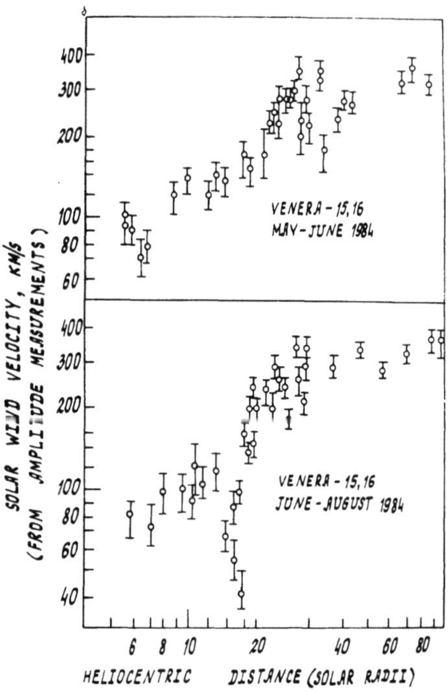

Fig. 5. Apparent radial velocities of the plasma irregularities derived from intensity fluctuations

surements. It is necessary to understand the origin of these discrepancies. This can be done by applying the different techniques to simultaneous observations. Such a possibility may be exist in 1996 when SOHO and Mars-94 will be in operation.

References

Armstrong, J.W., Woo, R.: 1981, *Astron. Astrophys.* **103**, 415

Armstrong, J.W., Coles, W.A., Kojima, M., Rickett, B.J.: 1986, in *The Sun and the Heliosphere in Three Dimensions*, ed. R.G. Marsden , Reidel, p. 59.

Bird, M.K., Edenhofer, P.: 1990, in *Physics of the Inner Heliosphere I*, eds. R. Schwenn, E. Marsch, Springer, p. 13.

Coles, W.A., Esser, R., Lovhang, U.-P., Markkanen, J.: 1991, *J. Geophys. Res.* **96**, 13849

Dryer, M., Cuperman, S.: 1972, in *Solar Wind Activity*, eds. P.S. McIntosh and M. Dryer, Cambridge, Mass.

Lotova, N.A.: 1988, *Solar Physics* **117**, 399

gilvie, K.W.:1978, in *A Close-Up of the Sun*, eds. M. Neugebauer and R.W. Davies, JPL Publication 78-70, p. 345.

Watanabe, T., Schwenn, R.: 1989, *Space Sci. Rev.* **51**, 147

Withbroe, G.: 1988, *Astrophys. J.* **325**, 442

MIX
Papier aus verantwortungsvollen Quellen
Paper from responsible sources
FSC® C105338

If you have any concerns about our products,
you can contact us on
ProductSafety@springernature.com

In case Publisher is established outside the EU,
the EU authorized representative is:
Springer Nature Customer Service Center GmbH
Europaplatz 3, 69115 Heidelberg, Germany

Printed by Libri Plureos GmbH
in Hamburg, Germany